CYTOSKELETON - STRUCTURE, DYNAMICS, FUNCTION AND DISEASE

Edited by **Jose C. Jimenez-Lopez**

Cytoskeleton - Structure, Dynamics, Function and Disease
http://dx.doi.org/10.5772/62622
Edited by Jose C. Jimenez-Lopez

Contributors

Francisco Javier Martin-Romero, Eulalia Pozo-Guisado, Aida M. Lopez-Guerrero, Carlos Pascual-Caro, Regina Pessoa-Pureur, Ariane Zamoner, Ivana Pajic-Lijakovic, Milan Milivojevic, Elsa Anes, Sukla Roychowdhury, Jorge Sierra-Fonseca, Doris Cerecedo, Stephane Gross, Galina Prokopchuk, Jacky Cosson, José Antonio Sánchez-Alcázar, Jesús Porcuna Doncel, Patricia De La Cruz Ojeda, Manuel Oropesa Ávila, Marina Villanueva Paz, Isabel De Lavera, Mario De La Mata, Mónica Álvarez Córdoba, Raquel Luzón Hidalgo, Juan Miguel Suarez Rivero, David Cotán, Yoshinobu Mineyuki, Miyuki Takeuchi, Andrew Staehelin, Jin Wang, Xuan Zhang, Zenglin Pei, Chunxia Ji, Xiaoyan Zhang, Jianqing Xu, Anca Hermenean, Aurel Ardelean, Fatima Pedrosa Domellöf, Juliana Vidal, Wanderley De Souza, Francisco Pelegri, Elaine L. Welch, Viji Draviam, Madeleine Hart, Duccio Conti, Roshan Shrestha, Asifa Islam, Naoka Tamura

Published by InTech
Janeza Trdine 9, 51000 Rijeka, Croatia

Publishing Process Manager Martina Usljebrka
Technical Editor SPi Global
Cover InTech Design team

First published May, 2017
Printed in Croatia
Legal deposit, Croatia: National and University Library in Zagreb

Additional hard copies can be obtained from orders@intechopen.com

Cytoskeleton - Structure, Dynamics, Function and Disease, Edited by Jose C. Jimenez-Lopez
p. cm.
Print ISBN 978-953-51-3169-4
Online ISBN 978-953-51-3170-0

For EU product safety concerns:
IN TECH d.o.o., Prolaz Marije Krucifikse Kozulić 3, 51000 Rijeka, Croatia,
info@intechopen.com or visit our website at intechopen.com.

Meet the editor

Dr. Jose C. Jimenez-Lopez finished his PhD degree in Plant Science in 2008 at the Spanish National Research Council (CSIC), which focused in the study of the actin-binding protein profilin in olive from functional and allergenic point of views. During this predoctoral period, he visited as an invited researcher different EU and US laboratories such as the Institute of Cellular and Molecular Botany (IZMB), University of Bonn, Germany (2005); the University of Vienna, Faculty of Life Sciences (2006); and the College of Sciences, Purdue University in the USA.
Soon after, he moved to the USA to start his postdoctoral research at Purdue University, the College of Sciences, until 2011 to work in the biochemical and cellular characterization of actin-binding proteins involved in plant actin cytoskeleton dynamics and signaling throughout membrane association.
In 2012, he was awarded with an EU research grant from the Marie Curie program to work until 2015 in the University of Western Australia and CSIC in a project related to the human health benefits of legume seed–derived foods and molecular aspects of allergy and cross allergenicity to legume seed proteins.
Currently, Jose is working at the Estación Experimental del Zaidín (EEZ), Spanish National Research Council (CSIC), as a senior research fellow thanks to a grant awarded from the "Ramon y Cajal" research program (MINECO), focusing in the functionality, health benefits, and allergy implications of proteins from reproductive tissues (pollen and seeds) in crop species (mainly legumes) of agro-industrial interest.

Contents

Preface IX

Section 1 Cytoskeleton: Structure and Dynamics 1

Chapter 1 **Reorganization of Vegetal Cortex Microtubules and Its Role in Axis Induction in the Early Vertebrate Embryo 3**
Elaine Welch and Francisco Pelegri

Chapter 2 **Actin-Microtubule Interaction in Plants 33**
Miyuki Takeuchi, L. Andrew Staehelin and Yoshinobu Mineyuki

Chapter 3 **Morphological and Functional Aspects of Cytoskeleton of Trypanosomatids 55**
Juliana Cunha Vidal and Wanderley de Souza

Chapter 4 **The Interplay between Cytoskeleton and Calcium Dynamics 73**
Francisco Javier Martin-Romero, Aida M. Lopez-Guerrero, Carlos Pascual-Caro and Eulalia Pozo-Guisado

Chapter 5 **Role of Band 3 in the Erythrocyte Membrane Structural Changes Under Isotonic and Hypotonic Conditions 89**
Ivana Pajic-Lijakovic and Milan Milivojevic

Chapter 6 **Dystrophin–Glycoprotein Complex in Blood Cells 105**
Doris Cerecedo

Chapter 7 **Biophysics of Fish Sperm Flagellar Movement: Present Knowledge and Original Directions 127**
Galina Prokopchuk and Jacky Cosson

Chapter 8 **Cytoskeleton Rearrangements during the Execution Phase of Apoptosis 151**
Jesús Porcuna Doncel, Patricia de la Cruz Ojeda, Manuel Oropesa-Ávila, Marina Villanueva Paz, Isabel De Lavera, Mario De La Mata, Mónica Álvarez Córdoba, Raquel Luzón Hidalgo, Juan Miguel Suarez Rivero, David Cotán and José Antonio Sánchez-Alcázar

Chapter 9 **How are Dynamic Microtubules Stably Tethered to Human Chromosomes? 171**
Duccio Conti, Madeleine Hart, Naoka Tamura, Roshan Shrestha, Asifa Islam and Viji M. Draviam

Section 2 Cytoskeleton: Function and Disease 195

Chapter 10 **Muscle Fibers Lacking Desmin in the Extraocular Muscles: A Paradigm Shift 197**
Fatima Pedrosa Domellöf

Chapter 11 **The Actomyosin Network and Cellular Motility: A S100A4 Regulatory View into the Process 205**
Stephane R. Gross

Chapter 12 **Intermediate Filaments as a Target of Signaling Mechanisms in Neurotoxicity 233**
Ariane Zamoner and Regina Pessoa-Pureur

Chapter 13 **Acting on Actin During Bacterial Infection 257**
Elsa Anes

Chapter 14 **Heterotrimeric G Proteins and the Regulation of Microtubule Assembly 279**
Sukla Roychowdhury and Jorge A. Sierra-Fonseca

Chapter 15 **Novel Insights into the Role of the Cytoskeleton in Cancer 299**
Xuan Zhang, Zenglin Pei, Chunxia Ji, Xiaoyan Zhang, Jianqing Xu and Jin Wang

Chapter 16 **Targeting the Cytoskeleton with Plant-Bioactive Compounds in Cancer Therapy 315**
Anca Hermenean and Aurel Ardelean

Preface

All cells of all domains of life (archaea, bacteria, and eukaryotes) contain constituents of the cytoskeleton, in which systems of different organisms are composed of similar proteins. However, the structure, function, and dynamic behavior of the cytoskeleton can be very different, depending on organism and cell type. Even within one cell, the cytoskeleton can change through association with other proteins and the previous history of the network.

In eukaryotes, the cytoskeletal matrix is a dynamic structure integrated in three main proteins (microtubules, intermediate filaments, and actin filaments), which are capable of a fast growth or disassembly depending on the cell's requirements at a certain period of time. Underlying all this complexity is the knowledge that the cell is alive and is constantly changing its properties, actively, as a consequence of many external factors, by integrating environment stimuli through different cell signaling pathways.

In recent years, it has been shown that bacteria contain a number of cytoskeletal structures and elements that include homologues of eukaryotic cytoskeletal proteins (actin, tubulin, and intermediate filament proteins) and a fourth group, the MinD-ParA group, which appears to be unique to bacteria.

The cytoskeleton is the overall name given to protein filaments and motor proteins in the cell, which form a massive and dynamic three-dimensional (3D) scaffolding structure, driving cell shape and cell organization, in determining the spatial arrangement of membrane receptors and thus in the development of cell polarity, movement, and cell stability. It plays a crucial role in many fundamental processes of cell growth and development such as chromosome segregation during cellular division and cytokinesis (the division of a mother cell into two daughter cells), cell expansion, and intracellular organization, also contributing to the organelles' functionality.

Furthermore, filaments can be cross-linked to other similar filaments and to membranes, having accessory and/or actin-binding proteins as intermediaries. This interlinking significantly increases rigidity, as the time that some filaments are used as trackways for motor proteins to transport molecules and cargo-containing vesicles around the cell and the uptake of extracellular material (endocytosis).

Many proteins of the cytoskeleton have been newly identified and have tentative functions assigned, and recently deficiencies were identified in disease states. In addition, components of the cytoskeleton form specialized structures, such as flagella, cilia, sperm, lamellipodia, and podosomes, allowing cells to move, cell organelles to be moved and positioned along the protein filaments using them as roadways rather like how a railway locomotive runs on rail tracks, and muscles to function.

The cytoskeleton is of fundamental importance in a wide range of cellular processes and that an appreciation of its features and functions impacts on a wide range of cell physiology, spanning research from fundamental cell biology to cancer pathology. We believe that knowledge of the cytoskeleton will be of benefit to a large number of physiologists, biochemists, and medical students as well as cell and molecular biologists.

Until recently, the emphasis in cytoskeleton research has been on identification of protein components and structural studies of the proteins and to stress functional aspects of the cytoskeleton and how this may be related to cell and tissue physiology and to disease.

This chapter presents the preface to cytoskeletal advances in structure, functional dynamics, and disease. It discusses protein structure-composition of the major filament systems (microfilaments, microtubules, and intermediate filaments) and major cytoplasmic components (actin and tubulin, the monomeric constituents of microfilaments and microtubules, are major cell proteins), which play important roles in cell function, and investigations into the functional role of the cytoskeleton currently represent a major area of cell biological research.

We would like to thank the authors for writing these interesting articles.

Jose C. Jimenez-Lopez, Ph.D.
Spanish National Research Council (CSIC)
Estacion Experimental del Zaidin (EEZ)
Department of Biochemistry, Cell and Molecular Biology of Plants,
Spain

Section 1

Cytoskeleton: Structure and Dynamics

Chapter 1

Reorganization of Vegetal Cortex Microtubules and Its Role in Axis Induction in the Early Vertebrate Embryo

Elaine Welch and Francisco Pelegri

Additional information is available at the end of the chapter

http://dx.doi.org/10.5772/66950

Abstract

In vertebrate species, induction of the embryonic axis is initiated by the transport of maternally supplied determinants, initially localized to the vegetal pole of the egg, toward the prospective organizer in the animal region. This transport process remains incompletely understood. Here, we review studies involving embryonic manipulations, visualization, and functional analysis of the cytoskeleton and loss- and gain-of-function conditions, which provide insights in this process. Transport of dorsal determinants requires cytoskeletal reorganization of a vegetal array of microtubules, microtubule motors, and an off-center movement of the vegetal cortex with respect to the inner egg core, a so-called cortical rotation. Additional mechanisms may be used in specific systems, such as a more general animally directed movement found in the teleost embryo. Initial polarity of the microtubule movement depends on early asymmetries, which are amplified by the movement of the outermost cortex. An interplay between microtubule organization and axis specification has also been reported in other animal species. Altogether, these studies show the importance of cytoskeletal dynamic changes, such as bundling, force-inducing motor activity, and regulated cytoskeletal growth, for the intracellular transport of maternally inherited factors to their site of action in the zygote.

Keywords: microtubules, dorsoventral axis, cortical rotation, zebrafish, *Xenopus*, embryo

1. Introduction

One of the main events that take place during vertebrate development is the establishment of the dorsoventral (DV) axis. This process has been studied in a variety of vertebrate species, in particular in the amphibian *Xenopus laevis* and the teleost fish *Danio rerio*. In these model systems, embryological manipulations show that the ligation of the vegetal pole of the freshly laid egg results in embryos that lack a primary (dorsal) axis (reviewed in Ref. [1]). However, the ligation of the same vegetal region after the second cell cycle does not have this effect. These manipulations allowed to infer the presence of dorsal determinants initially localized to the vegetal pole of the egg, which following fertilization are transported to a more animal region to specify prospective dorsal cells. These determinants, through mechanisms that have not been fully determined, result in the activation of the canonical Wnt/β-catenin signaling pathway, leading to dorsal gene expression and the induction of the dorsal organizer [2–4]. In *Xenopus*, the inferred transport of these determinants is coincident with the shift of the outer cortex, the "cortical rotation," relative to the entire cytoplasm, a shift that is readily apparent due to pigmentation patterns of granules in the cortex.

It has been shown that the process of transport of dorsal determinants is dependent on the microtubule cytoskeleton in the egg cortex, specifically on the reorganization of vegetal microtubules as long tracks of parallel bundles (**Figure 1**, left and center). In *Xenopus*, this array of aligned microtubule bundles extends the relatively long span from the vegetal pole to the prospective dorsal region near the animal pole, and visualization of particles, vesicles, and fluorescently labeled factors suggests that these tracks of microtubules may be acting as a

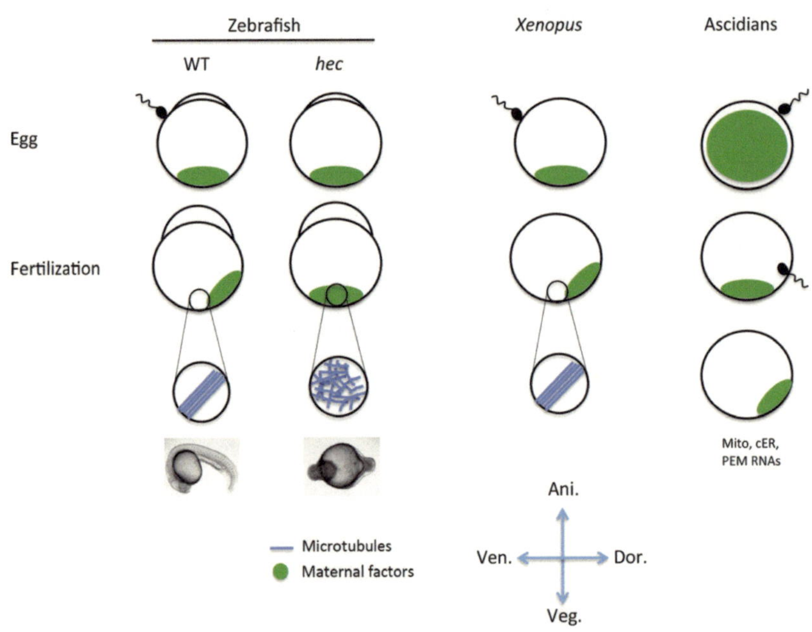

Figure 1. Schematic of early developmental processes in fish, amphibians, and ascidians. Prior to fertilization in zebrafish and *Xenopus* wild-type embryos, maternal factors are localized at the vegetal pole. Upon fertilization, they are transported to the dorsal region via a parallel array of vegetal microtubules. In zebrafish *hecate/grip2a* (*hec*) mutants, this vegetal microtubule array is compromised, preventing an initial early off-center dorsal shift of maternal factors, subsequently leading to a ventralized embryo (bottom row, second from left, compared to wild-type at left). A second phase of animally directed transport in zebrafish (not shown) appears to depend on a more general mechanism, independent of vegetal microtubule alignment [26]. In ascidian embryos, first the egg cortex and plasma membrane contract, resulting in the segregation of microfilaments, mitochondria, cER, postplasmic/PEM RNAs, and muscle-forming and endoderm-forming determinants toward the vegetal pole region. These components subsequently move toward the posterior pole through the attraction of a microtubule aster-based center.

substrate for long-range transport [5, 6]. Early zebrafish embryos do not exhibit an outwardly apparent cortical shift [7], and aligned vegetal microtubule tracks appear to span a more restricted area [6, 8], yet vegetal cortex microtubules may have similar transport functions as in *Xenopus*. Analysis of dynamic changes during microtubule reorganization in the context of the embryo has led to a model in which cortical rotation and microtubule-dependent transport are interdependent processes that together mediate the transport of dorsal determinants (see below) [5]. Forward and reverse genetic approaches in various systems, primarily zebrafish and *Xenopus*, have contributed to our understanding of these processes.

This chapter reviews events involved in the cytoskeletal reorganization required for the movement of determinants leading to axis induction. The outcome of microtubule reorganization in the early embryo is the induction of the dorsal axis, and we first briefly review this process in the zebrafish as well as the amphibian *X. laevis*.

2. Induction signals for axis specification

A primary event in the establishment of the dorsoventral axis in zebrafish and *Xenopus* is the translocation of the normally cytoplasmic protein β-catenin into the nuclei of dorsal blastomeres during cleavage stages (**Figure 2**) [2, 9, 10]. In the absence of Wnt signaling, levels of the cytoplasmic pool of β-catenin are reduced by the activity of glycogen synthase kinase-3 (GSK-3), which promotes β-catenin degradation (reviewed in Ref. [11]). Accordingly, enrichment of β-catenin in dorsal cell nuclei, as well as dorsal axis induction, is blocked by ectopic expression of GSK-3. Activation of the canonical Wnt signaling pathway, resulting in localized inhibition of GSK-3 on the future dorsal side of the embryo, is thought to promote the accumulation of cytoplasmic β-catenin in the prospective dorsal region. Accumulated β-catenin in turn translocates to the nucleus [3, 12], where it can influence gene expression (reviewed in Refs. [13, 14]).

A key mediator of Wnt signaling, when localized to the nuclei, β-catenin acts as a transcriptional effector to activate dorsal-specific genes such as *bozozok/dharma, nodal-related 1,* and *chordin* [13]. Products expressed from these dorsal genes along with those from their targets antagonize ventralizing signals such as bone morphogenetic proteins (BMPs), thus promoting dorsal cell fate specification (reviewed in Refs. [14, 15]). Failure of nuclear β-catenin to localize to the nuclei of dorsal blastomeres results in the ventralization of embryos [16].

The intricacies of the Wnt signaling pathway and its role in vertebrate axis induction have been determined through many studies, including functional manipulation of genes through ectopic expression and knockdown or expression of dominant-negative constructs (reviewed in Refs. [11, 17]). For example, overexpression of β-catenin induces a secondary axis in *Xenopus* embryos [18]. When β-catenin is overexpressed in zebrafish embryos, it is able to induce the expression of target genes such as *goosecoid* and *ntl* [19]. Similar results have been observed with the overexpression of some Wnt ligands [20], including overexpression of Wnt8 and Wnt8b in zebrafish embryos [19].

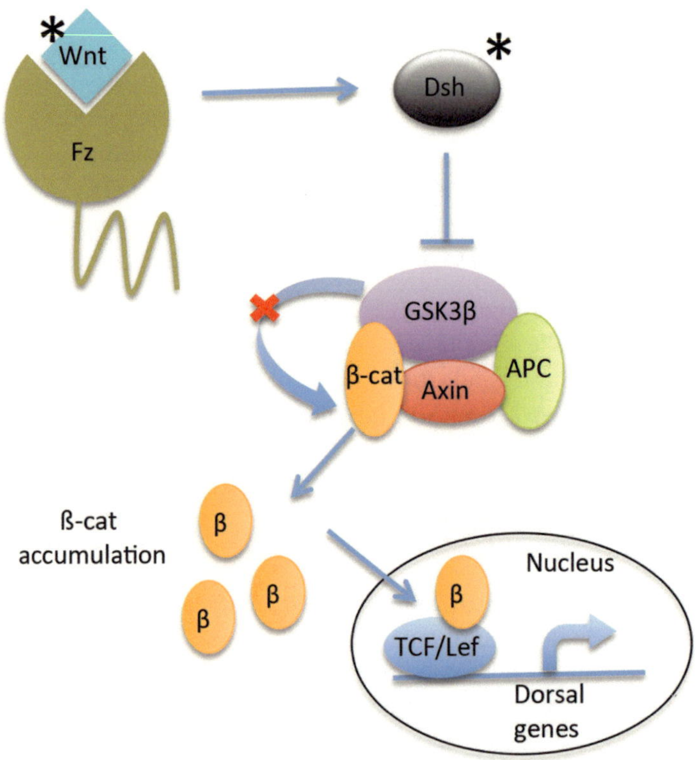

Figure 2. Simplified diagram of Wnt activity involved in axis specification. In the canonical Wnt signaling pathway, extracellular Wnt protein ligands signal through Frizzled transmembrane receptors to activate the cytoplasmic protein dishevelled (Dsh). Dsh in turn inhibits GSK-3 activity. GSK-3 is part of a complex that normally destabilizes β-catenin protein; hence, inhibition of GSK-3 activity results in β-catenin stabilization and its translocation into the nucleus. β-Catenin forms a complex with the transcription factor Tcf/Lef to activate dorsal-specific gene expression. Asterisk denotes factors thought to undergo a translocation to the prospective dorsal site through a process involving cortical rotation and/or microtubule-dependent transport.

Several mutations in zebrafish have allowed for the confirmation of an endogenous role for Wnt signaling pathway in axis induction in the early embryo. An identified recessive maternal-effect mutant in *ichabod* results in defective dorsal organizer formation and severe ventralization and shows impaired nuclear localization of maternal β-catenin protein [21]. This mutation was found to be closely linked to β-*catenin-2* (ctnnb2), a duplicate copy of the β-catenin gene located on a different linkage group from the previously characterized β-*catenin-1* [22]. It was shown that although the *ichabod* mutation does not functionally alter the β-*catenin-2* open reading frame, the level of maternal β-*catenin-2* transcript (but not that of the unlinked β-*catenin-1* gene) is substantially lower in *ichabod* mutant embryos. Reduction of β-*catenin-2* function in wild-type embryos by the injection of a gene-specific morpholino antisense oligonucleotide (MO) results in ventralized phenotypes [22], which are similar to those seen in *ichabod* mutant embryos. In contrast, MOs directed against β-*catenin-1* have no ventralizing effect on wild-type embryos. These data strongly suggest that the *ichabod* mutation corresponds to the β-*catenin-2* gene, providing genetic evidence for the role of this factor in axis induction. These results indicate that activation of Wnt signaling via the stabilization of β-catenin is essential for proper organization of the embryonic axis.

Activation of the Wnt/β-catenin signaling pathway, as well as its important role in the expression of dorsal genes, has been extensively studied in a number of cellular systems (reviewed in Refs. [11, 17]). However, the identity of the molecules thought to activate the pathway in early vertebrate embryos, referred to as dorsal determinants, remains to be fully elucidated. Wnt11 has been proposed to be a dorsal determinant in amphibian species [23]. In *Xenopus*, *wnt11* mRNA is located at the vegetal pole of the mature egg and, after fertilization, becomes enriched at the future dorsal side of the embryo. Thus, the localization of *wnt11* RNA exhibits the expected behavior of the inferred dorsal determinant, as predicted from transplantation of the dorsal-inducing activity. Additionally, it was shown that depletion of *wnt11* mRNA from oocytes results in embryos defective in dorsal axis induction [24]. Further studies have implicated ubiquitously present Wnt5 as acting together with Wnt11 in Wnt/β-catenin activation [23]. Studies in zebrafish have not implicated Wnt11 or Wnt5 function in axis induction. However, a role for zebrafish Wnt8a has been suggested in this process [25, 26]. Similarly to *Xenopus wnt11*, zebrafish *wnt8a* mRNA is localized to the vegetal pole of the egg and can be observed to translocate after fertilization toward the animally located blastomeres. These studies also indicate that, while Wnt/β-catenin pathway activation may be highly conserved in axis induction across the animal kingdom, there are variations in maternally based mechanisms leading to pathway activation.

Studies in *Xenopus* and zebrafish also showed that the transport of dorsal determinants, which results in the translocation of β-catenin to the nuclei of dorsal blastomeres, requires an array of parallel microtubules originating in the vegetal pole region [6, 27]. Miller and colleagues investigated the mechanisms responsible for the dorsal activation of the Wnt signaling pathway in *Xenopus* eggs and the subsequent specification of dorsal cell fates in the embryo. It was shown that dishevelled (Dsh) protein, a cytoplasmic component of the Wnt pathway that functions upstream of β-catenin [28], is associated with vesicle-like organelles that become enriched in the prospective dorsal side of the egg at the end of the first cell cycle and that the accumulation of Dsh persists through early cleavage stages [27]. Further experiments revealed that when embryos were UV irradiated at the vegetal hemisphere, the distribution of Dsh was blocked, which also blocked dorsal axis formation. Subsequently, when observing the subcellular localization of Dsh fused to GFP, it was revealed that during cortical rotation Dsh-GFP is translocated toward the future dorsal side via the vegetal cortex microtubule array [27]. Together, these data suggest a model in which dorsal-determining factors including *wnt* gene products and Dsh protein are transported via a microtubule-dependent pathway to the future dorsal side of the embryo, leading to the localized activation of the Wnt signaling pathway, the accumulation of β-catenin in dorsal blastomeres, and the induction of dorsal cell fates [27].

3. Transport of dorsal determinants in *Xenopus* and zebrafish

3.1. Molecular mechanism underlying cortical rotation

As mentioned above, embryological manipulations showed that, both in *Xenopus* and zebrafish, dorsal determinants are localized to the vegetal pole of the egg at the time of fertilization but have within several cell cycles moved to an animal region where they influence cell fate. Spatial changes that lead to these determinants translocating to the prospective dorsal region appear to be facilitated by two processes: the rotation of the zygote cortex with respect to the core during cortical rotation and the intracellular movement of factors (e.g., *wnt* RNA or Dsh-bound vesicles) along aligned vegetal microtubules. These are likely intertwined processes, as tracks of parallel microtubules appear to be required not only for the movement of vegetal factors to the prospective dorsal side but also for cortical rotation [29]. Treatment of the vegetal portion of embryos to prevent microtubule polymerization, such as exposure to nocodazole, cold shock, hydrostatic pressure, or UV irradiation [30, 31], shows that microtubules are required for cortical rotation in normal conditions [31]. In contrast, cytochalasin D, an inhibitor of actin polymerization, does not interfere with cortical rotation, indicating that microfilaments are not required for this process. Inhibition of protein synthesis with cycloheximide, known to have dramatic effects such as cell cycle arrest [32], also does not inhibit rotation, indicating that the control of cortical rotation is posttranslational and depends on preformed maternal proteins [32].

Failure of cortical rotation results in a ventralized mutant phenotype in the embryo. However, in embryos treated to inhibit microtubules, a cortical rotation can be artificially induced by gravity after immobilizing the embryo in a matrix and physically turning it 90°. This manipulation results in the formation of dorsal structures, albeit delayed [33]. Under these conditions, gravity leads to a rearrangement of the heavier yolk-containing core of the embryo relative to the cortex. This is thought to increase the proximity of vegetally localized cortical signals to internal regions in the more animally located prospective dorsal region. The ability of the entire cortex to move as a whole relative to the embryonic core contrasts with the visualization of moving particles along microtubule tracks. These observations suggest that both transport along cortical microtubules and a cortical shift relative to the embryonic core contribute to the redistribution of signals involved in axis induction during the early embryonic cell cycles. We subsequently address each of these processes.

3.2. Relocalization of RNA determinants during oogenesis and early embryogenesis

The mRNA for the putative zebrafish dorsal determinant *wnt8a* is localized to the Balbiani body during oogenesis. The zebrafish Balbiani body [34] is a mitochondria-rich subcellular structure in the forming oocyte shown to be essential for the creation of animal-vegetal polarity in the oocyte. This structure, thought to be homologous to the early messenger transport organizer (METRO) pathway of localization in *Xenopus* [35], constitutes a crucial component of a vegetally directed transport pathway that entraps mRNAs and other gene products necessary for patterning of the embryo and germ cell formation [34, 35]. Association of *wnt8a*

RNA with the Balbiani body leads to the localization of this RNA to the vegetal pole of the mature zebrafish oocyte [25]. Thus, fertilized embryos initiate development with *wnt8a* RNA localized to the vegetal pole. However, in wild-type embryos starting at 30 min, this mRNA experiences an asymmetric movement toward a more animal region that will become the prospective organizer region [25, 26].

In addition to *wnt8a*, genetic studies in zebrafish have allowed the identification of other maternally inherited factors involved in the transport of determinants essential for dorsal axis induction, such as *hecate/grip2a* mRNA and Tokkaebi/Syntabulin proteins. Molecular characterization of the three independent mutant alleles of the zebrafish maternal effect gene *hecate/grip2a* shows that loss of function for its product results in embryos with reduced dorsal gene expression and concomitant defects in forming dorso-anterior structures [26]. Similar effects are caused by a single mutation in *tokkaebi* [36]. Mutations in genes coding for either *hecate/grip2a* or *tokkaebi/syntabulin* do not interfere with vegetal pole localization of *wnt8a* RNA during oogenesis, but abolish the animally directed asymmetric movement of this RNA that normally occurs after fertilization [25, 36, 37]. Given the proposed role for Wnt8a as the dorsal determinant in zebrafish [25], the postfertilization defect in *wnt8a* RNA asymmetric movement in *hecate* and *tokkaebi* mutants explains axis induction defects observed in these mutants.

Positional cloning of *hecate* shows that this gene encodes glutamate receptor-interacting protein (Grip) 2a, a factor whose *Drosophila* homologue protein is associated with membrane vesicles in postsynaptic neuronal cells, where it acts in the reception of Wnt signals across the synapse [38]. Zebrafish Grip2a protein has four PDZ domains, which are known to interact with membrane-associated factors including members of the Wnt signaling pathway. Mutant alleles in this protein exhibit a range of phenotypes whose severity roughly correlates with the extent of unaffected protein, with the strongest allele causing a premature stop codon that truncates the Grip2a protein, removing all four PDZ domains [26]. The mutation in *tokkaebi* corresponds to syntabulin, which codes for a linker of the kinesin I motor protein [36], and acts as a linker molecule that attaches mitochondria to the kinesin-1 motor, thereby contributing to anterograde trafficking of mitochondria to neuronal processes [39]. The known roles for Grip and syntabulin in the transport of membranous organelles and signaling in neuronal types begin to draw similarities between microtubule-based transport of vesicles in neurons and the transport of dorsal determinants, also thought to at least partially associate with vesicles (as highlighted by Dsh-GFP movement [27]), in early vertebrate embryos.

Consistent with the effect of maternal-effect mutations in *hecate/grip2a* and *tokkaebi/syntabulin* on the formation of dorsal structures, products of these genes are localized in patterns that likely facilitate the movement of dorsal determinants [26, 36, 40]. In wild-type embryos, *grip2a* mRNA, like *wnt8a*, is localized via a Balbiani body-dependent mechanism to the vegetal pole of the oocyte and early embryo, and following egg activation and fertilization, the localization of this mRNA shifts off-center about 30° from the vegetal pole. During oogenesis, as in the case of *grip2a* RNA, *syntabulin* RNA becomes localized to the vegetal pole of the oocyte via a Balbiani body-dependent pathway, resulting in the localization of both syntabulin mRNA and protein to the vegetal pole of the egg. After fertilization, as in the case of *wnt8a* RNA and *grip2a* RNA, Syntabulin protein (but not its RNA) exhibits an off-center shift upon egg activation [36]. The off-center shift from the vegetal pole exhibited by *wnt8a* and *grip2a* mRNAs and Syntabulin protein roughly corresponds to a 30° arc offset from the vegetal pole that contains an aligned set of arrayed microtubules in the zebrafish embryo and which has been observed to contain moving subcellular particles [8]. Thus, the movement of these mRNAs roughly corresponds to a region in the teleost early embryo thought to undergo mass movements toward the future dorsal side, reminiscent of the amphibian cortical rotation. The coordinated asymmetric movement of vegetally localized products such as *wnt8a* RNA, *grip2a* RNA, and Syntabulin protein is consistent with the observed mass transport of particles in the vegetal cortex [8], although they may also reflect specialized transport mechanisms involving microtubule tracks, Syntabulin-mediated motor movement, and *wnt8* RNA- and *grip2a* RNA-containing RNPs.

Genetic analysis indicates that the Hecate/Grip2a and Tokkaebi/Syntabulin products are required for the off-center, asymmetric shift of vegetally localized determinants that follows fertilization. *hecate/grip2a* mutants show defects in this off-center movement for vegetally localized products such as *wnt8a* RNA and Syntabulin protein, as well as *grip2a* RNA itself [26]. Mutations in *tokkaebi/syntabulin* also result in defects in *wnt8a* RNA and Syntabulin protein asymmetric movement [36]. However, in both of these mutants, localization of dorsal factors (*wnt8a* RNA, *hecate/grip2a* RNA, and *tokkaebi* products) during oogenesis remains unaffected. Localization of these factors during oogenesis is instead dependent on the function of buckyball [25, 26, 36], a novel protein required for Balbiani body formation [34, 41]. Thus, localization of dorsal factors to the vegetal pole of the oocyte relies on a Balbiani body-dependent pathway, and the asymmetric movement of these factors after fertilization, which is required for axis induction, depends on the subsequent action of *hecate* and *tokkaebi*. As discussed below, these functions rely on microtubule-dependent reorganization and transport processes.

Additional studies have shown that, as in *Xenopus*, zebrafish vegetal cortex microtubules become reorganized into parallel bundles (**Figure 3**) [6, 8]. The studies paint a picture of translocation of dorsal axis determinants that is remarkably similar to that of the known *Xenopus* cortical rotation. However, transport of dorsal determinants in zebrafish appears to use a dual system, in which microtubule alignment initiates an off-center shift, and other cytoskeletal processes mediate long-range transport (see below). In spite of observed differences, these studies show that microtubule-dependent transport of dorsal determinants plays an essential role in canonical Wnt pathway activation and dorsal axis determination in teleost embryos, as in amphibians.

Interestingly, the RNA for the *grip2* homologue in *Xenopus, XGRIP2* is, like its zebrafish homologue *grip2a*, localized to the mitochondrial cloud (the Balbiani body in zebrafish) during *Xenopus* oocyte development and subsequently to the vegetal pole of the mature oocyte. However, in contrast to zebrafish *grip2a* RNA, *Xenopus XGRIP2* RNA does not have an apparent role in axis induction, and after fertilization its RNA becomes localized to germplasm masses that coalesce in the embryo (see below) [42–44].

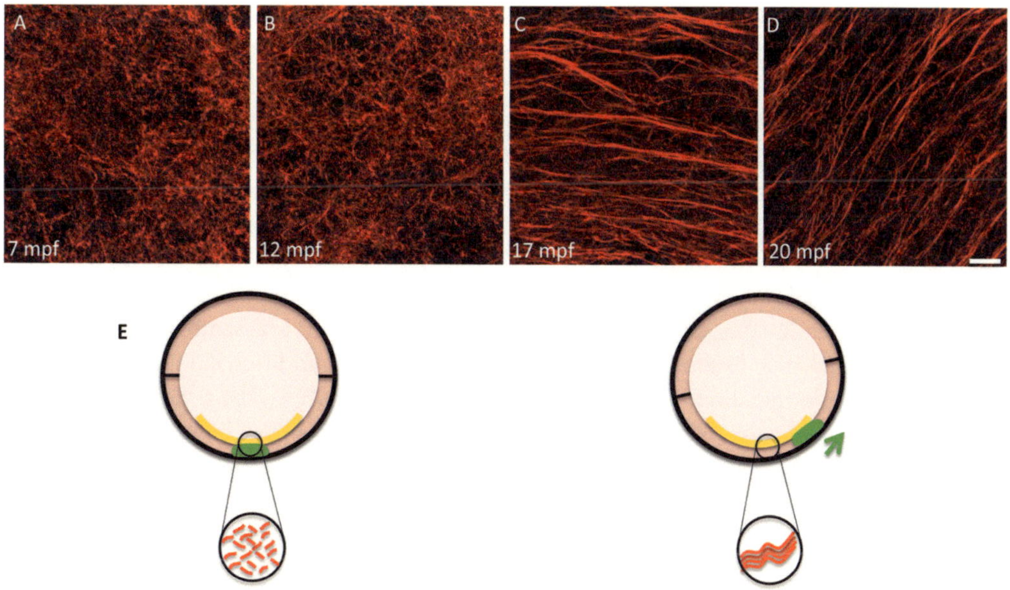

Figure 3. Alignment of microtubules at the vegetal cortex in wild-type zebrafish embryos. (A–B) Between 7 and 12 min postfertilization (mpf), microtubules at the vegetal cortex start to become reorganized to form parallel bundles. (C–D) Around 17 mpf, microtubules become organized into parallel bundles. This organization facilitates the movement of dorsal determinants from the base of the vegetal pole to the dorsal region. Scale bar in D represents two microns for all panels. (E) Diagrammatic representation of dorsal determinants (green) with respect to more internally located determinants, such as vegetally localized germplasm determinants in the zebrafish (orange), depicting microtubule reorganization (red lines), before (left) and after (right) cortical rotation.

Altogether, these studies indicate key roles for RNA localization pathways during oogenesis leading to the localization of factors required for axis induction to the vegetal pole of the egg. Initially localized to the vegetal pole through the action of the mitochondrial cloud during oogenesis, after fertilization and egg activation these factors exhibit an off-center shift dependent on the function of vegetally localized factors, such as Grip2a and the kinesin motor adaptor protein Syntabulin.

3.3. Reorganization of microtubules during cortical rotation

At least in the case of *Xenopus*, it is clear that the rotation of the cortex facilitates the relocation of dorsalizing factors from the vegetal pole to the presumptive future dorsal side or to a more animal (equatorial in the case of *Xenopus*) region, where they act to initiate gene expression programs corresponding to the body axis, at a signaling center known as Spemann's organizer [45]. Following fertilization in the *Xenopus* egg, the cortex rotates an average of 30° within the first cell cycle, relative to the inner cytoplasm [29, 46], a rotation mediated by an array of aligned microtubules beneath the vegetal cortex [47]. At the same time, these microtubules become aligned in a parallel arrangement with plus ends directed toward the direction of cortical translocation [48], a reorganization that coincides with the initiation of cortical rotation [30].

In the early *Xenopus* embryo, microtubule nucleation occurs deep within the animal hemisphere [49] by the sperm-derived centriole near the site of sperm entry. These microtubules extend through the cytoplasm toward the vegetal pole, where they contribute to the formation of the vegetal microtubule array (**Figure 4**, top left) [50]. Thus, in *Xenopus*, rotation (and dorsal site formation) typically occurs away from the sperm site of entry. On the other hand, the orientation of the vegetal microtubule array can occur in potentially any direction with respect to the cleavage site [5, 29], so that it is unlikely that there is an intrinsic preexisting dorsal asymmetry in the egg with respect to the site of cellular cleavage, itself determined by the orientation of the spindle [51, 52]. Studies have also shown that cortical rotation can occur toward the sperm entry point in specific cases, such as when the sites of meiotic spindle assembly and polar body extrusion are oppositely located [53], suggesting the existence of additional unknown variables that influence the orientation of the vegetal microtubule array. In *Xenopus*, cortical rotation is halted right before the first cellular cleavage occurs [47], when the microtubules of the vegetal microtubule array are depolymerized under the influence of M-phase-promoting factor [54].

A morphologically apparent cortical rotation, observed through changes in the position of an outer cortex relative to an inner core as observed in *Xenopus*, is not readily apparent in the zebrafish embryo. However, studies have indicated the existence of processes in the zebrafish embryo that share similarities with the amphibian cortical rotation. Early studies showed that fluorescent polystyrene beads injected at the vegetal pole were transported

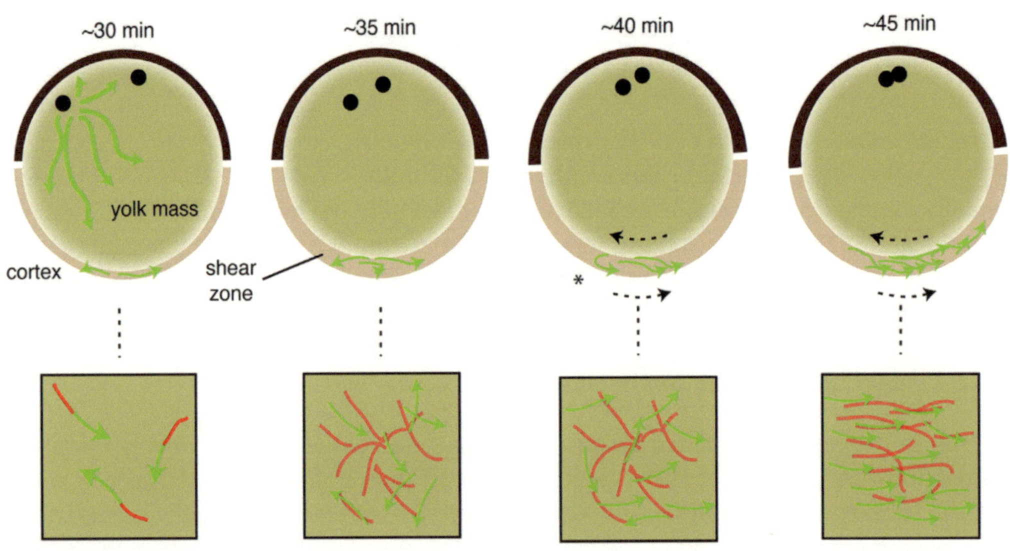

Figure 4. Microtubule dynamics during cortical rotation. In the *Xenopus* embryo, microtubule polymerization is initiated approximately 30 min after fertilization at the vegetal cortex, when astral microtubules derived from sperm components at the animal pole reach the vegetal cortex. Microtubule polymerization also occurs at the vegetal pole (growing microtubule (plus) ends indicated by green arrows and preformed microtubules by red lines). Relative movement between the yolk cell and the cortex (dashed arrows), initiated by the asymmetry conferred by the sperm-derived asters, facilitates the alignment of both growing and preformed microtubules in the direction of movement. Microtubule alignment in turn contributes to cortical movement. Microtubules oriented toward the dorsal side, the direction in which the cortex rotates (Reprinted from Ref. [58], with permission).

http://dx.doi.org/10.5772/66950

animally along microtubule-based cortical tracks in a microtubule-dependent manner [6] and that this movement had temporal dynamics and functional requirements similar to that of the movement of putative dorsal determinants as defined by embryological manipulations. However, this movement was shown to occur by visualizing injected fluorescent beads, as opposed to an entire cortex, consistent with translocation along microtubule arrays but not necessarily a shift of the outer cortex analogous to a cortical rotation. A cortical rotation process in the zebrafish was later suggested by the observation of coordinated movement of optically visible particles in the vegetal cortex, and that long-term tracking of these particles occurs toward the presumed dorsal side, as expected from a cortical rotation [8].

A cortical rotation-like process is also consistent with differences in the changes in RNP particle distribution at different cortical depths, as visualized by fluorescent in situ hybridization, since RNPs located at the outermost cortex undergo a spatial shift with respect to more internally located RNPs (**Figure 3E**) [55]. To understand the basis of transport for differentially localizing factors at the zebrafish vegetal-most embryonic cortex, double fluorescence in situ hybridization (FISH) was used to detect pairs of RNAs for factors involved in axis induction (*wnt8a* and *grip2a*) and RNAs for vegetally localized germ cell specification factors (*dazl*). Localization of these three factors occurs in different RNPs at the vegetal cortex. Moreover, RNAs for dorsal factors, *wnt8a* and *grip2a*, are enriched in the outermost layer of the cortex, whereas RNPs for the primordial germ cell determining factor *dazl* are present in more internal regions [55]. Although domains containing RNPs for these two sets of vegetally localized factors are both centered at the vegetal pole in the egg, upon fertilization the domain containing the outer cortex RNPs, coding for dorsal induction factors, shifts relative to the more internal domain containing germplasm determinant RNPs. RNPs in the outer cortex have a function in axis determination and need to experience a relative shift to generate an asymmetry in the embryo, facilitated by the cortical rotation-like movement. These observations further add to the finding of bulk particle movement at the zebrafish embryo vegetal cortex [8] and are consistent with a cortical rotation-like process in the early zebrafish embryo. As in amphibians, this teleost cortical rotation-like process may be involved in generating an asymmetry in the location and function of dorsal determinants.

Thus, both in amphibians and teleost, an array of aligned microtubules is associated with the movement of RNA molecules and the vegetal cortex itself with respect to the inner egg core, which altogether mediates the transport of dorsal determinants toward the prospective dorsal site.

3.4. Long-range vs. short-range transport

In both *Xenopus* and zebrafish, the process of cortical rotation appears to be an important part of the mechanism that directs dorsal determinants to their final destination at the animal pole. However, zebrafish and *Xenopus* embryos display some differences in mechanism of animally directed transport. In the *Xenopus* embryo, aligned tracks of microtubules appear to span most if not all of the space between the vegetal pole and the prospective dorsal region. In zebrafish, in contrast, transport with an end point in blastomeres at the animal pole of the embryo appears to depend on two sequential steps: an initial short-range transport of vegetal localizing factors generating a slight off-center shift toward the animal pole, followed by animally-directed transport via a more general mechanism. The first, off-center asymmetry, is revealed by changes in the distribution of RNAs such as *wnt8a* and *grip2a* in a process that appears to correspond to a cortical rotation-like event. As in *Xenopus*, the initial cortical rotation-like event in zebrafish depends on the alignment of microtubules in parallel bundles at the vegetal cortex. The microtubule reorganization into parallel bundles in turn is dependent on the function of Grip2a (**Figure 1**, left). Short-range shift in vegetal signals is affected in homozygous *hecate/grip2a* mutant embryos, evidenced by defective off-center shift of RNAs such as *wnt8a* and other factors [26].

The second step involves a long-range transport along the mediolateral region of the embryo to the base of the blastomeres by a mechanism that is neither restricted to the dorsal side nor dependent on Grip2a function [6, 26]. The presence of such a second transport mechanism can be inferred by the observation that *hecate/grip2a* mutants do not exhibit a defect in the long-range animally directed translocation of vegetally injected beads, indicating that animally directed movement occurring in mediolateral regions is independent of *hecate* function. Indeed, injection of beads in opposite sides of the embryo indicates that animally directed travel along the mediolateral region of the yolk cortex occurs in both injected sides, implying that, as opposed to *Xenopus*, the entirety of the zebrafish mediolateral cortex, and not only the prospective dorsal region, is competent for long-range movement [26]. It is possible that the second step in zebrafish depends on a more general transport mechanism associated with animally directed transport in meroblastic embryos, through which other factors with a function unrelated to dorsal axis induction, such as vegetally localized germplasm RNAs [56], need to travel animally toward the forming blastodisc. Thus, both *Xenopus* and zebrafish experience animally directed movement of dorsal determinants facilitated by a microtubule-dependent cortical rotation-like process. However, the *Xenopus* embryo uses a mechanism in which cortical rotation and microtubule alignment into parallel tracks together implement long-range movement of dorsal determinants through an apparently seamless mechanism. In teleost embryos, on the other hand, embryonic-scale differences along the dorsoventral axis are generated by the sequential action of a short-range off-center movement mediated by less expansive vegetal microtubule array, which is subsequently amplified by a more general animally directed system.

3.5. Other factors involved in vegetal microtubule reorganization

Additional factors have been identified to be important for dorsal axis induction. A mutation in the maternal-effect mutant *brom bones*, which has a nonsense mutation in the gene *hnRNP I*, shows egg activation defects, disorganized vegetal microtubule array formation, and subsequently defects in axis formation [16]. Additionally, these mutant embryos display egg activation defects as evidenced by failure of cortical granule exocytosis and chorion expansion. In zebrafish, cortical granule exocytosis is one of the first cellular responses to egg activation and is initiated by a wave of elevated cytoplasmic calcium that is impaired in *brom bones* mutants [16]. It is possible that the defect in vegetal microtubule alignment in *brom bones* is similarly based on the calcium release defects after egg activation, which is required for vegetal microtubule array formation [8].

http://dx.doi.org/10.5772/66950

Studies have also revealed that an ubiquitin ligase, *tripartite motif-containing 36* (*trim36*), is required for vegetal microtubule reorganization mediating axis induction. *Xenopus trim36* is maternally expressed, with mRNA enrichment at the Balbiani body in stage 1 oocytes and localization to the vegetal cortex of stage VI oocytes. *trim36* mRNA is also detectable in the germplasm of fertilized eggs and cleavage-stage embryos [57]. Embryos depleted of *trim36* function by injection of antisense oligos into oocytes exhibit defects in vegetal microtubule reorganization and cortical rotation, leading to reduced organizer formation and severe embryo ventralization at later stages [57]. As expected, injection of *wnt11* mRNA rescues this effect, confirming that Trim36 functions upstream of Wnt/β-catenin activation. Recent studies have shown that Trim36 attenuates the growth of plus ends of vegetal microtubules during array formation (see below) [58], indicating a role for this factor, possibly through the mediation of protein degradation, in the regulation of microtubule dynamics essential for array formation.

The mRNA for *dead end*, which codes for an RNA-binding protein initially shown to be essential for the development of the germ line [59–68], has been shown to have a role in vegetal microtubule array formation. In *Xenopus*, the mRNA for *dead end1*, like that of *trim36*, is localized to the vegetal pole of the oocyte [61]. Early embryos depleted of dead end exhibit an unexpected defect in the formation of arrays of parallel vegetal microtubules and consequently axis specification [69]. This requirement appears to depend on the function of Dead end protein to directly bind *trim36* mRNA and anchor it to the oocyte vegetal pole, likely increasing Trim36 protein local concentration in this region [46].

Thus, a variety of factors are required for the reorganization of vegetal cortex microtubules leading to dorsal determinant transport. In some cases, these factors are important for general processes essential for the microtubule reorganization, such as in the case of *hnRNP I* and dependent calcium signaling. In other cases, these factors begin to delineate a pathway for microtubule reorganization, as in the case of *dead end* and *trim36*, involved in the regulation of vegetal microtubule growth.

3.6. Mechanism of microtubule alignment during cortical rotation

Even though it has been shown that microtubule-dependent cortical rotation is important for axis formation, the molecular mechanisms underlying the organization and orientation of cortical microtubule have not been fully elucidated. The process of cortical rotation is highly conserved, and it likely requires the embryo to use a significant amount of energy. Weaver and Kimelman [70] asked the question that if dorsal determinants can travel along microtubules, then what is the purpose of the cortical rotation? As described above, cortical rotation might directly contribute to the overall animally directed movement of the dorsalizing activity. However, studies have also suggested that cortical rotation might serve to facilitate aligning the polymerizing microtubules into parallel bundles and orienting their plus ends toward the dorsal side. One favored model for the orientation of the microtubule array is a positive feedback mechanism where initial random asymmetry in microtubule growth is amplified by continuous movement of the cortex [31, 58].

Microtubules that form the vegetal microtubule array appear to arise from several sources [70]. Some are nucleated by the centriole of the sperm, which acts as a minus-end microtubule-organizing center, others extend toward the periphery from unknown sources deep in the cytoplasm and bend into the vegetal shear zone, and, finally, some arrays appear to polymerize spontaneously in the vegetal shear zone [49, 50]. As the vegetal microtubule array begins to form, it becomes progressively stabilized by movement of the cortex during cortical rotation, which provides an amplifying loop for microtubule alignment [58]. The precise manner by which this cortical movement contributes to microtubule alignment and stabilization is not fully understood. Suggested mechanisms, described below, include a combing process mediated by cortically anchored kinesin-related proteins [54, 70] or the stabilization of microtubules by membrane compartments such as the endoplasmic reticulum and vesicles [58].

Vegetal microtubules originally appear with their plus ends in a random orientation yet subsequently become aligned in parallel arrays with plus ends directed toward the dorsal side (**Figure 4**) (reviewed in Ref. [70]; see also Ref. [58]). Marrari and colleagues suggested how microtubules could become aligned through cortical motor proteins and the process of the cortical rotation [54] (reviewed in Ref. [70]). They proposed that cortically anchored plus-end-directed motor proteins, such as kinesins, move toward microtubule plus ends, generating a cortical displacement with respect to the inner core [47, 54, 71]. The attachment of plus ends to the moving cortex mediates aligning of microtubules in the same direction. Thus, the movement and action of these kinesin-related proteins could potentially align the microtubules as well as generate the pulling force that is needed to translocate the cortex relative to the cytoplasm [54, 70]. This positive feedback loop also allows amplifying an original small asymmetry into the observed prominent array of parallel microtubule bundles.

Marrari and colleagues also investigated the role of kinesin and dynein motors in the formation of the cortical microtubule array as well as their role in the translocation of the vegetal cortex [47, 54, 71]. The function of kinesin was inhibited using an antibody against a highly conserved peptide of the kinesin motor domain, LAGSE. Anti-LAGSE antibodies block spindle elongation in semi-in vitro systems [47, 54, 71, 72] and successfully interfere with kinesin function [47, 54, 71]. The function of dynein was inhibited by microinjection of p50/dynamitin beneath the vegetal cortex [54]. In *Xenopus* egg extracts as well as cells, excess dynamitin inhibits processes dependent on dynein function by disrupting the dynactin complex [73].

Inhibition of kinesin-related function results not only in expected defects in mitosis and cell cleavage but also in disruptions in the array of vegetal microtubules and cortical rotation [71]. On the other hand, inhibition of dynein causes an inward shift in the distribution of microtubules with respect to the cortex, indicating that dynein functions to move microtubules outward, into the vegetal subcortical layer [47]. Moreover, these experiments showed that the formation of the vegetal microtubule array (and therefore cortical rotation) is sensitive to dynein inhibition prior to array formation, but that cortical rotation remains sensitive to inhibition of kinesin function throughout the normal period of rotation [47]. Together, these data suggest that kinesin and dynein motors have different functions during cortical rotation (**Figure 5**) [47]. In this model, dynein motors anchored to internal

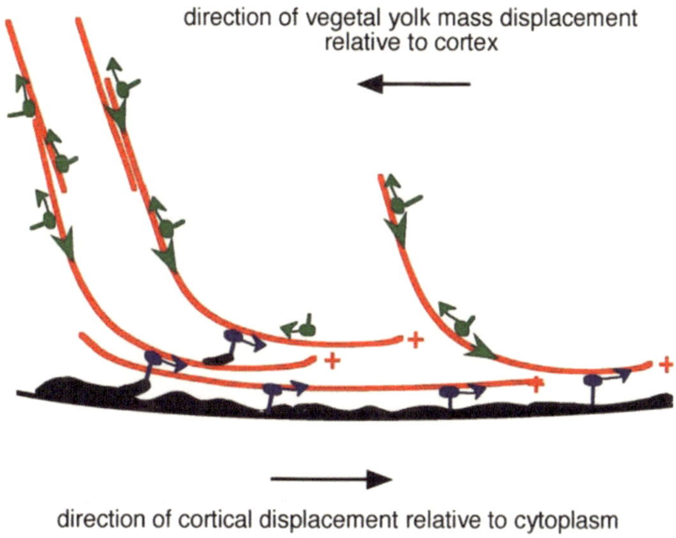

Figure 5. Proposed role of microtubule-dependent motors on the rotation of the vegetal cortex, as suggested by inhibitor studies [47, 54, 71]. A pushing force from the minus-end-directed microtubule motor dynein (green; green arrows show direction of motor movement relative to microtubules) helps translocate microtubules (depicted in red) from the inner cytoplasm outward onto the cortical surface (green arrowheads indicate direction of microtubule movement). Plus-end-directed microtubule motors such as kinesins (blue) anchor microtubules to the cortex and facilitate cortical movement relative to the yolk mass (blue arrows show direction of motor movement relative to microtubules). (Reprinted from Ref. [71], with permission. The original image has been rotated horizontally for a better comparison to others in this chapter.)

elements generate an outward force to facilitate bringing microtubules from the inner egg core region to the vegetal cortex. Kinesins, on the other hand, are thought to act by tethering microtubule plus ends to the cortex, thus generating a pulling force on microtubule arrays, mediating the rotation of the cortex itself, and favoring further parallel alignment of microtubules within the array. It is important to note that, after the vegetal microtubule array has formed, further microtubule alignment and cortical rotation can occur independent of dynein function, but motors of the kinesin-related protein family are needed for the movement of the cortex [47]. Thus, kinesin motor function appears to be essential for *Xenopus* cortical rotation, whereas the role of dynein appears to be more indirect. Altogether, these data suggest that both motor proteins interact early in the process of vegetal microtubule array, followed by a period in which kinesin-dependent translocation is sufficient to generate cortical movement.

Olson and colleagues performed experiments that would characterize microtubule plus-end dynamics in *Xenopus* oocytes and eggs, identified changes in microtubule stability and plus-end flux during the oocyte to egg transition, and characterized behaviors that are present at the onset of cortical rotation (**Figure 4**) [58]. They showed that the initial phase of microtubule assembly is between 25 and 35 min post egg activation. During this time, microtubules are short and dynamic with a low initial density that increases rapidly [58]. In the second phase of assembly, microtubules polymerize rapidly from sites within the vegetal cortex. Microtubules became thinner or less bundled, and the entire network appears to sink deep into the cytoplasm. At this time the microtubule array is referred to as exhibiting a "fine-combed" appearance, which is thought to be the result of the continual action of cortical kinesin-related proteins that straighten microtubules as the cortex moves along them [47, 58, 70]. At the same time, at approximately 36 min post activation, the cortical shift in relation to the egg core becomes apparent [58]. These studies also reveal that microtubule-directed growth, occurring after the initial cortical microtubule alignment, has an important contribution to the formation of the vegetal array of parallel microtubules, which powers cortical rotation.

It was previously noted that the direction of microtubule polymerization in cultured cells depends on the arrangement of elongated tubes of endoplasmic reticulum [74]. Endoplasmic reticulum, vesicles, and tubes possess kinesin-like microtubule-associated proteins that associate with microtubules during transport and elongation, and it is possible that similar membrane organelles are attached to the vegetal cortex and facilitate kinesin-mediated anchoring of microtubules during cortical rotation [31]. A precedent for this is the association of cortical ER with aligned microtubules in early ascidian embryos (see below) [75]. Further studies will be required to address a potential role for membrane organelle attachment in *Xenopus* vegetal microtubule array formation and cortical rotation, such as membrane organelle sliding between membrane organelles and microtubules, or associations of ER extensions with growing microtubule tips [76].

Studies in zebrafish are consistent with mechanisms for cortical microtubule array formation and alignment as detailed in amphibians, including the presence of early internal microtubules, increase in cortical microtubule polymerization concomitant with microtubule alignment and bulk movement of the cortex, and the aligned orientation of microtubule plus ends toward the prospective dorsal site [8].

Altogether, these studies suggest that the formation of the vegetal microtubule array is dependent on the orchestration of various influences, including dynein-dependent outward translocation of existing microtubules, kinesin-dependent vegetal anchoring of cortical microtubules, and microtubule polymerization at the vegetal cortex. Vegetal microtubule and cortical rotation are interdependent and enhance each other, resulting in the alignment of preexisting and new microtubules and allowing dorsal determinant transport.

4. Cortical rotation and cytoskeletal dynamics in invertebrate and protovertebrate systems

As described above, a cortical rotation process has been described in amphibians, and a related process proposed in teleosts. However, other studies have described processes of cytoskeletal reorganization that serve a similar purpose as the cortical rotation, namely, the early distribution of cellular determinants that will help pattern the egg or embryonic axis. We briefly discuss three such examples below, in ascidians (a chordate protovertebrate), the nematode *Caenorhabditis elegans*, and the dipteran *Drosophila melanogaster*, highlighting similarities with cortical rotation-like processes in lower vertebrates. For a more in-depth description of these processes, the reader is referred to Refs. [77–81] .

4.1. Ascidians

In ascidians, gastrulation and neurulation involve cellular rearrangements that are comparable to those in vertebrates, with the exception that ascidians are composed of just a few hundred cells, while vertebrate embryos contain thousands of cells [82]. In fact, the very first classical evidence that localized determinants control cell fate specification was found in ascidians [82, 83].

The ascidian egg undergoes dramatic cytoplasmic and cortical reorganizations between fertilization and the beginning of the first cleavage, a process that has been referred to as ooplasmic segregation [83–85]. Ascidian ooplasmic segregation occurs in two major phases (**Figure 1**, right). The first phase occurs shortly after fertilization. The first consequence of fertilization is that a calcium wave is initiated from the site where the sperm and egg fuse [86]. Upon fertilization, the sperm activates the stage IV oocyte, which was arrested in metaphase I of meiosis, resulting in the contraction of the egg cortex and the plasma membrane as a wave that travels across the egg in the animal to vegetal direction. It was suggested early on that an oocyte actomyosin cortical network can only contract in a general animal to vegetal direction regardless of the sperm entry site, because of it being less dense around the animal pole, in a basket-like arrangement [86, 87]. This animal-to-vegetal contraction in turn causes the segregation of cortical and subcortical components including microfilaments, mitochondria, and the cortical endoplasmic reticulum (cER) [77, 88, 89].

Unfertilized eggs after the first phase of ooplasmic segregation are radially symmetrical along the animal-vegetal (A-V) axis. This symmetry is broken in the second phase of reorganization after the movement of cortical and subcortical components from the vegetal pole toward the posterior pole occurs, generating an anteroposterior asymmetry, and eggs become bilaterally symmetrical [77]. In this second ooplasmic segregation phase, a number of cellular organelles such as the ER and mitochondria are brought toward the future posterior pole [90]. These organelles also anchor specific RNAs, termed postplasmic/PEM, which are important for muscle determination and the specification of the posterior cell fate, in particular the germ line [91]. Other factors involved in endoderm formation and gastrulation do not move toward the future posterior pole and instead expand their distribution to the vegetal hemisphere (see **Figure 1**) [77]. Reminiscent of asymmetry development in *Xenopus*, it has been suggested that also in the ascidian egg, reorganization of plus-end-directed motors attached to the ER could provide the major force to move the vegetal cortex dorsally to a more equatorial location [48, 92].

Ascidian embryonic polarity is directed by a posteriorly located centrosome, introduced through sperm entry in this region [77, 93, 94]. In contrast to the first phase which is driven by microfilaments, and where the sperm triggers a cortical contraction [88], the second phase is mediated by anchoring one of the centrosomes of the bipolar spindle to the vegetal posterior cortex, resulting in the posterior asymmetric localization of germ line-determining components. Spindle pole posterior anchoring also results in the eccentric, posteriorly located placement of the spindle, which in turn (because of the influence of the spindle midzone on furrow induction) [51, 52], results in asymmetric division [75, 77]. In this manner, the embryo generates sets of smaller posterior cells fated to become the germ line.

Thus, in both *Xenopus* and ascidians, microtubule-dependent function results in the redistribution of embryonic determinants just before the onset of embryonic mitoses, the posterior-specifying cytoplasmic components such as the cER-mRNA and myoplasm domains being displaced posteriorly in ascidians and dorsalizing factors being translocated toward the future dorsal side in *Xenopus* [89, 95].

4.2. *Caenorhabditis elegans*

In the nematode *C. elegans*, a role for PAR proteins in anterior-posterior (AP) axis specification is well documented [96]. In contrast, dorsal-ventral (DV) patterning in this system is less understood. It was recently reported that the so-called cytokinetic midbody remnant (MBR), a thus-far poorly studied organelle, acts as a polarity cue to define the *C. elegans* DV axis [97]. The MBR is an organelle that forms from the cytokinetic midbody when the fully constricted actomyosin furrow embraces the condensed material of the spindle midzone [98, 99]. To understand the role of the MBR in DV axis specification, Singh and Pohl [100] analyzed the pattern of segregation and the movements of the MBR during the first divisions of the *C. elegans* embryo. The AP axis of the *C. elegans* embryo is established by the asymmetric distribution of PAR proteins during the P0 division producing an anterior AB and a posterior P1 blastomere. Subsequently, the DV axis is established in the transition from the two-cell to the four-cell stage [101]. During prophase of the second cell division in the P1 cell, a 90° rotation of the nucleus-centrosome complex relative to the AP axis takes place, and is regarded as a key event in DV axis formation [102, 103]. It was not clear as to what generates this movement, which has long been a point of interest. The authors showed that the MBR was displaced toward the ventral side of the embryo and that it acts as a positional cue for mitotic spindle rotation in the P1 cell, thereby establishing DV axis patterning. Importantly, the authors demonstrated that ventral displacement of the MBR is directed by myosin II cortical flow [97, 100]. In this system, again microtubules together with coordinated actomyosin regulation are important for symmetry-breaking events in the embryo.

4.3. *Drosophila melanogaster*

In *D. melanogaster*, the transition from a round to an elongated egg is driven by the rearrangement of the polar arrays of microtubules [80, 81], a process that is again facilitated by the actomyosin cytoskeleton [81]. As in *C. elegans* and ascidians, this reorganization results in the segregation of cell determinants to the posterior pole of the egg, except that in the case of *Drosophila*, these changes occur during oogenesis and not early embryogenesis.

Altogether, these studies show that the microtubule cytoskeleton, and in some cases the actomyosin cortex, is used to generate axis asymmetry in various organisms, although the precise details of the interactions, and whether microtubules act as tracks that mediate transport or attraction centers, are specific to different species [97, 104].

5. Relationship between axis induction and germ cell specification

As mentioned above, in addition to dorsal determinants, anuran and teleost embryos contain other vegetally localized factors, particularly RNAs that become associated with the germplasm. The germplasm, also referred to as nuage, is a maternally inherited cytoplasmic structure containing RNPs present in some animal species. Through a mechanism referred to a preformation, inherited germplasm determines the germ cell fate [105]. Evidence for preformation mechanism for PGC induction in anurans was originally shown by the inheritance of electron-dense cytoplasm, corresponding to germplasm, into the primordial germ cells of this organism [106]. This electron-dense cytoplasm was later shown to contain specialized mRNAs involved in germ cell specification [107]. Similarly, RNAs involved in germ cell development in zebrafish, such as for the gene *vasa* [108] and subsequently other mRNAs [56, 59, 109], were shown to localize in electron-dense particles and become segregated to primordial germ cells.

Maternally inherited germplasm in *Xenopus* and zebrafish contains shared sets of factors for primordial germ cell specification, such as *deleted in azoospermia-like* (*dazl*) and *Xcat2/nanos*. Zebrafish and *Xenopus* additionally share similarities in the way in which germplasm masses are assembled and segregated, including the gradual condensation of germplasm masses from smaller particles, the formation of four germ masses, and their asymmetric segregation during cell division in the cleavage stages [110, 111].

Recent studies in these systems have begun to suggest a functional connection between axis induction and germ cell determination. As described above, during oogenesis both dorsal determinants are transported to the vegetal pole of the egg through the mitochondrial cloud in *Xenopus* and its equivalent structure, the Balbiani body, in zebrafish [111, 112]. Moreover, during early embryogenesis, genes acting in dorsal induction functionally overlap and share localization patterns with genes involved in germ cell determination. For example, the germplasm component *dead end*, which has been well characterized as a germplasm-specific transcript both in *Xenopus* [61] and zebrafish [59] and is known to function in germ cell migration and survival, has been shown in *Xenopus* to have an unexpected role in axis induction [46]. *Xdead end* RNA localizes to the vegetal pole in oocytes beginning at the early stage III to stage VI, when it becomes transported to the vegetal pole via the late RNA transport pathway [61]. It has recently been shown that maternal XDead end plays a role in vegetal microtubule reorganization required for dorsal axis induction [46]. When XDead end function is disrupted, the expression of dorsal-specific genes is reduced, and embryos become ventralized, due to the disruption in vegetal microtubule reorganization [46]. As mentioned above, this requirement appears to be due to a role for XDead end function in the vegetal cortex anchoring of the RNA for the Trim36 ubiquitin ligase [46], itself needed for growth regulation of the vegetal microtubule array [58].

Conversely, factors known to be involved in dorsal axis induction also function in germ cell development. One example is maternal Syntabulin, which as mentioned above is important for vegetal microtubule array reorganization and axis induction in both zebrafish and *Xenopus* [36, 113]. Recently, *syntabulin* mRNA has been shown to localize in *Xenopus*-cleaving embryos to clusters near the cleavage furrow on the vegetal hemisphere of the early embryo, consistent with germplasm localization and colocalization with *Xpat* RNA, a germ cell marker, during later stages [113]. *Xenopus* Syntabulin is also expressed in scattered cells localized along the posterior endoderm, presumably primordial germ cells [113]. These data suggest that, in addition to a role in DV patterning, Syntabulin may have a role in germ cell development.

Similarly, *grip2a,* which as mentioned above is required for vegetal cortex microtubule reorganization in zebrafish [26], has gene homologues involved in germ cell development in the *Xenopus* embryo [43, 44], suggesting a potential scenario in which an ancestral *grip* gene had a role in both processes. Altogether, these findings suggest that there is functional overlap between factors involved in germplasm segregation and axis induction. Whether this functional overlap is caused by evolutionary history or convergent evolution remains to be determined.

It is important to note, as stated above, that there is a difference with respect to cortical depth between the factors that are localized to the vegetal pole. Those that are important for microtubule reorganization, and thereby patterning the embryonic axis, namely, *grip2a* and *wnt8a,* are located toward the outermost region of the cortex. This allows them to be transported from the vegetal to the prospective dorsal region of the egg and embryo through a cortical rotation-like process. Those factors that are important for germ cell specification, such as *dazl* RNA, are localized deeper within the embryo and are transported via the actin cytoskeleton to the animal pole, where they become localized to the aggregating zebrafish germplasm [55]. Thus, RNA localization at the cortex reflects transport mechanisms consistent with the function of the localized product.

These set of studies highlight commonalities between processes and factors involved in axis induction and germ cell specification. Factors such as Dead end, Grip2, and Syntabulin may form a core gene set with a current or ancestral function in both axis induction and germ cell determination.

6. Conclusion: challenges and future directions

The cytoskeleton plays an essential role in axis specification, through its role mediating the movement of maternal factors within the early zygote. Studies have shown that the reorganization of the microtubule cytoskeleton is important for the transport of factors from the vegetal pole of the embryo to the future dorsal side in both zebrafish and *Xenopus,* in a process associated with the shift of the outermost cortical layer of the embryo—a cortical rotation. This cytoskeletal reorganization allows for the asymmetric transport of localized dorsal determinants, involved in the specification of the main embryonic axis. Precise mechanisms for microtubule reorganization remain incompletely understood, although are known to involve microtubule dependent motors and a positive feedback loop in which an early asymmetry and microtubule alignment triggers the rotation of the cortex, which in turn amplifies and stabilizes the incipient cytoskeletal rearrangement. Anurans and teleosts show similarities in the use of microtubule arrays and a cortical rotation-like mechanism, although they also exhibit differences in the spatial extent implemented by these coordinated processes, variations that may be related to the different cleavage type of these embryos. Components of the germplasm, which also become localized to the vegetal pole of the fertilized embryo, may escape cortical rotation by virtue of differential localization in more internal regions of the embryo. A comparison of early cytoplasmic segregation events in other species, such as ascidians, nematodes, and dipterans, highlights the importance of microtubule- and other cytoskeletal-dependent processes in the generation of early asymmetries in the embryos. Further studies will allow better understanding for mechanisms of microtubule generation, bundling, and alignment that drive the movement of cellular determinants in the early vertebrate embryo and their relation to similar processes in other animal lineages.

Acknowledgements

We would like to thank the members of the Pelegri Lab for feedback and encouragement. This work was supported by NIH grants 5RO1GM065303 and 3RO1GM65303-10A1S1.Elaine L. Welch received additional support from NIH Training Grant 2 T32 GM007133 to the Laboratory of Genetics.

Author details

Elaine Welch and Francisco Pelegri*

*Address all correspondence to: fjpelegri@wisc.edu

Laboratory of Genetics, University of Wisconsin-Madison, Madison, WI, USA

References

[1] Sive HL. The frog prince-ss: A molecular formula for dorsoventral patterning in *Xenopus*. Genes Dev. 1993;7: 1–12. doi:10.1101/gad.7.1.1

[2] Schneider S, Steinbeisser H, Warga RM, Hausen P. B-catenin translocation into nuclei demarcates the dorsalizing centers in frog and fish embryos. Mech Dev. 1996;57: 191–198. doi:10.1016/0925-4773(96)00546-1

[3] Larabell CA, Torres M, Rowning BA, Yost C, Miller JR, Wu M, et al. Establishment of the dorso-ventral axis in *Xenopus* embryos is presaged by early asymmetries in β-catenin that are modulated by the Wnt signaling pathway. J Cell Biol. 1997;136: 1123–1136. doi:10.1083/jcb.136.5.1123

[4] Petersen CP, Reddien PW. Wnt signaling and the polarity of the primary body axis. Cell. 2009;139: 1056–1068. doi:10.1016/j.cell.2009.11.035

[5] Houston DW. Cortical rotation and messenger RNA localization in *Xenopus* axis formation. Wiley Interdiscip Rev Dev Biol. 2012;1: 371–388. doi:10.1002/wdev.29

[6] Jesuthasan S, Strähle U. Dynamic microtubules and specification of the zebrafish embryonic axis. Curr Biol. 1997;7: 31–42. doi:10.1016/S0960-9822(06)00025-X

[7] Kimmel CB, Ballard WW, Kimmel SR, Ullmann B, Schilling TF. Stages of embryonic development of the zebrafish. Dev Dyn. 1995;203: 253–310. doi:10.1002/aja.1002030302

[8] Tran LD, Hino H, Quach H, Lim S, Shindo A, Mimori-Kiyosue Y, et al. Dynamic microtubules at the vegetal cortex predict the embryonic axis in zebrafish. Development. 2012;139: 3644–3652. doi:10.1242/dev.082362

[9] Fagotto F, Funayama N, Glück U, Gumbiner BM. Binding to cadherins antagonizes the signaling activity of β-catenin during axis formation in *Xenopus*. J Cell Biol. 1996;132: 1105–1114. doi:10.1083/jcb.132.6.1105

[10] Heasman J, Ginsberg D, Geiger B, Goldstone K, Pratt T, Yoshida-Noro C, et al. A functional test for maternally inherited cadherin in *Xenopus* shows its importance in cell adhesion at the blastula stage. Development. 1994;120: 49–57.

[11] Clevers H. Wnt/beta-catenin signaling in development and disease. Cell. 2006;127: 469-480. doi:10.1016/j.cell.2006.10.018

[12] Yost C, Torres M, Miller JR, Huang E, Kimelman D, Moon RT. The axis-inducing activity, stability, and subcellular distribution of beta-catenin is regulated in *Xenopus* embryos by glycogen synthase kinase 3. Genes and Dev. 1996;10: 1443-1454. doi:10.1101/gad.10.12.1443

[13] Schier AF, Talbot WS. Molecular genetics of axis formation in zebrafish. Annu Rev Genet. 2005;39: 561–613. doi:10.1146/annurev.genet.37.110801.143752

[14] Langdon YG, Mullins MC. Maternal and zygotic control of zebrafish dorsoventral axial patterning. Annu Rev Genet. 2011;45: 357–377. doi:10.1146/annurev-genet-110410-132517

[15] Little SC, Mullins MC. Extracellular modulation of BMP activity in patterning the dorsoventral axis. Birth Defects Res Part C – Embryo Today Rev. 2006;78: 224–242. doi:10.1002/bdrc.20079

[16] Mei W, Lee KW, Marlow FL, Miller AL, Mullins MC. hnRNP I is required to generate the Ca2+ signal that causes egg activation in zebrafish. Development. 2009;136: 3007–3017. doi:10.1242/dev.037879

[17] Clevers H, Nusse R. Wnt/β-catenin signaling and disease. Cell. 2012;149: 1192-1205. doi:10.1016/j.cell.2012.05.012

[18] Funayama N, Fagotto F, McCrea P, Gumbiner BM. Embryonic axis induction by the armadillo repeat domain of β-catenin : Evidence for intracellular signaling. Journal of Cell Biology 1995;128: 959–968.

[19] Kelly GM, Erezyilmaz DF, Moon RT. Induction of a secondary embryonic axis in zebrafish occurs following the overexpression of β-catenin. Mech Dev. 1995;53: 261–273. doi:10.1016/0925-4773(95)00442-4

[20] Du SJ, Purcell SM, Christian JL, McGrew LL, Moon RT. Identification of distinct classes and functional domains of Wnts through expression of wild-type and chimeric proteins in *Xenopus* embryos. Mol Cell Biol. 1995;15: 2625–34. doi:10.1128/MCB.15.5.2625

[21] Kelly C, Chin AJ, Leatherman JL, Kozlowski DJ, Weinberg ES. Maternally controlled (beta)-catenin-mediated signaling is required for organizer formation in the zebrafish. Development. 2000;127: 3899–3911.

[22] Bellipanni G, Varga M, Maegawa S, Imai Y, Kelly C, Myers AP, et al. Essential and opposing roles of zebrafish beta-catenins in the formation of dorsal axial structures and neurectoderm. Development. 2006;133: 1299–1309. doi:10.1242/dev.02295

[23] Cha S, Tadjuidje E, White J, Wells J, Mayhew C, Wylie C. Wnt11/5a complex formation caused by tyrosine sulfation increases canonical signaling activity. Curr Biol. 2009;19: 1573–1580. doi:10.1016/j.cub.2009.07.062

[24] Tao Q, Yokota C, Puck H, Kofron M, Birsoy B, Yan D, et al. Maternal Wnt11 activates the canonical Wnt signaling pathway required for axis formation in *Xenopus* embryos. Cell. 2005;120: 857–871. doi:10.1016/j.cell.2005.01.013

[25] Lu F, Thisse C, Thisse B. Identification and mechanism of regulation of the zebrafish dorsal determinant. Proc Natl Acad Sci U S A. 2011;108: 15876–15880. doi:10.1073/pnas.1106801108/-/DCSupplemental.www.pnas.org/cgi/doi/10.1073/pnas.1106801108

[26] Ge X, Grotjahn D, Welch E, Lyman-Gingerich J, Holguin C, Dimitrova E, et al. Hecate/Grip2a acts to reorganize the cytoskeleton in the symmetry-breaking event of embryonic axis induction. PLoS Genet. 2014;10: e1004422. doi:10.1371/journal.pgen.1004422

[27] Miller JR, Rowning BA, Larabell CA, Yang-snyder JA, Bates RL, Moon RT. Establishment of the dorsal – ventral axis in *Xenopus* embryos coincides. J Cell Biol. 1999;146: 427–437.

[28] Nordermeer J, Klingensmith J, Perrimon N, Nusse R. Dishevelled and armadillo act in the wingless sinaling pathway in *Drosophila*. Nature. 1994;367: 80–83. doi:10.1038/367080a0

[29] Moon RT, Kimelman D. From cortical rotation to organizer gene expression: Toward a molecular explanation of axis specification in *Xenopus*. BioEssays. 1998;20: 536–545. doi:10.1002/(SICI)1521-1878(199807)20:7<536::AID-BIES4>3.0.CO;2-I

[30] Elinson RP, Rowning B. A transient array of parallel microtubules in frog eggs: Potential tracks for a cytoplasmic rotation that specifies the dorso-ventral axis. Dev Biol. 1988;128: 185–197. doi:10.1016/0012-1606(88)90281-3

[31] Gerhart J, Danilchik M, Doniach T, Roberts S, Rowning B, Stewart R. Cortical rotation of the *Xenopus* egg: Consequences for the anteroposterior pattern of embryonic dorsal development. Development. 1989;107(Suppl): 37–51. Available: http://www.ncbi.nlm.nih.gov/pubmed/2699856

[32] Gerhardt J, Wu M, Kirschner MW. Cell cycle dynamics of an M-phase specific cytoplasmic factor in *Xenopus laevis* oocytes and eggs. J Cell Biol. 1984;98: 1247–1255.

[33] Scharf SR, Gerhart JC. Determination of the dorsal-ventral axis in eggs of *Xenopus laevis*: Complete rescue of uv-impaired eggs by oblique orientation before first cleavage. Dev Biol. 1980;79: 181–198. doi:10.1016/0012-1606(80)90082-2

[34] Marlow FL, Mullins MC. Bucky ball functions in Balbiani body assembly and animal-vegetal polarity in the oocyte and follicle cell layer in zebrafish. Dev Biol. 2008;321: 40–50. doi:10.1016/j.ydbio.2008.05.557

[35] Kloc M, Etkin LD. RNA localization mechanisms in oocytes. J Cell Sci. 2005;118: 269–282. doi:10.1242/jcs.01637

[36] Nojima H, Rothhämel S, Shimizu T, Kim C-H, Yonemura S, Marlow FL, et al. Syntabulin, a motor protein linker, controls dorsal determination. Development. 2010;137: 923–933. doi:10.1242/dev.046425

[37] Ge X, Grotjahn D, Welch E, Lyman-gingerich J, Holguin C, Dimitrova E, et al. Hecate/Grip2a acts to reorganize the cytoskeleton in the symmetry-breaking event of embryonic axis induction. PloS Genetics 2014;10: e1004422. doi:10.1371/journal.pgen.1004422

[38] Ataman B, Ashley J, Gorczyca D, Gorczyca M, Mathew D, Wichmann C, et al. Nuclear trafficking of *Drosophila* Frizzled-2 during synapse development requires the PDZ protein dGRIP. Proc Natl Acad Sci U S A. 2006;103: 7841–7846. doi:10.1073/pnas.0600387103

[39] Cai Q, Gerwin C, Sheng ZH. Syntabulin-mediated anterograde transport of mitochondria along neuronal processes. J Cell Biol. 2005;170: 959–969. doi:10.1083/jcb.200506042

[40] Nojima H, Shimizu T, Kim CH, Yabe T, Bae YK, Muraoka O, et al. Genetic evidence for involvement of maternally derived Wnt canonical signaling in dorsal determination in zebrafish. Mech Dev. 2004;121: 371–386. doi:10.1016/j.mod.2004.02.003

[41] Bontems F, Stein A, Marlow F, Lyautey J, Gupta T, Mullins MC, et al. Bucky ball organizes germ plasm assembly in zebrafish. Curr Biol. 2009;19: 414–422. doi:10.1016/j.cub.2009.01.038

[42] Kaneshiro K, Miyauchi M, Tanigawa Y, Ikenishi K, Komiya T. The mRNA coding for *Xenopus* glutamate receptor interacting protein 2 (XGRIP2) is maternally transcribed, transported through the late pathway and localized to the germ plasm. Biochem Biophys Res Commun. 2007;355: 902–906. doi:10.1016/j.bbrc.2007.02.059

[43] Tarbashevich K, Koebernick K, Pieler T. XGRIP2.1 is encoded by a vegetally localizing, maternal mRNA and functions in germ cell development and anteroposterior PGC positioning in *Xenopus laevis*. Dev Biol. 2007;311: 554–565. doi:10.1016/j.ydbio.2007.09.012

[44] Kirilenko P, Weierud FK, Zorn AM, Woodland HR. The efficiency of *Xenopus* primordial germ cell migration depends on the germplasm mRNA encoding the PDZ domain protein Grip2. Differentiation. International Society of Differentiation; 2008;76: 392–403. doi:10.1111/j.1432-0436.2007.00229.x

[45] Elinson RP, Holowacz T. Specifying the dorsoanterior axis in frogs: 70 years since Spemann and Mangold. Curr Top Dev Biol. 1995;30: 253–285. doi:10.1016/S0070-2153(08)60569-4

[46] Mei W, Jin Z, Lai F, Schwend T, Houston DW, King ML, et al. Maternal Dead-End1 is required for vegetal cortical microtubule assembly during *Xenopus* axis specification. Development. 2013;140: 2334–2344. doi:10.1242/dev.094748

[47] Marrari Y, Rouvière C, Houliston E. Complementary roles for dynein and kinesins in the *Xenopus* egg cortical rotation. Dev Biol. 2004;271: 38–48. doi:10.1016/j.ydbio.2004.03.018

[48] Houliston E, Elinson RP. Evidence for the involvement of microtubules, ER, and kinesin in the cortical rotation of fertilized frog eggs. J Cell Biol. 1991;114: 1017–1028. doi:10.1083/jcb.114.5.1017

[49] Schroeder MM, Gard DL. Organization and regulation of cortical microtubules during the first cell cycle of *Xenopus* eggs. Development. 1992;114: 699–709.

[50] Houliston E, Elinson RP. Patterns of microtubule polymerization relating to cortical rotation in *Xenopus laevis* eggs. Development. 1991;112: 107–117. Available: http://www.ncbi.nlm.nih.gov/pubmed/1769322

[51] Rappaport R. Experiments concerning the cleavage stimulus in sand dollar eggs. Journal of Experimental Zoology 1961;148: 81–89. doi:10.1002/jez.1401480107

[52] Oegema K, Mitchison TJ. Rappaport rules: Cleavage furrow induction in animal cells. Proc Natl Acad Sci U S A. 1997;94: 4817–4820. doi:10.1073/pnas.94.10.4817

[53] Brown EE, Margelot KM, Danilchik MV. Provisional bilateral symmetry in *Xenopus* eggs is established during maturation. Zygote 1994;2: 213–220.

[54] Marrari Y, Clarke EJ, Rouvière C, Houliston E. Analysis of microtubule movement on isolated *Xenopus* egg cortices provides evidence that the cortical rotation involves dynein as well as Kinesin Related Proteins and is regulated by local microtubule polymerisation. Dev Biol. 2003;257: 55–70. doi:10.1016/S0012-1606(03)00057-5

[55] Welch E, Pelegri F. Cortical depth and differential transport of vegetally localized dorsal and germ line determinants in the zebrafish embryo. Bioarchitecture. 2015;5: 13–26. doi:10.1080/19490992.2015.1080891

[56] Theusch EV, Brown KJ, Pelegri F. Separate pathways of RNA recruitment lead to the compartmentalization of the zebrafish germ plasm. Dev Biol. 2006;292: 129–141. doi:10.1016/j.ydbio.2005.12.045

[57] Cuykendall TN, Houston DW. Vegetally localized *Xenopus* trim36 regulates cortical rotation and dorsal axis formation. Development. 2009;136: 3057–65. doi:10.1242/dev.036855

[58] Olson DJ, Oh D, Houston DW. The dynamics of plus end polarization and microtubule assembly during *Xenopus* cortical rotation. Dev Biol. 2015;401: 249–263. doi:10.1016/j.ydbio.2015.01.028

[59] Weidinger G, Stebler J, Slanchev K, Dumastrei K, Wise C, Lovel-Badge R, et al. Dead end a novel vertebrate germ plasm component is required fir zebrafish primordial germ cell migration and survival. Curr Biol. 2003;13: 1429–1434. doi:10.1016/S0960-9822(03)00537-2

[60] Youngren KK, Coveney D, Peng X, Bhattacharya C, Schmidt LS, Nickerson ML, et al. The Ter mutation in the dead end gene causes germ cell loss and testicular germ cell tumours. Nature. 2005;435: 360–4. doi:10.1038/nature03595

[61] Horvay K, Claußen M, Katzer M, Landgrebe J, Pieler T. *Xenopus* Dead end mRNA is a localized maternal determinant that serves a conserved function in germ cell development. Dev Biol. 2006;291: 1–11. doi:10.1016/j.ydbio.2005.06.013

[62] Aramaki S, Sato F, Kato T, Soh T, Kato Y, Hattori MA. Molecular cloning and expression of dead end homologue in chicken primordial germ cells. Cell Tissue Res. 2007;330: 45–52. doi:10.1007/s00441-007-0435-1

[63] Kedde M, Strasser MJ, Boldajipour B, Vrielink JAFO, Slanchev K, le Sage C, et al. RNA-binding protein Dnd1 inhibits microRNA access to target mRNA. Cell. 2007;131: 1273–1286. doi:10.1016/j.cell.2007.11.034

[64] Lam MYJ, Heaney JD, Youngren KK, Kawasoe JH, Nadeau JH. Trans-generational epistasis between Dnd1Ter and other modifier genes controls susceptibility to testicular germ cell tumors. Hum Mol Genet. 2007;16: 2233–2240. doi:10.1093/hmg/ddm175

[65] Zhu R, Bhattacharya C, Matin A. The role of dead-end in germ-cell tumor development. Ann N Y Acad Sci. 2007;1120: 181–186. doi:10.1196/annals.1411.006

[66] Koebernick K, Loeber J, Arthur PK, Tarbashevich K, Pieler T. Elr-type proteins protect *Xenopus* Dead end mRNA from miR-18-mediated clearance in the soma. Proc Natl Acad Sci U S A. 2010;107: 16148–16153. doi:10.1073/pnas.1004401107

[67] Cook MS, Coveney D, Batchvarov I, Nadeau JH, Capel B. BAX-mediated cell death affects early germ cell loss and incidence of testicular teratomas in Dnd1 Ter/Ter mice. Dev Biol. 2009;328: 377–383. doi:10.1016/j.ydbio.2009.01.041

[68] Cook MS, Munger SC, Nadeau JH, Capel B. Regulation of male germ cell cycle arrest and differentiation by DND1 is modulated by genetic background. Development. 2011;138: 23–32. doi:10.1242/dev.057000

[69] Mei W, Jin Z, Lai F, Schwend T, Houston DW, King ML. Maternal Dead-End1 is required for vegetal cortical microtubule assembly during *Xenopus* axis specification. Development. 2013;140: 2334–2344. doi:10.1242/dev.094748

[70] Weaver C, Kimelman D. Move it or lose it: Axis specification in *Xenopus*. Development.2004;131: 3491–9. doi:10.1242/dev.01284

[71] Marrari Y, Terasaki M, Arrowsmith V, Houliston E. Local inhibition of cortical rotation in *Xenopus* eggs by an anti-KRP antibody. Dev Biol. 2000;224: 250–262. doi:10.1006/dbio.2000.9773

[72] Hogan CJ, Wein H, Wordeman L, Scholey JM, Sawin KE, Cande WZ. Inhibition of anaphase spindle elongation in vitro by a peptide antibody that recognizes kinesin motor domain. Proc Natl Acad Sci U S A. 1993;90: 6611–5. Available: http://www.pubmedcentral.nih.gov/articlerender.fcgi?artid=46982&tool=pmcentrez&rendertype=abstract

[73] Echeverri CJ, Paschal BM, Vaughan KT, Vallee RB. Molecular characterization of the 50-kD subunit of dynactin reveals function for the complex in chromosome alignment and spindle organization during mitosis. J Cell Biol. 1996;132: 617–633. doi:10.1083/jcb.132.4.617

[74] Terasaki M, Chen LB, Fujiwara K. Microtubules and the endoplasmic reticulum are highly interdependent structures. J Cell Biol. 1986;103: 1557–1568. doi:10.1083/jcb.103.4.1557

[75] Sawada T, Schatten G. Microtubules in ascidian eggs during meiosis, fertilization, and mitosis. Cell Motil Cytoskeleton. 1988;9: 219–30. doi:10.1002/cm.970090304

[76] Gurel PS, Hatch AL, Higgs HN. Connecting the cytoskeleton to the endoplasmic reticulum and Golgi. Curr Biol. 2014;24: R660–R672. doi:10.1016/j.cub.2014.05.033

[77] Nishida H. Specification of embryonic axis and mosaic development in ascidians. Dev Dyn. 2005;233: 1177–1193. doi:10.1002/dvdy.20469

[78] Hyman AA. Centrosome movement in the early divisions of *Caenorhabditis elegans*: A cortical site determining centrosome position. J Cell Biol. 1989;109: 1185–1193.

[79] Hird S, White J. Cortical and cytoplasmic flow in early *C. elegans* embryos. J Cell Biol. 1993;121: 1343–1355. Available: http://jcb.rupress.org/content/121/6/1343.long

[80] Viktorinová I, Dahmann C. Microtubule polarity predicts direction of egg chamber rotation in *Drosophila*. Curr Biol. 2013;23: 1472–1477. doi:10.1016/j.cub.2013.06.014

[81] Haigo SL, Bilder D. Global tissue revolutions in a morphogenetic movement controlling elongation. Science. 2011;331: 1071–1074. doi:10.1126/science.1199424

[82] Corbo JC, Di Gregorio A, Levine M. The Ascidian as a model organism in developmental and evolutionary biology. Cell. 2001;106: 535–538. doi:10.1016/S0092-8674(01)00481-0

[83] Conklin EG. The Organization and Cell-Lineage of the Ascidian Egg. Proceedings of the Academy of Natural Sciences of Philadelphia. 1905;13: 1–119. doi:10.1126/science.23.583.340

[84] Sardet C, Dru P, Prodon F. Maternal determinants and mRNAs in the cortex of ascidian oocytes, zygotes and embryos. Biology of the cell. 2005. pp. 35–49. doi:10.1042/BC20040126

[85] Ishii H, Shirai T, Makino C, Nishikata T. Mitochondrial inhibitor sodium azide inhibits the reorganization of mitochondria-rich cytoplasm and the establishment of the anteroposterior axis in ascidian embryo. Dev Growth Differ. 2014;56: 175–188. doi:10.1111/dgd.12117

[86] Sardet C, Paix A, Dru P, Chenevert J. From oocyte to 16-cell stage : Cytoplasmic and cortical reorganizations that pattern the ascidian embryo. Developmental Dynamics 2007 ;236: 1716–1731. doi:10.1002/dvdy.21136

[87] Sardet C, Speksnijder J, Terasaki M, Chang P, Biologie U De, Marine C, et al. Polarity of the ascidian egg cortex before fertilization. Development 1992;237: 221–237.

[88] Roegiers F, Djediat C, Dumollard R, Rouvière C, Sardet C. Phases of cytoplasmic and cortical reorganizations of the ascidian zygote between fertilization and first division. Development. 1999;126: 3101–3117.

[89] Prodon F, Dru P, Roegiers F, Sardet C. Polarity of the ascidian egg cortex and relocalization of cER and mRNAs in the early embryo. J Cell Sci. 2005;118: 2393–2404. doi:10.1242/jcs.02366

[90] Nishida H. Cell-fate specification by localized cytoplasmic determinants and cell interaction in ascidian embryo. Int Rev Cytol. 1997;176: 245–306. doi:10.1126/scisignal.2001449.Engineering

[91] Prodon F, Yamada L, Shirae-Kurabayashi M, Nakamura Y, Sasakura Y. Postplasmic/PEM RNAs: A class of localized maternal mRNAs with multiple roles in cell polarity and development in ascidian embryos. Dev Dyn. 2007;236: 1698–1715. doi:10.1002/dvdy.21109

[92] Houliston E, Le Guellec R, Kress M, Philippe M, Le Guellec K. The kinesin-related protein Eg5 associates with both interphase and spindle microtubules during *Xenopus* early development. Dev Biol. 1994;164: 147–159. Available: http://dx.doi.org.ezproxy.library.wisc.edu/10.1006/dbio.1994.1187

[93] Sardet C, Speksnijder J, Inoue S, Jaffe L. Fertilization and ooplasmic movements in the ascidian egg. Development. 1989;105: 237–249. doi:10.1142/9789812790866_0049

[94] Roegiers F, McDougall A, Sardet C. The sperm entry point defines the orientation of the calcium-induced contraction wave that directs the first phase of cytoplasmic reorganization in the ascidian egg. Development. 1995;121: 3457–3466.

[95] Weaver C, Farr GH, Pan W, Rowning BA, Wang J, Mao J, et al. GBP binds kinesin light chain and translocates during cortical rotation in *Xenopus* eggs. Development. 2003;130: 5425–5436. doi:10.1242/dev.00737

[96] Ja K, Ra N. Asymmetric cell division: Recent developments and their implications for tumour biology. Dividing cellular asymmetry : Asymmetric cell division and its implications for stem cells and cancer . Mechanisms of asymmetric stem cell division. Nat Rev Mol Cell Biol. 2010;11: 849–860. doi:10.1038/nrm3010.

[97] Singh D, Pohl C. A function for the midbody remnant in embryonic patterning. Commun Integr Biol. 2014;7: 1–3. doi:10.4161/cib.28533

[98] Fededa JP, Gerlich DW. Molecular control of animal cell cytokinesis. Nat Cell Biol. Nature Publishing Group; 2012;14: 440–7. doi:10.1038/ncb2482

[99] Green RA, Mayers JR, Wang S, Lewellyn L, Desai A, Audhya A, et al. The midbody ring scaffolds the abscission machinery in the absence of midbody microtubules. J Cell Biol. 2013;203: 505–520. doi:10.1083/jcb.201306036

[100] Singh D, Pohl C. Coupling of rotational cortical flow, asymmetric midbody positioning, and spindle rotation mediates dorsoventral axis formation in *C. elegans*. Dev Cell. 2014: 253–267. doi:10.1016/j.devcel.2014.01.002

[101] Pinheiro D, Bellaïche Y. Making the most of the midbody remnant: Specification of the dorsal-ventral axis. Dev Cell. 2014;28: 219–220. doi:10.1016/j.devcel.2014.01.026

[102] Hyman AA, White J. Determination of cell division axes in the early embryogenesis of *Caenorhabditis elegans*. J Cell Biol. 1987;105: 2123–2135. doi:10.1083/jcb.105.5.2123

[103] Keating HH, White JG. Centrosome dynamics in early embryos of *Caenorhabditis elegans*. J Cell Sci. 1998;111: 3027–3033.

[104] Singh D, Pohl C. Coupling of rotational cortical flow, asymmetric midbody positioning, and spindle rotation mediates dorsoventral axis formation in *C. elegans*. Dev Cell. 2014;28: 253–267. doi:10.1016/j.devcel.2014.01.002

[105] Wylie C. Germ cells. CurrOpinGenetDev. 2000;10: 410–413. doi:10.1016/S0092-8674(00)80557-7

[106] Extavour CG, Akam M. Mechanisms of germ cell specification across the metazoans: Epigenesis and preformation. Development. 2003;130: 5869–5884. doi:10.1242/dev.00804

[107] Elinson RP, Sabo MC, Fisher C, Yamaguchi T, Orii H, Nath K. Germ plasm in *Eleutherodactylus coqui*, a direct developing frog with large eggs. EvoDevo. 2011;2: 20. doi:10.1186/2041-9139-2-20

[108] Yoon C, Kawakami K, Hopkins N. Zebrafish vasa homologue RNA is localized to the cleavage planes of 2- and 4-cell-stage embryos and is expressed in the primordial germ cells. Development. 1997;124: 3157–3165. Available: http://www.ncbi.nlm.nih.gov/pubmed/9272956

[109] Köprunner M, Thisse C, Thisse B, Raz E. A zebrafish nanos-related gene is essential for the development of primordial germ cells. Genes Dev. 2001;15: 2877–2885. doi:10.1101/gad.212401

[110] Pelegri F, Knaut H, Maischein HM, Schulte-Merker S, Nüsslein-Volhard C. A mutation in the zebrafish maternal-effect gene nebel affects furrow formation and vasa RNA localization. Curr Biol. 1999;9: 1431–1440. doi:10.1016/S0960-9822(00)80112-8

[111] Eno C, Pelegri F. Germ cell determinant transmission, segregation, and function in the zebrafish embryo. In: Carreira RP, editor. Intech. 2016. pp.115-142. Available: 10.5772/62207

[112] Minakhina S, Steward R. Axes formation and RNA localization. Curr Opin Genet Dev. 2005;15: 416–421. doi:10.1016/j.gde.2005.06.006

[113] Colozza G, De Robertis EM. Maternal syntabulin is required for dorsal axis formation and is a germ plasm component in Xenopus. Differentiation. 2014;88: 17–26. doi:10.1016/j.diff.2014.03.002

Chapter 2

Actin-Microtubule Interaction in Plants

Miyuki Takeuchi, L. Andrew Staehelin and
Yoshinobu Mineyuki

Additional information is available at the end of the chapter

http://dx.doi.org/10.5772/66930

Abstract

Interactions between actins and microtubules play an important role in many fundamental cellular processes in eukaryotes. Although several studies have shown actins and microtubules to be involved in specific cellular activities, little is known about how actins and microtubules contribute together to a given process. Preprophase band formation, which plays an essential role in plant division site determination, is a cellular process that lends itself to studies of actin-microtubule interactions and how they contribute to important cellular functions. Recently, we have analyzed microtubule-associated microfilaments during preprophase band formation in onion cotyledon epidermal cells using a combination of high-pressure freezing/freeze substitution and electron tomography. Quantitative analysis of our electron tomography data showed that relatively short single microfilaments form bridges between two adjacent microtubules in the process of narrowing of the preprophase microtubule band. Two types of microtubule-microfilament-microtubule connections are observed, and these microfilament-microtubule interactions suggest a direct role of F-actins in microtubule bundling. Based on these observations, we discuss how different actin-microtubule linkers might contribute to preprophase band narrowing and to other changes in microtubule organization in plant cells.

Keywords: actin-microtubule interaction, electron tomography, microtubule bundling, plant morphogenesis, preprophase band

1. Introduction

Actin filaments (F-actins) and microtubules (MTs) are major components of the cytoskeleton of eukaryotic cells. Each involved in fundamental cellular processes, such as cell division, directional cell expansion, organelle movement and signal transduction. Studies of cross-talk between F-actin and MT networks have led to the conclusion that actin-MT interactions are essential for the regulation of these cytoskeletal networks in eukaryotes [1]. In plants, actin has been postulated to play critical roles in MT-mediated cytomorphogenesis. Plant shape is determined by the spatial and temporal regulation of cell division and cell expansion. During interphase, directional cell expansion is controlled by interphase cortical MTs, whereas during cell division, arrays of MTs, such as preprophase bands (PPBs), spindles and phragmoplasts contribute to division site determination, chromosome movement and cell plate formation, respectively. Although electron microscope (EM) observations, studies with fluorescence probes, pharmacological studies and molecular genetic studies have provided insights into the possible cross-talk between actin and MT-systems, little is known about how actin controls MT organization during morphogenesis [2]. The PPB is a plant-specific cytokinetic apparatus that establishes the future site of cell division, and the evidence for actin-MT cross-talk has been obtained from fluorescent microscope studies [3]. Recently, we have succeeded in observing how microfilaments (MFs) interact with MTs and thereby cause MT bundling using electron tomography analysis of cryofixed cells [4]. In this chapter, we review how interactions between F-actins and MTs control plant cell expansion and cell division. We then explain why electron tomography is an excellent tool for the analysis of the 3D architecture of MF-MT interactions during PPB development. Finally, we discuss candidate molecules that might mediate these actin-MT interactions.

2. Evidence for the involvement of actin-MT interactions in plants

A list of studies supporting the hypothesis that actin-MT interactions play important roles in plant morphogenesis is shown in **Tables 1** and **2**. In electron micrographs, F-actins give rise to ~6 nm in diameter MFs and are seen in close proximity to and aligned with MTs. Such MFs are linked to the MTs through cross-bridging structures (see papers marked EM in **Tables 1** and **2**). Fluorescent microscopy provides several methods for studying F-actins and MTs. Most frequently, F-actins and MTs are visualized by means of specific antibodies with attached fluorescent tags (see papers marked IF in **Tables 1** and **2**). Fluorescent-labeled phalloidins are also used to detect F-actins (see papers marked Ph in **Tables 1** and **2**). Microinjection of fluorescent-labeled cytoskeletal proteins or phalloidin has been used to examine cytoskeletal dynamics in living cells (see papers marked Inj (Ph/) in **Tables 1** and **2**). Live cell studies of expressed chimera peptides with fluorescent protein-tags have also yielded important results (see papers marked e.g. live GFP-AtFim1/ in **Tables 1** and **2**). An alternative is to use cytoskeleton-modifying drugs to investigate cytoskeletal involvement in cellular processes in different types of cells (**Tables 1B**, **C** and **2B**). For example, to determine whether actins exert a role in a MF-mediated function of a cell, the disruption or stabilization on the organization of MTs, and *vice versa* can be determined. Genetic modifications in combination with cytoskeleton-altering drugs provide yet another tool for investigating MF-MT interactions (**Tables 1B**, **C** and **2B**)

2.1. Directional cell expansion

The shape of plant cells is defined by the mechanical properties of their cell walls, and cortical MTs help determine plant cell shape by controlling the direction of deposition of cell

http://dx.doi.org/10.5772/66930

References	Plant	Cell type	Imaging method
(A) Colocalization of actins with MTs/ MFs associated with MTs			
Franke et al. [5]	*Liliumlongiflorum*	Pollen tube	EM (chem)
Seagull and Heath [6]	*Raphanussativus*	Root hair	EM (chem)
Hardham et al. [7]	*Graptopetalumparaguayense*	Leaf	EM (chem)
Tiwari et al. [8]	*Triticum* *Agrostis* *Phleum*	Leaf trichome Leaf epidermal cell Root tip cell	EM (cryo) EM (chem)
Emons [9]	*Equisetum hyemale* *Limnobiumstoloniferum*	Root hair Root hair	EM (cryo)
Lancelle et al. [10]	*Nicotianaalata*	Pollen tube	EM (cryo)
Kobayashi et al. [11]	*Zinnia elegans*	Suspension culture cell (tracheary element)	Ph/ IF
Pierson et al. [12]	*Nicotianatabacum* *Liliumlongiflorum*	Pollen tube Pollen tube	Ph/IF
Sonobe and Shibaoka [13]	*Nicotianatabacum*	Suspension culture cell	Ph/IF
Ding et al. [14]	*Nicotianatabacum*	Root tip cell	EM (cryo)
Kengen and Derksen [15]	*Nicotianaplumbaginifolia*	Protoplast	EM (DC)
Lancelle and Hepler [16]	*Nicotianaalata*	Pollen tube	EM (cryo)
Cleary et al. [17]	*Selaginellakraussiana*	Differentiating guard cell	Ph/MT
Tominaga et al. [18]	*Hydrocharisdubia*	Root hair	Ph/IF, EM (chem)
Murata et al. [19]	*Nicotianatabacum* *Allium cepa*	Root tip cell Cotyledon epidermal cell	EM (cryo)
Sampathkumar et al. [20]	*Arabidopsis thaliana*	Epidermal cell of root	Live (GFP-FABD/ mCherry-TUA5)
(B) F-actin disruption/stabilization causes alternation in MT organization			
Kobayashi et al. [11]	*Zinnia elegans*	Suspension culture cell (tracheary element)	Ph/IF
Seagull [21]	*Gossypiumhirsutum*	Cotton fiber	IF/IF
Kadota and Wada [22]	*Adiantumcapillus-veneris*	Protonemal cell	Ph/IF
Hasezawa et al. [23]	*Nicotianatabacum*	Suspension culture cell	Ph/IF
Takesue and Shibaoka [24]	*Vignaangularis*	Epidermal cell of epicotyl	Ph/IF
Blancaflor [25]	*Zea mays*	Cortical cell of root	Ph/IF
Schwab et al. [26]	*Arabidopsis thaliana (dis2)*	Trichome	Live (-/GFP-MAP4)
Saedler et al. [27]	*Arabidopsis thaliana (dis2)*	Trichome	Live (CFP-mTalin/ YFP-MAP4)
Timmers et al. [28]	*Medicagotruncatula*	Root hair	Live (GFP-FABD2/ GFP-MBD,YFP-EB1)
Sainsbury et al. [29]	*Allium porrum*	Leaf epidermall cell	Live (-/GFP-MBD)

References	Plant	Cell type	Imaging method
(C) MT disruption/stabilization causes alteration in actin organization			
Kobayashi et al. [11]	*Zinnia elegans*	Suspension culture cell (tracheary element)	Ph/ IF
Chu et al. [30]	*Secalecereale*	Root tip cell	Ph/ IF
Tominaga et al. [18]	*Hydrocharisdubia*	Root hair	Ph/ IF
Collings et al. [31]	*Nicotianatabacum*	Suspension culture cell	IF/ IF
Collings and Wasterneys [32]	*Arabidopsis thaliana*	Root epidermal cell	IF/ IF
Collings et al. [33]	*Arabidopsis thaliana (mor1)*	Root epidermal cell	IF/ IF
Smertenko et al. [34]	*Nicotianatabacum* *Arabidopsis thaliana*	Suspension culture cell Root	Live (GFP-Lifeact/-)
Shevchenko [35]	*Arabidopsis thaliana*	Root transition zone	Liv (GFP-ABD2/-)

EM (chem), electron microscopy with chemical fixed materials;EM (cryo), electron microscopy with cryo-fixed materials; EM (DC), electron microscopy of dry cleaved sample; IF, immunofluorescence method; Ph, phalloidin labelled with fluorescent dyes; Live, live imaging

Table 1. Actin-MT cross-talks in cell expansion.

References	Plant	Cell type	Imaging method
(A) F-actins (or MFs) in the PPB			
Kakimoto and Shibaoka [36]	*Nicotianatabacum*	Suspension culture cell	Ph/IF
Palevitz [37]	*Allium cepa*	Root tip cell	Ph/IF
Traas et al. [38]	*Daucuscarota*	Suspension culture cell	Ph/-
Lloyd and Traas [39]	*Daucuscarota*	Suspension culture cell	Ph/-
McCurdy et al. [40]	*Triticumaestivum*	Root tip cell	IF/-
Palevitz[41]	*Allium cepa*	Root tip cell	Ph/IF
Katsuta et al. [42]	*Nicotianatabacum*	Suspension culture cell	Ph/IF
McCurdy and Gunning [43]	*Tririćumaesrivum*	Root tip cell	IF/IF
Mineyuki and Palevitz [44]	*Allium cepa*	Cotyledon epidermal cell	Ph/IF
Ding et al. [45]	*Nicotianatabacum*	Root tip cell	EM (cryo)
Cleary et al. [46]	*Tradescantiavirginiana*	Stamen hair cell	Inj (Ph/tubulin)
Eleftheriou and Palevitz [47]	*Allium cepa*	Root tip cell	IF/IF
Liu and Palevitz [48]	*Allium cepa* *Tradescantiavirginiana*	Root tip cell	IF/IF
Panteris et al. [49]	*Adiantumcapillus-veneris*	Root tip cell	Ph/IF
Cleary [50]	*Tradescantiavirginiana*	Stomatal complexes	Inj (Ph/-)
Cleary and Mathesius [51]	*Tradescantiavirginiana*	Leaf epidermal cell	Ph/IF
Baluska et al. [52]	*Zea mays*	Root tip cell	IF/-
Zachariadis et al. [53]	*Pinusbrutia*	Root tip cell	Ph/IF
Sano et al. [54]	*Nicotianatabacum*	Suspension culture cell	Live (GFP-ABD2/-)

References	Plant	Cell type	Imaging method
Li et al. [55]	*Arabidopsis thaliana*	Suspension culture cell	Ph/IF
Higaki et al. [56]	*Nicotianatabacum*	Suspension culture cell	Live (GFP-ABD2/ -)
Panteris et al. [57]	*Zea mays, Triticumturgidum*	Leaf stomatal complexes	Ph/ IF
Takeuchi et al. [58]	*Allium cepa*	Root tip cell	IF/ IF
Takeuchi et al. [4]	*Allium cepa*	Cotyledon epidermal cell	EM (cryo)
(B) F-actin disruption causes alternation in MT organization in the PPB			
Mineyuki and Palevitz [44]	*Allium cepa*	Cotyledon epidermal cell	Ph/ IF
Eleftheriou and Palevitz [47]	*Allium cepa*	Root tip cell	IF/ IF
Granger and Cyr [59]	*Nicotianatabacum*	Suspension culture cell	Live (-/ GFP-MBD)
Li et al., [55]	*Arabidopsis thaliana*	Suspension culture	IF/ IF
Kojo et al. [60]	*Nicotianatabacum*	Suspension culture cell	Live (-/GFP-tubulin)
Takeuchi et al. [58]	*Allium cepa*	Root tip cell	IF/ IF
(C) Actins in spindle			
Forer and Jackson, Forer et al. [61–63]	*Haemanthuskatherinae*	Endosperm	EM (chem)
Seagull et al. [64]	*Medicago sativa*	Suspension culture cell	Ph/IF
Traas et al. [38]	*Daucuscarota*	Suspension culture cell	Ph/-
Traas et al. [65]	*Splanummelongena*	Miotic cell	Ph/IF
Czaban and Forer [66]	*Haemanthuskatherinae*	Endosperm	Ph/IF
(D) Actins in phragmoplast			
Clayton and Lloyd [67]	*Allium cepa*	Root tip cell	Ph/IF
Gunning and Wick [68]	*Tradescantiavirginiana*	Stamen hair cell	Ph/IF
Schmit et al. [69]	*Haemanthuskatherinae*	Endosperm	Ph/IF
Kakimoto and Shibaoka [36]	*Nicotianatabacum*	Suspension culture cell	Ph/IF
Traas et al. [38]	*Daucuscarota*	Suspension culture cell	Ph/-
Palevitz [70]	*Allium cepa*	Root tip cell	Ph
Schmit and Lambert [71]	*Haemanthuskatherinae*	Endosperm	Ph, IF/ IF, EM (chem)
Lloyd and Traas [39]	*Daucuscarota*	Suspension culture cell	Ph/IF
Kakimoto and Shibaoka [72]	*Nicotianatabacum*	Suspension culture cell	EM (chem)
Traas et al. [65]	*Splanummelongena*	Miotic cell	Ph/ IF
Schmit and Lambert [73]	*Haemanthuskatherinae*	Endosperm	Inj (Ph/-)
Zhang et al. [74]	*Tradescantiavirginiana*	Stamen hair cell	Inj (Ph/tubulin)
Collings et al. [75]	*Allium cepa*	Root tip cell	IF/IF
Collings and Wasteneys [32]	*Arabidopsis thaliana*	Root tip cell	IF/IF

EM(chem), electron microscopy with chemical fixed materials; EM(cryo), electron microscopy with cryo-fixed materials;Live, live imaging; IF, immunofluorescence method; Ph, phalloidin labeled with fluorescent dyes; Inj, microinjection of fluorescent-labeled probes

Table 2. Actin-MT cross-talks in cell division.

wall fibrils [76–78]. In pollen tubes and root hair cells, cortical MTs run parallel to the cell axis and a similar F-actin pattern has been observed by EM images (see papers marked EM in **Table 1A**). Transversely aligned cortical MTs are observed in the elongating cells of roots, suspension culture cells and fern protonemata. Transversely arranged F-actins have also been observed in these cells (**Table 1A**) as well as MFs associated with cortical MTs (see papers marked EM in **Table 1A**). In dry-cleaved preparations of tobacco protoplasts, Kengen and Derksen [15] were able to follow MT-associated MFs over distances up to 1.46 μm. The presence of MTs connected to MFs via cross-bridging molecules [10, 14] suggests that the organization of such MTs is influenced by the direct interaction with F-actins. The fact that the short F-actins can slide along MTs as seen in live cell imaging studies using fluorescent-labeled MTs and F-actins suggests an involvement of motor proteins in this cross-bridging process [20]. Pharmacological studies suggest that F-actins either play a role in cortical MT organization, or that the MTs can regulate the organization of actin (**Table 1B**, **C**).

In addition to the direct physical actin-MT interactions described in the previous paragraph, there are other MT systems in which different types of actin-MT interactions have been observed. For example, during the formation of tracheary elements in cultured cells of *Zinnia*, F-actins are formed between reticulate arrays of MT bundles. The fact that the disruption of the F-actin aggregates by cytochalasin affects the MT organization in this system suggests that the F-actin aggregates are involved in MT organization [11, 79]. Coordination of MT and actin networks is also required for morphogenesis of cells showing complex expansion patterns such as leaf pavement cells and trichomes [80–83]. Actins and MTs have been shown to serve distinct roles during the formation of the intricately shaped leaf mesophyll and epidermal cells. Thus, the cortical MT bundles appear to determine the sites of deposition of cell wall molecules that prevent cell wall expansion, whereas actin patches regulate the formation of the cell lobes [80]. Formation of separate MT and actin domains appears to be coordinated by the activity of RopGTPase [82].

During trichome morphogenesis, control of MT dynamics by actin has been inferred from studies of *distorted2* (*dis2*) mutants defective in the actin-related ARP2/3 complex. The ARP2/3 complex regulates actin polymerization and the *dis2* mutation gives rise to distorted trichomes. The mutant trichomes exhibit changes in MT organization that is similar to those seen in cytochalasin-treated cells [26, 27]. The *tortifolia2* (*tor2*) mutant has a mutation in α-tubulin4 that causes aberrant cortical MT dynamics and overbranching of the trichomes. The double mutant of *tor2* with *dis1*, another ARP2/3 complex mutant, shows complete loss of anisotropic growth, and MT organization in the mutant is severely disturbed in comparison with the respective single mutants. Based on these observations, Sambade [84] proposed that cortical MTs have two different functions, actin-dependent and actin-independent functions during trichome differentiation.

2.2. Cell division

Spindle: Forer's group has suggested that F-actins are present in the spindles of *Haemanthus* endosperm based on EM observations of chemically fixed cells decorated with heavy meromyosin as well as on fluorescent microscope observations using rhodamine phalloidin [61–63, 66]. In contrast, Schmit and co-workers [69, 71, 73] using live *Hemanthus* endosperms injected with fluorescent phalloidin probes or fixed *Hemanthus* endosperms stained with fluorescent phalloidin, reported that the concentration of F-actins is low inside spindles and that instead they form a cage-like structure around the spindle. The actin cage has also been observed in *Tradescantia* stamen hair cells [46]. Disruption of the actin cages around spindles alters both spindle positioning and spindle orientation [60, 85].

Phragmoplast: Actins have been shown to be present in phragmoplasts by means of fluorescent dye-labeled phalloidin [36–38, 67, 68]. However, both EM and fluorescent microscope observations have shown that the organization of phragmoplast actin differs from that of the MTs [32, 71, 75, 85]. A high density of short F-actins is seen in the equatorial plane oriented parallel to the MTs. F-actins identified by heavy meromyosin decoration have also been observed near the forming cell plate, and the heavy meromyosin arrowheads on the F-actins point away from the cell plate [72].

PPB: The presence of actins in PPBs has been documented in a variety of tissues through fluorescence microscopy techniques (**Table 2A**). Single MFs associated with PPBs in high-pressure frozen cells have also been visualized by EM [4, 45]. The development of PPBs is an interesting process in terms of actin-MT interactions, because actins are associated with MTs during the early developmental stages but are absent from later stages. This is illustrated in **Figure 1**. PPBs develop from broad bands of MTs in the G2 phase of the cell cycle. Cells exhibiting an early broad PPB also possess a broad cortical F-actin array (Interphase in **Figure 1**),

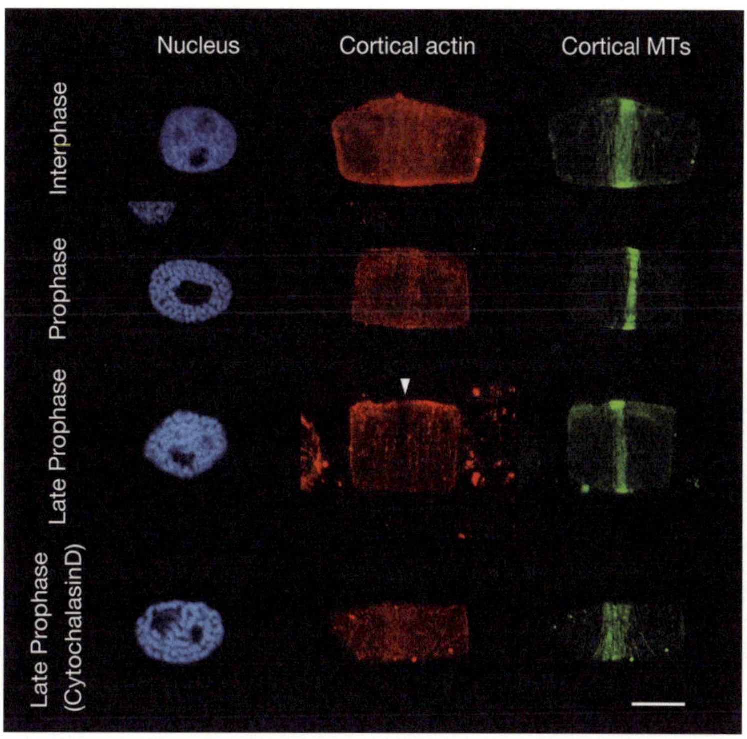

Figure 1. Actin-MT cross-talk during PPB development in root tip cells of onion seedlings. Cells were triple labeled for actin (red), tubulin (green) and DNA (blue). Interphase, the broad PPB MT and actin. Prophase, the narrow PPB MT and actin. Late prophase, a narrow PPB with ADZ (arrowhead). Late prophase (Cytochalasin D), a late prophase cell treated with 20 μM cytochalasin D for 30 min. Scale bar: 10 μm. Modified from Ref. [58].

and in the subsequent period, the actin band narrows in parallel with the narrowing of the PPB MT band (Prophase in **Figure 1**). However, when the PPB MTs reach their narrowest configuration in late prophase (Late prophase in **Figure 1**), the fluorescent actin signals start to disappear from the cell cortex region occupied by the MT band, giving rise to the actin-depleted zone (ADZ) [46, 48]. The ADZ forms prior to the breakdown of the nuclear envelope and persists throughout mitosis [46, 48]. It is known that F-actin disrupters, such as cytochalasins and latrunculins, not only prevent narrowing of the PPB MT but also rewidening of narrowed MT band (see references in **Table 2B**). The effect of cytochalasin D on PPB actin is unique. When cytochalasin D is applied to late prophase cells, PPB MT widening occurs and actins stay in the broadened PPB instead of disappearing from the PPB region (Late prophase (cytochalasin D) in **Figure 1**).

3. Evidences for MF-MT association revealed by electron tomography

3.1. Visualization of single F-actins in electron micrographs

Bundled MFs in plant cells were first observed in streaming *Nitella* cells by Nagai and Rebhun [86]. Subsequently, the MFs were identified as F-actins by heavy meromyosin decoration [87]. Since then, MFs have been observed in a variety of plant cells [6–8, 10, 14, 16, 45, 88–91]. Murata et al. [19] optimized the condition to preserve and visualize MFs in the cortical cytoplasm of plant cells. They applied high-pressure freezing/freeze-substitution followed by post-fixation with OsO_4 and uranyl acetate to achieved stable preservation and high contrast of the fine MFs in both tobacco root tips and onion cotyledons (**Figure 2**). A single, ~1-μm-long MF preserved by this method is illustrated in **Figure 2B**.

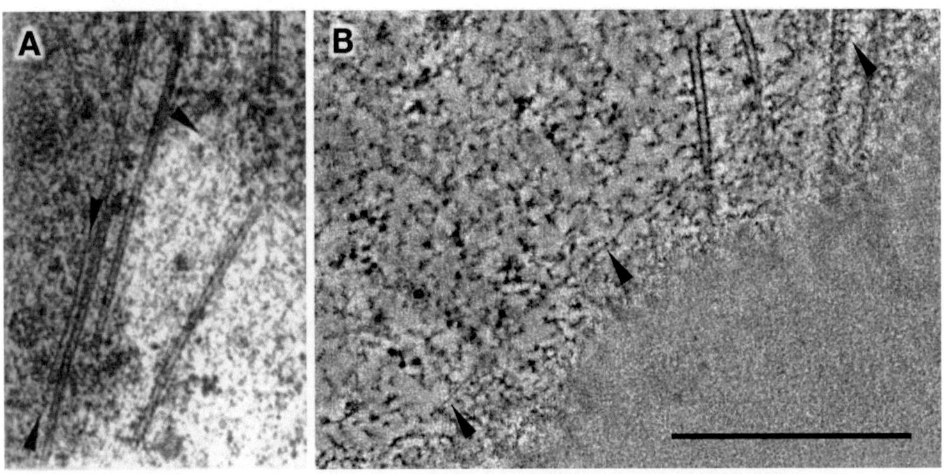

Figure 2. MFs in the high pressure frozen/freeze-substituted plant cell cortex. (A) A tobacco root tip cell. (B) An epidermal cell of an onion cotyledon. Arrowheads, MFs. Bar = 500 nm. (A) Modified from Ref. [19], (B) modified from Ref. [4].

3.2. Spatial relationship between MFs and MTs during PPB development

Electron tomography is a powerful method for visualizing and quantitatively analyzing the ultrastructural features of cells in 3D. The 3D volume is reconstructed from a series of 2D-EM images of successively tilted samples. It enables researchers to analyze the 3D organization of filamentous structures in cells and has been applied to study of F-actin networks in lamellipodia [92] and MTs in hyphae [93]. Electron tomography has also been applied to cell division studies in plants, most notably to obtain quantitative information on the organization of cortical MTs as well as cell plate forming structures [94–96].

PPB development in epidermal cells of onion cotyledons is well-characterized [96, 97], and this experimental system has proven advantageous for elucidating the events associated with PPB formation. In particular, it has enabled us to characterize the organization of MFs in the PPBs of onion cotyledon epidermal cells by means of electron tomography of high pressure frozen/freeze-substituted tissues. For these studies, a basal part of the cotyledon was cut and immediately high-pressure frozen, then samples were freeze-substituted and embedded in Spurr's resin [19]. Tomograms were generated from the tilted image series of 250-nm-thick serial sections of PPB-containing cells, and the tomograms were analyzed with IMOD [98]. The tomograms obtained by these means covered a ~4 μm^2 large area of the cell cortex and enabled us to obtain quantitative information on the distribution of coated and non-coated vesicles in PPBs, the number and length of PPB MFs, as well as on the spatial relationship between individual fibers of MFs and MTs in the PPBs [4, 96]. During the course of these studies, we made approximately 50 serial sections for each cell, which enabled us to determine both the nuclear stage and the corresponding developmental stage of the PPB for each cell used for the electron tomography analysis. The results are reported in Ref. [4] and are briefly summarized in **Figure 3**.

Tomographic models of **Figure 3A–E** show changes in MT and MF organization during this process. As reported previously [44], most cortical MTs (magenta lines) are not arrange transversely in interphase cells of the basal region of the onion cotyledon epidermis and very few MT-MF interactions are seen (**Figure 3A**). At the onset of PPB formation, the MTs become organized into loose, irregular arrays transversely oriented to the longitudinal axis of the cells and the first MT-associated MFs (yellow lines) start to appear (**Figure 3B**). During this MT reorganization, groups of two or three MTs initiate the formation of MT bundles by moving closer together and becoming more aligned (**Figure 3C**). Then, pairs of closely aligned MTs serve as templates for the assembly of increasingly large MT clusters (**Figure 3D**). A majority of the MFs become aligned with and many become bound to the MTs. This spatial relationship is maintained until the MT band narrows. In late prophase, the density of the MFs declines within the PPB region (bracketed area) compared to the area outside of the PPB where the cortical MFs are randomly oriented (**Figure 3E, F**). This stage appears to correspond to late prophase during which the ADZ is formed (**Figure 1**). As the behavior of MFs is similar to those reported in F-actins by fluorescent microscopy, the MFs described here seem to correspond to F-actins. MFs in PPBs were single, relatively short (168 ± 14 nm) filaments. In contrast, the longer (>500 nm) MFs were seen only in the cortex of interphase cells and were not bound to MTs (**Figure 2B**). This suggests that single short F-actins play an essential role in the formation of actin-MT interactions during PPB formation.

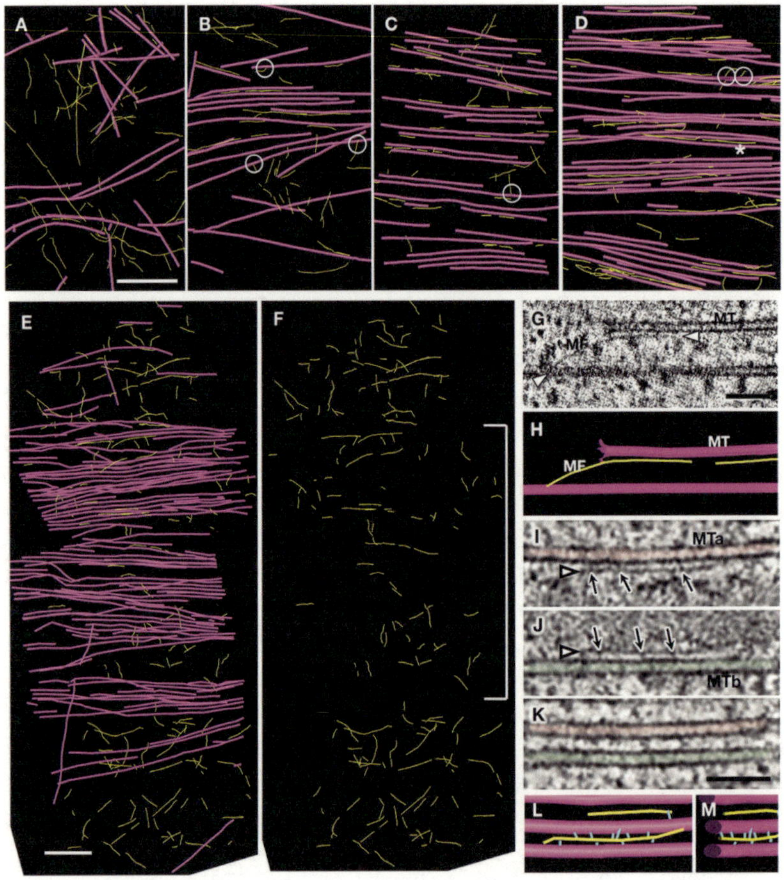

Figure 3. Changes in MT-MF interactions during PPB formation of onion cotyledon epidermal cells. All images are adapted from Figures 1, 3 and 4 in Ref. [4]. (A)–(F) Electron tomographic models showing the distribution of MFs (yellow lines) and MTs (magenta lines) in the cortical cytoplasm of tangentially sectioned cells. Circles, Contact sites between a MF end and a MT. Asterisk, A MF running along two MTs that is linked with these MTs. Bar = 500 nm. (A) Interphase cell with randomly oriented cortical MTs and MFs. (B) Interphase cell with some transversely oriented MTs. (C) Early prophase cell with a broad PPB. (D) Prophase cell containing a narrow PPB. (E) Late prophase cell with a narrow PPB. (F) Same model as shown in (E) but without the MT images. White bracket, the PPB region. (G)–(M) Images of MFs from tomograms of tangentially sectioned cell cortex regions. Bars = 100 nm. (G) A tomographic slice image showing a MF (arrowheads) associated with a cortical MT. (H) A tomographic model showing the MF and MTs illustrated in (G). (I)–(K) Three tomographic slices from a small volume of a PPB. (I) and (J) The single MF (arrowhead) is bound to two adjacent MTs, MTa (I) and MTb (J), by cross-bridges (arrows). (L) and (M) Tomographic models showing arrangement of the two MTs and one MF illustrated in (I)–(K).

Detailed analyses of the PPB tomograms revealed several interesting features of the MF-MT interactions. Single short MFs whose one end attaches to the surface of a MT are seen when PPB formation starts (circles in **Figure 3**). Subsequently, single MFs are seen running along MTs as they become attached to the MTs by cross-linkers. Two types of MF-MT cross-linkers can be distinguished. Short cross-bridges (~14 nm long) appear in early prophase and slightly longer ones (~17 nm long) during the later stages of PPB maturation [4]. However, the number of MFs forming bridges between MTs is typically low during the early stages of PPB formation but increases as the development of the PPBs progresses. Similarly, during the earliest stages of PPB formation, the length of the majority of MFs is 50–100 nm. As the PPB matures, the length of the MFs increases suggesting that they are elongated while attached to the MTs. Two types of single MFs connecting adjacent MTs are seen. One type of MF forms bridging structures between MTs (**Figure 3G**, **H**), whereas the other consists of MFs running between two adjacent MTs, with close connections to both MTs. Interestingly, the linkers to the two MTs are not in the same plane (**Figure 3I–M**). When the PPBs mature, the number of MT-associated MFs decreases, whereas the number of MT-MT interactions increases. The MT-MT bridges form tight ladder-like connections between the MTs.

Based on these observations, we have developed a hypothesis for the role of short F-actins on MT bundling during PPB formation ([4], **Figure 4**). Actin-MT interactions in PPBs start when one end of a short F-actin fragment (~70 nm) attaches to a MT (**Figure 4A**). The F-actin fragment then becomes aligned along the MT by linkers (**Figure 4B**) and starts elongating (**Figure 4C**). Elongation of the MT-associated MFs enables them to connect two MTs that are separated by spaces that are too big to be bridged by MT-MT linker proteins (**Figure 4D**). As the zippering up of the MF-linked MTs continues, the MTs come closer together with the MFs becoming sandwiched between the two parallel MTs to which they are connected through alternating cross-links (**Figure 4E**). The MF mediated cross-links are replaced by direct MT-MT bridging structures (shown in Figure 7 in Ref. [4]) and the MFs disappear from the dense MT arrays. Based on their length, the MT-MT linkers can be divided into two groups. Long linkers (**Figure 4F**) are dominant during the early stages of PPB formation and are replaced by the short ones as the MT band narrows (**Figure 4G**). Although the molecules that form the linkers have yet to be positively identified, several MT associated proteins have been localized to PPBs [99, 100].

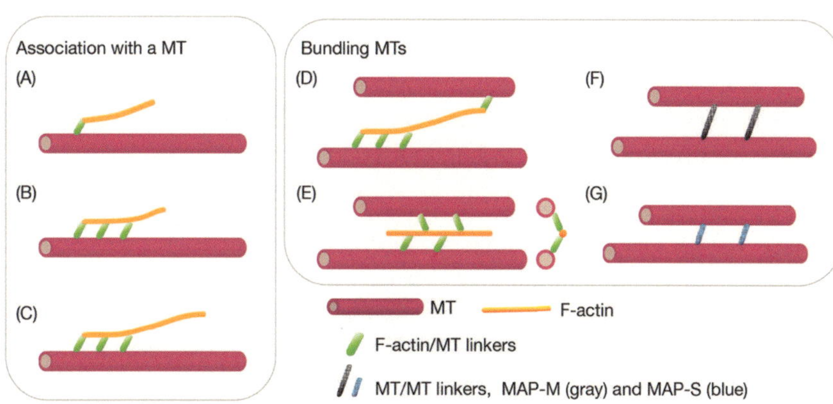

Figure 4. A model showing how single F-actins bind to MTs and contribute to MT bundling during the early stages of PPB formation. (A)–(C) Development of MT-associated MFs. (D) and (E) Two types of F-actin-MT connections between two adjacent MTs. (F) and (G) Two types of MT-MT connection by MAPs. Adapted from Ref. [4].

4. Molecular candidates involved in actin-MT interaction

The interaction between actins and MTs is mediated by linkers that connect the MFs to the PPB MTs. Here, we discuss several candidate proteins capable of binding to actins and to MTs and thereby control their organization and functions. The list includes several kinesins, formins and MT-associated proteins [101, 102].

Kinesins: One group of candidate proteins capable of forming actin-MT bridges belongs to the kinesin-14 family of proteins with actin-binding and calponin-homology domains (KCHs) reviewed in Ref. [103]. KCHs have been identified as plant-specific kinesins and belong to the minus end-directed kinesin subfamily. Members of the KCH protein family capable of forming actin-MT cross-bridges have been identified in a variety of plant species and tissues, in cotton fibers [104, 105], rice coleoptiles [106] and tobacco cells [106]. Two kinesin-14 family proteins, KingG [107] and NtKCH [108] localize to PPBs. Localization of the KingG also suggest its movement towards the minus-ends of MTs in vivo [107]. Although their function in the PPB is unknown, the KCHs may function to align F-actin to MTs and hereby contribute to the formation of MT bundles.

Formins: In plants, formins constitute a large family of proteins. Their primary function is to serve as regulators of the actin cytoskeleton. Since some plant formins have the ability to nucleate F-actins, they are considered candidate proteins for initiating the assembly of PPB F-actins [109–112]. Furthermore, considering that some formins have been shown to also possess MT-binding activity [113], they may play a role in generating MT-associated F-actins as described by Sampathkumar et al. [20]. The class II formin isoform AFH14 of *Arabidopsis* is a candidate protein for generating links between actin and MTs in PPBs, since it has been localized to PPBs. AFH14 can bind directly to and bundle actin and MTs via its FH1FH2 domain, which is a highly conserved site in formin family proteins. In addition, it shows cross-linking activity between F-actin and MT [114]. Another class II formin, the rice formin FH5 has been shown to interact with actin and MT [115, 116]. Binding of the *Arabidopsis* AtFH4 to MTs is mediated by the GOE domain, a conserved domain in the class I subfamily of *Arabidopsis* formins [117]. AtFH1, which is the main housekeeping formin in *Arabidopsis*, has also been suggested to participate in actin-MT cross-talk. Mutations of the *AtFH1* affect root cell expansion, root hair morphogenesis, and cytoskeleton structures and alter the dynamics of actin and MTs even as it lacks known MT-binding motifs [118].

Actin-related protein-2/3 (ARP2/3) complex: The ARP2/3 complex is another regulator of the F-actin in plants that can initiate actin polymerization as well as control F-actin organization and thereby cell shape. In contrast to formin that promotes the nucleation of unbranched filaments, ARP2/3 functions as an initiator of new F-actins that branch off of existing filaments [119–121]. Mutants defected in ARPC2, a subunit of ARP2/3 complex, displays aberrant trichomes, and the organization of both actin and cortical MTs is disturbed in the mutants [26, 27]. Control of the actin and MT cytoskeletal networks mediated by the ARP2/3 complex was also implied in a study of the SCAR2 mutant [122]. Recently, Havelkova et al. [123] reported that ARPC2 binds directly to MTs, suggesting a new mechanism of actin-MT interaction mediated by the ARP2/3 complex.

Other candidates: Phospholipase D (PLD) belongs to a superfamily of signaling enzymes that are associated with the plasma membrane. A PLD from tobacco has been postulated to mediate MT-plasma membrane linkages [124] and PLD activation correlated with MT reorganization [125]. PLDδ is detected in the periphery of *Arabidopsis* suspension culture cells, and it binds both to MTs and actin in vitro [126]. The PLDδ may play a role in initiating cytoskeleton remodeling. A pollen-specific MT-associated protein (MAP), SB401 in *Solanum* localizes to the cortical cytoplasm of pollen tubes where it binds to and bundles MTs and F-actins [127, 128] and they are possibly involved in F-actin and MT organization. The CLIP-associated protein (CLASP) is a MT-plus end directed motor protein. It is involved in both cell division and cell expansion, and in the organization of the cortical MTs [129, 130]. Although its interaction with actin in plants is unclear [129], human CLASP molecules have been reported to function as actin/ MT cross-linkers in interphase cells [131].

5. Conclusion

How actin-MT interactions contribute to cellular activities in plants is poorly understood. To address this question, we have employed electron tomography to examine the 3D relationship between MFs and MTs during PPB formation. At the onset, short MFs form bridges between adjacent MTs thereby starting the process of PPB narrowing. The narrowing process initially involves two types of MF-MT linkers. During later stages, the actin-mediated cross-links disappear and are replaced by two types of MT-MT linkers that act sequentially to complete MT bundling. The focus is now on identifying the different types of cross-linkers.

Author details

Miyuki Takeuchi[1], L. Andrew Staehelin[2] and Yoshinobu Mineyuki[3]*

*Address all correspondence to: mineyuki@sci.u-hyogo.ac.jp

1 Graduate School of Agricultural and Life Sciences, The University of Tokyo, Tokyo, Japan

2 MCD Biology, University of Colorado, Boulder, CO, USA

3 Graduate School of Life Science, University of Hyogo, Himeji, Hyogo, Japan

References

[1] Rodriguez OC, Schaefer AW, Mandato CA, Forscher P, Bement WM, Waterman-Storer CM: Conserved microtubule-actin interactions in cell movement and morphogenesis. Nat Cell Biol. 2003;**5**:599–609.

[2] Collings DA. Crossed-wires: interactions and cross-talk between the microtubule and microfilament networks in plants. In: Nick P, editor. Plant microtubules. Plant cell monographs. 11. Berlin, Heidelberg: Springer; 2008. p. 47–79.

[3] Mineyuki Y: The preprophase band of microtubules: its function as a cytokinetic apparatus in higher plants. Internationa review of cytology. 1999;**187**:1–49.

[4] Takeuchi M, Karahara I, Kajimura N, Takaoka A, Murata K, Misaki K, Yonemura S, Staehelin LA, Mineyuki Y: Single microfilaments mediate the early steps of microtubule bundling during preprophase band formation in onion cotyledon epidermal cells. Mol Biol Cell. 2016;**27**:1809–1820.

[5] Franke WW, Herth W, Vanderwoude WJ, Morre DJ: Tubular and filamentous structures in pollen tubes. Planta. 1972;**105**:317–341.

[6] Seagull RW, Heath IB: The effects of tannic acid on the *in vivo* preservation of microfilaments. Eur J Cell Biol. 1979;**20**:184–188.

[7] Hardham AR, Green PB, Lang JM: Reorganization of cortical microtubules and cellulose deposition during leaf formation in *Graptopetalum*. Planta. 1980;**149**:181–195.

[8] Tiwari SC, Wick SM, Williamson RE, Gunning BES: Cytoskeleton and integration of cellular function in cells of higher-plants. J Cell Biol. 1984;**99**:S63–S69.

[9] Emons AMC: The Cytoskeleton and secretory vesicles in root hairs of *Equisetum* and *Limnobium* and cytoplasmic streaming in root hairs of *Equisetum*. Ann Bot-London. 1987;**60**:625–632.

[10] Lancelle SA, Cresti M, Hepler PK: Ultrastructure of the cytoskeleton in freeze-substituted pollen tubes of *Nicotiana alata*. Protoplasma. 1987;**140**:141–150.

[11] Kobayashi H, Fukuda H, Shibaoka H: Interrelation between the spatial disposition of actin-filaments and microtubules during the differentiation of tracheary elements in cultured *Zinnia*cells. Protoplasma. 1988;**143**:29–37.

[12] Pierson ES, Kengen HMP, Derksen J: Microtubules and actin filaments co-localize in pollen tubes on *Nicotiana tabacum* L. and *Lilium longflorum*. Protoplasma. 1989;**150**:75–77.

[13] Sonobe S, Shibaoka H: Cortical fine actin filaments in higher plant cells visualized by rhodamine-phalloidin after pretreatment with m-maleimidobenzoyl N-hydroxysuccinimide ester. Protoplasma. 1989;**148**:80–86.

[14] Ding B, Turgeon R, Partharathy MV: Microfilament organization and distribution in freeze substituted tobacco plant tissues. Protoplasma. 1991;**165**:96–105.

[15] Kengen HMP, Derksen J: Organization of microtubules and microfilaments in protoplasts from suspension cells of*Nicotiana plumbaginifolia* - a quantitative-analysis. Acta Bot Neerl. 1991;**40**:29–40.

[16] Lancelle SA, Hepler PK: Association of actin with cortical microtubules revealed by immunogold localization in *Nicotiana* pollen tubes. Protoplasma. 1991;**165**:167–172.

[17] Cleary AL, Brown RC, Lemmon BE: Organisation of microtubules and actin filaments in the cortex of differentiating *Selaginella* guard cells. Protoplasma. 1993;**177**:37–44.

[18] Tominaga M, Morita K, Sonobe S, Yokota E, Shimmen T: Microtubules regulate the organization of actin filements at the cortical region in root hair cells of *Hydrocharis*. Protoplasma. 1997;**199**:83–92.

[19] Murata T, Karahara I, Kozuka T, Thomas HG, Jr., Staehelin LA, Mineyuki Y: Improved method for visualizing coated pits, microfilaments, and microtubules in cryofixed and freeze-substituted plant cells. J Electron Microsc (Tokyo). 2002;**51**:133–136.

[20] Sampathkumar A, Lindeboom JJ, Debolt S, Gutierrez R, Ehrhardt DW, Ketelaar T, Persson S: Live cell imaging reveals structural associations between the actin and microtubule cytoskeleton in *Arabidopsis*. Plant Cell. 2011;**23**:2302–2313.

[21] Seagull RW: The effects of microtubule and microfilament disrupting agents on cytoskeletal arrays and wall deposition in developing cotton fibers. Protoplasma. 1990; **159**:44–59.

[22] Kadota A, Wada M: The circular arrangement of cortical microtubules around the bubapex of tip-growing fern protonemata is sensitive to cytochalasin B. Plant Cell Physiol. 1992;**33**:99–102.

[23] Hasezawa S, Sano T, Nagata T: The role of microfilaments in the organization and orientation of microtubules during the cell cycle transition from M phase to G1 phase in tobacco BY-2 cells. Protoplasma. 1998;**202**:105–114.

[24] Takesue K, Shibaoka H: The cyclic reorientation of cortical microtubules in epidermal cells of azuki bean epicotyls. Planta. 1998;**205**:539–546.

[25] Blancaflor EB: Cortical actin filaments potentially interact with cortical microtubules in regulating polarity of cell expansion in primary roots of maize (*Zea mays* L.). J Plant Growth Regul. 2000;**19**(4):406–414.

[26] Schwab B, Mathur J, Saedler RR, Schwarz H, Frey B, Scheidegger C, Hulskamp M: Regulation of cell expansion by the DISTORTED genes in *Arabidopsis thaliana*. Mol Genet Genomics. 2003;**269**:350–360.

[27] Saedler R, Mathur N, Srinivas BP, Kernebeck B, Hulskamp M, Mathur J: Actin control over microtubules suggested by DISTORTED2 encoding the *Arabidopsis* ARPC2 subunit homolog. Plant Cell Physiol. 2004;**45**:813–822.

[28] Timmers AC, Vallotton P, Heym C, Menzel D: Microtubule dynamics in root hairs of *Medicago truncatula*. Eur J Cell Biol. 2007;**86**:69–83.

[29] Sainsbury F, Collings DA, Mackun K, Gardiner J, Harper JD, Marc J: Developmental reorientation of transverse cortical microtubules to longitudinal directions: a role for actomyosin-based streaming and partial microtubule-membrane detachment. Plant J. 2008;**56**:116–131.

[30] Chu B, Kerr GP, Carter JV: Stabilizing microtubules with taxol increases microfilament stability during freezing of rye root-tips. plant cell and environment. 1993;**16**:883–889.

[31] Collings DA, Asada T, Allen NS, Shibaoka H: Plasma membrane-associated actin in bright yellow 2 tobacco cells. Evidence for interaction with microtubules. Plant Physiol. 1998;**118**:917–928.

[32] Collings DA, Wasteneys GO: Actin microfilament and microtubule distribution patterns in the expanding root of *Arabidopsis thaliana*. Can J Bot. 2005;**83**(6):579–590.

[33] Collings DA, Lill AW, Himmelspach R, Wasteneys GO: Hypersensitivity to cytoskeletal antagonists demonstrates microtubule-microfilament cross-talk in the control of root elongation in *Arabidopsis thaliana*. New Phytol. 2006;**170**:275–290.

[34] Smertenko AP, Deeks MJ, Hussey PJ: Strategies of actin reorganisation in plant cells. J Cell Sci. 2010;**123**:3019–3028.

[35] Shevchenko G: Actin microfilament organization in the transition zone of *Arabidopsis*-ABD2-GFP roots under clinorotation. Microgravity Sci Tec. 2012;**24**:427–433.

[36] Kakimoto T, Shibaoka H: Actin filaments and mocrotubules in the preprophase band and phragmoplast of Tobacco cells. Protoplasma. 1987;**140**:151–156.

[37] Palevitz BA: Actin in the preprophase band. J Cell Biol. 1987;**104**:1515–1519.

[38] Traas JA, Doonan JH, Rawlins DJ, Shaw PJ, Watts J, Lloyd CW: An actin network is present in the cytoplasm throughout the cell cycle of carrot cells and associates with the dividing nucleus. J Cell Biol. 1987;**105**:387–395.

[39] Lloyd CW, Traas JA: The role of F-actin in determining the division plane of carrot suspension cells. Drug studies. Development. 1988;**102**:211–221.

[40] McCurdy DW, Sammut M, Gunning BES: Immunofluorescent visualization of arrays of transverse cortical actin microfilaments in wheat root-tip cells. Protoplasma. 1988; **147**:204–206.

[41] Palevitz BA: Cytochalasin-induced reorganization of actin in *Allium* root cells. Cell Motil Cytoskeleton. 1988;**9**:283–298.

[42] Katsuta J, Hashiguchi Y, Shibaoka H: The Role of the cytoskeleton in positioning of the nucleus in premitotictobaccoBY-2 Cells. J Cell Sci. 1990;**95**:413–422.

[43] McCurdy DW, Gunning ES: Reorganizaztion of cortical actin microfilaments and microtubules at preprophase and mitosis in wheat root-tip cells: a double immunofluorescence study. Cell motil Cytosleteton. 1990;**15**:76–87.

[44] Mineyuki Y, Palevitz BA: Relationship between preprophase band organization, F-actin and the division site in *Allium*. J Cell Sci. 1990;**97**:283–295.

[45] Ding B, Turgeon R, Partharathy MV: Microfilaments in the preprophase band of freeze substituted tobacco root cells. Protoplasma. 1991;**165**:209–211.

[46] Cleary AL, Gunning BES, Wasteneys GO, Hepler PK: Microtubule and F-actin dynamics at the division site in living tradescantia stamen hair-cells. J Cell Sci. 1992;**103**:977–988.

[47] Eleftheriou EP, Palevitz BA: The effect of cytochalasin D on preprophase band organization in root tip cells of *Allium*. J Cell Sci. 1992;**103**:989–998.

[48] Liu B, Palevitz BA: Organization of cortical microfilaments in dividing root cells. Cell Motili Cytoskelton. 1992;**23**:252–264.

[49] Panteris E, Apostolakos P, Galatis B: The organization of F-actin in root tip cells of *Adiantu capillus veneris* througout the cell cycle. Protoplasma. 1992;**170**:128–137.

[50] Cleary AL: F-Actin Redistributions at the division site in living *Tradescantia* stomatal complexes as revealed by microinjection of rhodamine-phalloidin. Protoplasma. 1995;**185**:152–165.

[51] Cleary AL, Mathesius U: Rearrangements of F-actin during stomatogenesis visualised by confocal microscopy in fixed and permeabilised *Tradescantia* leaf epidermis. Bot Acta. 1996;**109**:15–24.

[52] Baluska F, Vitha S, Barlow PW, Volkmann D: Rearrangements of F-actin arrays in growing cells of intact maize root apex tissues. European J Cell Biol. 1997;**72**:113–121.

[53] Zachariadis M, Quader H, Galatis B, Apostolakos P: Endoplasmic reticulum preprophase band in dividing root-tip cells of *Pinus brutia*. Planta. 2001;**213**:824–827.

[54] Sano T, Higaki T, Oda Y, Hayashi T, Hasezawa S: Appearance of actin microfilament 'twin peaks' in mitosis and their function in cell plate formation, as visualized in tobacco BY-2 cells expressing GFP-fimbrin. Plant J. 2005;**44**:595–605.

[55] Li CL, Chen ZL, Yuan M: Actomyosin is involved in the organization of the microtubule preprophase band in *Arabidopsis* suspension cultured cells. J Integrative Plant Biol. 2006;**48**:53–61.

[56] Higaki T, Sano T, Hasezawa S: Actin microfilament dynamics and actin side-binding proteins in plants. Curr Opin Plant Biol. 2007;**10**:549–556.

[57] Panteris E, Galatis B, Quader H, Apostolakos P: Cortical actin filament organization in developing and functioning stomatal complexes of *Zea mays* and *Triticum turgidum*. Cell Motil Cytoskeleton. 2007;**64**:531–548.

[58] Takeuchi M, Mineyuki Y. Plate 6.13 Actin-microtubule interaction during preprophase band formation in onion root tips visualized by immuno-fluorescence microscopy. In: Noguchi T, Kawano S, Tsukaya H, Matsunaga S, Sakai A, Karahara I, et al., editors. Atlas of plant cell structure. Tokyo: Springer Verlag Tokyo; 2014.

[59] Granger CL, Cyr RJ: Use of abnormal preprophase bands to decipher division plane determination. J Cell Sci. 2001;**114**:599–607.

[60] Kojo KH, Higaki T, Kutsuna N, Yoshida Y, Yasuhara H, Hasezawa S: Roles of cortical actin microfilament patterning in division plane orientation in plants. Plant Cell Physiol. 2013;**54**:1491–1503.

[61] Forer A, Jackson WT: Actin-filaments in endosperm mitotic spindles in a higher plant, *Haemanthus-Katerinae* Baker. Cytobiologie. 1976;**12**:199–214.

[62] Forer A, Jackson WT: Actin in spindles of *Haemanthus katherinae* endosperm. I. General results using various glycerination methods. J Cell Sci. 1979;**37**:323–347.

[63] Forer A, Jackson WT, Engberg A: Actin in spindles of *Haemanthus katherinae* endosperm. II. Distribution of actin in chromosomal spindle fibres, determined by analysis of serial sections. J Cell Sci. 1979;**37**:349–371.

[64] Seagull RW, Falconer MM, Weerdenburg CA: Microfilaments: dynamic arrays in higher plant cells. J Cell Biol. 1987;**104**:995–1004.

[65] Traas JA, Burgain S, Devaulx RD: The Organization of the cytoskeleton during meiosis in eggplant (*Solanum-Melongena* (L)). J Cell Sci. 1989;**92**:541–550.

[66] Czaban BB, Forer A: Rhodamine-labeled phalloidin stains components in the chromosomal spindle fibers of crane-fly spermatocytes and *Haemanthus* endosperm cells. Biochem Cell Biol. 1992;**70**(8):664–676.

[67] Clayton L, Lloyd CW: Actin organization during the cell cycle in meristematic plant cells. Actin is present in the cytokinetic phragmoplast. Exp Cell Res. 1985;**156**:231–238.

[68] Gunning BE, Wick SM: Preprophase bands, phragmoplasts, and spatial control of cytokinesis. J Cell Sci Suppl. 1985;**2**:157–179.

[69] Schmit AC, Vantard M, Lambert AM. Microtubules and F-actin rearrangement during the initiation of mitosis in acentriolar higher plant cells. In: lshikawa H, Hatano S, Sato H, editors. Cell Motility: mechanism and regulation. Tokyo: University of Tokyo Press; 1985. p. 415–433.

[70] Palevitz BA: Accumulation of F-actin during cytokinesis in*Allium*. Correlation with microtubule distribution and the effects of drugs. Protoplasma. 1987;**141**:24–32.

[71] Schmit AC, Lambert AM: Characterization and dynamics of cytoplasmic F-actin in higher plant endosperm cells during interphase, mitosis, and cytokinesis. J Cell Biol. 1987;**105**:2157–2166.

[72] Kakimoto T, Shibaoka H: Cytoskeletal ultrastructure of phragmoplast -nuclei complexes isolated from cultured tobacco cells. In: Tazawa M, editor. Cell dynamics 2: molecular aspects of cell motility cytoskeleton in cellular structure and activity. Wien: Springer; 1988. p. 95–103.

[73] Schmit AC, Lambert AM: Microinjected fluorescent phalloidin *in vivo* reveals the F-actin dynamics and assembly in higher plant mitotic cells. Plant Cell. 1990;**2**:129–138.

[74] Zhang D, Wadsworth P, Hepler PK: Dynamics of microfilaments are similar, but distinct from microtubules during cytokinesis in living, dividing plant cells. Cell Motil Cytoskeleton. 1993;**24**:151–155.

[75] Collings DA, Harper JDI, Vaughn KC: The association of peroxisomes with the developing cell plate in dividing onion root cells depends on actin microfilaments and myosin. Planta. 2003;**218**:204–216.

[76] Lloyd C, Chan J: Microtubules and the shape of plants to come. Nat Rev Mol Cell Biol. 2004;**5**:13–22.

[77] Paredez AR, Somerville CR, Ehrhardt DW: Visualization of cellulose synthase demonstrates functional association with microtubules. Science. 2006;**312**:1491–1495.

[78] Oda Y, Fukuda H: Rho of plant GTPase signaling regulates the behavior of *Arabidopsis* kinesin-13A to establish secondary cell wall patterns. Plant Cell. 2013;**25**:4439–4450.

[79] Kobayashi H, Fukuda H, Shibaoka H: Reorganization of actin filaments associated with the differentiation of tracheary elements in*Zinnia* mesophyll cells. Protoplasma. 1987;**138**:69–71.

[80] Panteris E, Galatis B: The morphogenesis of lobed plant cells in the mesophyll and epidermis: organization and distinct roles of cortical microtubules and actin filaments. New Phytol. 2005;**167**:721–732.

[81] Smith LG, Oppenheimer DG: Spatial control of cell expansion by the plant cytoskeleton. Annu Rev Cell Dev Biol. 2005;**21**:271–295.

[82] Kotzer AM, Wasteneys GO: Mechanisms behind the puzzle: microtubule–microfilament cross-talk in pavement cell formation. Can J Bot. 2006;**84**:594–603.

[83] Yanagisawa M, Desyatova AS, Belteton SA, Mallery EL, Turner JA, Szymanski DB: Patterning mechanisms of cytoskeletal and cell wall systems during leaf trichome morphogenesis. Nat Plants. 2015;**1**:15014.

[84] Sambade A, Findlay K, Schaffner AR, Lloyd CW, Buschmann H: Actin-dependent and independent functions of cortical microtubules in the differentiation of *Arabidopsis* leaf trichomes. Plant Cell. 2014;**26**:1629–1644.

[85] Schmit A-C, Lambert A-M: Plant actin filament and microtubule interactions during anaphase - telophase transition: effects of antagonist drugs. Biol Cell. 1988;**64**:309–319.

[86] Nagai R, Rebhun LI: Cytoplasmic microfilaments in streaming *Nitella* cells. J Ultrastruct Res. 1966;**14**:571–589.

[87] Palevitz BA, Ash JF, Hepler PK: Actin in the green alga, *Nitella*. Proc Natl Acad Sci U S A. 1974;**71**:363–366.

[88] Heath IB, Seagull RW. Oriented cellulose fibrils and the cytoskeleton: a critical conparison of models In: Lloyd CW, editor. The cytoskeleton and plant growth and development. London, New York: Academic Press; 1982. p. 163–182.

[89] Kersey YM, Hepler PK, Palevitz BA, Wessells NK: Polarity of actin-filaments in *Characean*algae. P Natl Acad Sci USA. 1976;**73**:165–167.

[90] Kersey YM, Wessells NK: Localization of actin-filaments in internodal cells of *Characean*algae. J Cell Biol. 1976;**68**:264–275.

[91] Tiwari SC, Polito VS: Organization of the cytoskeleton in pollen tubes of P*yrus communis*. Protoplasma. 1988;**147**:100–112.

[92] Urban E, Jacob S, Nemethova M, Resch GP, Small JV: Electron tomography reveals unbranched networks of actin filaments in lamellipodia. Nat Cell Biol. 2010;**12**:429–435.

[93] Gibeaux R, Lang C, Politi AZ, Jaspersen SL, Philippsen P, Antony C: Electron tomography of the microtubule cytoskeleton in multinucleated hyphae of *Ashbya gossypii*. J Cell Sci. 2012;**125**:5830–5839.

[94] Segui-Simarro JM, Austin JR, 2nd, White EA, Staehelin LA: Electron tomographic analysis of somatic cell plate formation in meristematic cells of *Arabidopsis* preserved by high-pressure freezing. Plant Cell. 2004;**16**:836–856.

[95] Austin JR, 2nd, Segui-Simarro JM, Staehelin LA: Quantitative analysis of changes in spatial distribution and plus-end geometry of microtubules involved in plant-cell cytokinesis. J Cell Sci. 2005;**118**:3895–3903.

[96] Karahara I, Suda J, Tahara H, Yokota E, Shimmen T, Misaki K, Yonemura S, Staehelin LA, Mineyuki Y: The preprophase band is a localized center of clathrin-mediated endocytosis in late prophase cells of the onion cotyledon epidermis. Plant J. 2009; **57**:819–831.

[97] Mineyuki Y, Marc J, Palevitz BA: Development of the preprophase band from random cytoplasmic microtubules in guard mother cells of *Allium cepa*. Planta. 1989;**178**:291–296.

[98] Kremer JR, Mastronarde DN, McIntosh JR: Computer visualization of three-dimensional image data using IMOD. J Struc Biol. 1996;**116**:71–76.

[99] McMichael CM, Bednarek SY: Cytoskeletal and membrane dynamics during higher plant cytokinesis. New Phytol. 2013;**197**:1039–1057.

[100] Hamada T: Microtubule organization and microtubule-associated proteins in plant cells. Int Rev Cell Mol Biol. 2014;**312**:1–52.

[101] Krtkova J, Benakova M, Schwarzerova K: Multifunctional microtubule-associated proteins in plants. Front Plant Sci. 2016;**7**:474.

[102] Petrasek J, Schwarzerova K: Actin and microtubule cytoskeleton interactions. Curr Opin Plant Biol. 2009;**12**:728–734.

[103] Schneider R, Persson S: Connecting two arrays: The emerging role of actin-microtubule cross-linking motor proteins. Front Plant Sci. 2015;**6**:415

[104] Preuss ML, Kovar DR, Lee YR, Staiger CJ, Delmer DP, Liu B: A plant-specific kinesin binds to actin microfilaments and interacts with cortical microtubules in cotton fibers. Plant Physiol. 2004;**136**:3945–3955.

[105] Xu T, Qu Z, Yang X, Qin X, Xiong J, Wang Y, Ren D, Liu G: A cotton kinesin GhKCH2 interacts with both microtubules and microfilaments. Biochem J. 2009;**42**:171–180.

[106] Frey N, Klotz J, Nick P: Dynamic Bridges-acalponin-domain kinesin from rice links actin filaments and microtubules in both cycling and non-cycling cells. Plant Cell Physiol. 2009;**50**:1493–1506.

[107] Buschmann H, Green P, Sambade A, Doonan JH, Lloyd CW: Cytoskeletal dynamics in interphase, mitosis and cytokinesis analysed through Agrobacterium-mediated transient transformation of tobacco BY-2 cells. New Phytol. 2011;**190**:258–267.

[108] Klotz J, Nick P: A novel actin-microtubule cross-linking kinesin, NtKCH, functions in cell expansion and division. New Phytol. 2012;**193**:576–589.

[109] Banno H, Chua N-H: Characterization of the *Arabidopsis* formin-like protein AFH1 and its interacting protein. Plant Cell Physiol. 2000;**41**:617–626.

[110] Blanchoin L, Staiger CJ: Plant formins: diverse isoforms and unique molecular mechanism. Biochim Biophys Acta. 2010;**1803**:201–206.

[111] Cheung AY, Wu HM: Overexpression of an *Arabidopsis* formin stimulates supernumerary actin cable formation from pollen tube cell membrane. Plant Cell. 2004;**16**:257–269.

[112] Favery B, Chelysheva LA, Lebris M, Jammes F, Marmagne A, De Almeida-Engler J, Lecomte P, Vaury C, Arkowitz RA, Abad P: *Arabidopsis* formin AtFH6 is a plasma membrane-associated protein upregulated in giant cells induced by parasitic nematodes. Plant Cell. 2004;**16**:2529–2540.

[113] Wang J, Xue X, Ren H: New insights into the role of plant formins: regulating the organization of the actin and microtubule cytoskeleton. Protoplasma. 2012;**249**:S101–107.

[114] Li Y, Shen Y, Cai C, Zhong C, Zhu L, Yuan M, Ren H: The type II *Arabidopsis* formin14 interacts with microtubules and microfilaments to regulate cell division. Plant Cell. 2010;**22**:2710–2726.

[115] Yang W, Ren S, Zhang X, Gao M, Ye S, Qi Y, Zheng Y, Wang J, Zeng L, Li Q, Huang S, He Z: BENT UPPERMOST INTERNODE1 encodes the class II formin FH5 crucial for actin organization and rice development. Plant Cell. 2011;**23**:661–680.

[116] Zhang Z, Zhang Y, Tan H, Wang Y, Li G, Liang W, Yuan Z, Hu J, Ren H, Zhang D: RICE MORPHOLOGY DETERMINANT encodes the type II formin FH5 and regulates rice morphogenesis. Plant Cell. 2011;**23**:681–700.

[117] Deeks MJ, Fendrych M, Smertenko A, Bell KS, Oparka K, Cvrckova F, Zarsky V, Hussey PJ: The plant formin AtFH4 interacts with both actin and microtubules, and contains a newly identified microtubule-binding domain. J Cell Sci. 2010;**123**:1209–1215.

[118] Rosero A, Zarsky V, Cvrckova F: AtFH1 formin mutation affects actin filament and microtubule dynamics in *Arabidopsis thaliana*. J Exp Bot. 2013;**64**:585–597.

[119] Goley ED, Welch MD: The ARP2/3 complex: an actin nucleator comes of age. Nat Rev Mol Cell Biol. 2006;**7**:713–726.

[120] Deeks MJ, Hussey PJ: Arp2/3 and SCAR: plants move to the fore. Nat Rev Mol Cell Biol. 2005;**6**:954–964.

[121] Yanagisawa M, Zhang C, Szymanski DB: ARP2/3-dependent growth in the plant kingdom: SCARs for life. Front Plant Sci. 2013;**4**:166.

[122] Zhang X, Dyachok J, Krishnakumar S, Smith LG, Oppenheimer DG: IRREGULAR TRICHOME BRANCH1 in *Arabidopsis* encodes a plant homolog of the actin-related protein2/3 complex activator Scar/WAVE that regulates actin and microtubule organization. Plant Cell. 2005;**17**:2314–2326.

[123] Havelkova L, Nanda G, Martinek J, Bellinvia E, Sikorova L, Slajcherova K, Seifertova D, Fischer L, Fiserova J, Petrasek J, Schwarzerova K: Arp2/3 complex subunit ARPC2 binds to microtubules. Plant Sci. 2015;**241**:96–108.

[124] Gardiner JC, Harper JD, Weerakoon ND, Collings DA, Ritchie S, Gilroy S, Cyr RJ, Marc J: A 90-kD phospholipase D from tobacco binds to microtubules and the plasma membrane. Plant Cell. 2001;**13**:2143–2158.

[125] Dhonukshe P, Laxalt AM, Goedhart J, Gadella TW, Munnik T: Phospholipase D activation correlates with microtubule reorganization in living plant cells. Plant Cell. 2003;**15**:2666–2679.

[126] Ho AYY, Day DA, Brown MH, Marc J: *Arabidopsis* phospholipase Dδ as an initiator of cytoskeleton-mediated signalling to fundamental cellular processes. Funct Plant Biol. 2009;**36**:190–198.

[127] Huang S, Jin L, Du J, Li H, Zhao Q, Ou G, Ao G, Yuan M: SB401, a pollen-specific protein from *Solanum berthaultii*, binds to and bundles microtubules and F-actin. Plant J. 2007;**51**:406–418.

[128] Liu BQ, Jin L, Zhu L, Li J, Huang S, Yuan M: Phosphorylation of microtubule-associated protein SB401 from *Solanum berthaultii* regulates its effect on microtubules. J Integr Plant Biol. 2009;**51**:235–242.

[129] Ambrose JC, Shoji T, Kotzer AM, Pighin JA, Wasteneys GO: The *Arabidopsis* CLASP gene encodes a microtubule-associated protein involved in cell expansion and division. Plant Cell. 2007;**19**:2763–2775.

[130] Ambrose JC, Wasteneys GO: CLASP modulates microtubule-cortex interaction during self-organization of acentrosomal microtubules. Mol Biol Cell. 2008;**19**:4730–4737.

[131] Tsvetkov AS, Samsonov A, Akhmanova A, Galjart N, Popov SV: Microtubule-binding proteins CLASP1 and CLASP2 interact with actin filaments. Cell Motil Cytoskeleton. 2007;**64**:519–530.

Chapter 3

Morphological and Functional Aspects of Cytoskeleton of Trypanosomatids

Juliana Cunha Vidal and Wanderley de Souza

Additional information is available at the end of the chapter

http://dx.doi.org/10.5772/66859

Abstract

Trypanosomatidae are protozoans that include monogenetic parasites, such as the *Blastocrithidia* and *Herpetomonas* genera, as well as digenetic parasites, such as the *Trypanosoma* and *Leishmania* genera. Their life cycles alternate between insect vectors and mammalian hosts. The parasite's life cycle involves symmetrical division and different transitional developmental stages. In trypanosomatids, the cytoskeleton is composed of subpellicular microtubules organized in a highly ordered array of stable microtubules located beneath the plasma membrane, the paraflagellar rod, which is a lattice-like structure attached alongside the flagellar axoneme and a cytostome-cytopharynx. The complex life cycle, the extremely precise cytoskeletal organization and the single copy structures present in trypanosomatids provide interesting models for cell biology studies. The introduction of molecular biology, FIB/SEM (focused ion beam scanning electron microscopy) and electron microscopy tomography approaches and classical methods, such as negative staining, chemical fixation and ultrafast cryofixation have led to the determination of the three-dimensional (3D) structural organization of the cells. In this chapter, we highlight the recent findings on Trypanosomatidae cytoskeleton emphasizing their structural organization and the functional role of proteins involved in the biogenesis and duplication of cytoskeletal structures. The principal finding of this review is that all approaches listed above enhance our knowledge of trypanosomatids biology showing that cytoskeleton elements are essential to several important events throughout the protozoan life cycle.

Keywords: trypanosomatids, cytoskeleton, ultrastructure, microscopy, three-dimensional reconstruction

1. Introduction

Trypanosomatids are uniflagellated protozoan parasites belonging to the Kinetoplastid order. They are the etiological agents of several diseases [1] and exhibit particular features that differentiate them significantly from their mammalian host [2]. First, they have subpellicular microtubules (SPMT), which are a network of organized stable microtubules that are closely associated with the plasma membrane and to each other forming a corset that confers rigidity to the cell body and help to determine the shape of the cell. Second, trypanosomatids have the microtubule quartet (MtQ) of the flagellar pocket (FP) that encircles the flagellar pocket in a helicoidal pattern [3]. Third, they have microtubule sets of cytostome-cytopharynx that forms a gutter in this funnel invagination [4]. Fourth, trypanosomatids have a flagellum attachment zone (FAZ) where the flagellum emerges from the flagellar pocket and remains attached to the cell body [5]. Finally, they have a paraflagellar rod (PFR), a lattice-like structure that runs parallel to the axoneme from the flagellar pocket to the flagellar tip [6].

Recent studies using techniques, such as electron tomography and focused ion beam-scanning electron microscopy, that allow three-dimensional reconstruction of whole protozoan, allowed for the accurate understanding of their subcellular morphology. High-resolution microscopy studies have provided detailed cellular information regarding protein localization and phenotype after cytoskeleton-protein depletion aiming at the elucidation of protein function during life cycle of trypanosomatids. Using FIB/SEM, one can examine a large number of whole cells at the same time, which enables qualitative and quantitative studies about different morphological aspects of cytoskeleton elements during the life cycle of trypanosomatids. This chapter aims to provide an overview of the topographical relationship among the cytoskeletal elements throughout the protozoan life cycle.

2. Subpellicular microtubules

The shape of cells in trypanosomatids is defined by SPMT. SPMT are a cage of stable microtubules located underneath the plasma membrane and composed of α/β tubulin [7]. High-resolution field emission scanning electron microscopy (FESEM) revealed a helicoidal pattern of SPMT in the promastigote form of nonpathogenic *Herpetomonas megaseliae*; whereas in the procyclic trypomastigote form of *Trypanosoma brucei*, it appears as a straight pattern [8] (**Figure 1A–C**).

In contrast to the microtubules of mammalian cells, SPMT are resistant to low temperatures and drugs that usually promote microtubule depolymerization [10]; thus, SPMT contribute to the stabilization of the trypanosomatids shape. Transversal ultrathin sections of *Leptomonas samueli* fixed with glutaraldehyde and tannic acid showed that trypanosomatids' SPMT are formed by 13 typical protofilaments [11]. Immunolabeling studies using *T. brucei* revealed that the region of tubulin polymerization is localized at the posterior region of the procyclic forms, which is consistent with the identification of the plus end of microtubules at this region [12].

These SPMT are cross-linked to each other by short 6 nm thick filaments and to the plasma membrane [9, 13] (**Figure 1D**). Molecular studies using *T. brucei* indicated the presence of microtubule-associated proteins (MAPs) linking SPMT to each other [14]. These links are particularly strong and are not disrupted by cell lysis. It is likely that they play a role in the rigidity of the trypanosomes' cell bodies [15]. Moreover, a connection between SPMT and the endoplasmic reticulum was observed in *Leishmania amazonensis* [16]. The transmission electron microscopy image of thin sections showed that the distance between the SPMT of

http://dx.doi.org/10.5772/66859

trypanosomatids is 44 nm and approximately 15 nm between SPMT and the plasma membrane. The number of microtubules varies by the region of the cell body. In *Trypanosoma cruzi* trypomastigotes, the posterior region where the Golgi apparatus is located has approximately 120 microtubules. At the cell extremities, approximately 40 microtubules were counted. In the intracellular dividing amastigote, the highest number of microtubules, 222, was observed. These data suggest that newly formed microtubules are inserted between preexisting microtubules during the cell division, because the number of microtubules remained constant [17]. A study on the biogenesis of subpellicular microtubules in *T. brucei* indicated that new microtubules are incorporated primarily in a region adjacent to the new FAZ and between the new and the old FAZs during cell growth [18]. The addition of new microtubules into the FAZ region facilitates the segregation of the basal bodies and consequently, mitochondrial genome segregation. Furthermore, this incorporation promotes the formation of the membrane fold in preparation for cell division [12, 19, 20].

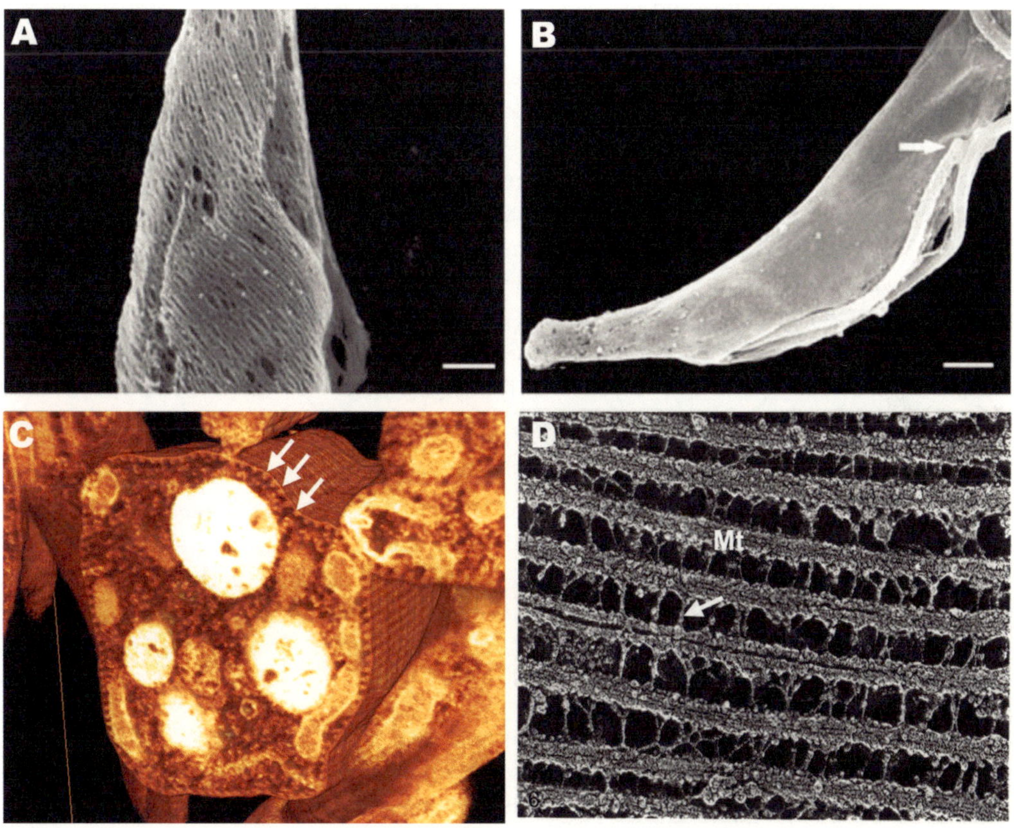

Figure 1. Ultrastructural pattern of subpellicular microtubules (SPMT). (A–B) Field emission scanning electron micrographs of whole-extracted *H. megaseliae* and *T. brucei*, respectively, showing the SPMT. (A) The helical pattern of *H. megaseliae* SPMT and (B) The straight pattern of *T. brucei* SPMT [8]. (C) Several images of *T. cruzi* epimastigotes obtained by focused ion beam scanning electron microscopy (FIB-SEM) were joined in order to emphasizing the SPMT cage (arrows) underneath the plasma membrane [unpublished data from Wanderley de Souza]. (D) Promastigotes of *H. megaseliae* were extracted using Triton-X-100, revealing the parallel array of SPMT. Additionally, the fibrils connecting SPMT to each other were observed (arrow) [9]. Scale bars: (A) 3.5 μm and (B) 8 μm.

Due to the rigidity of the SPMT cage beneath the cell membrane, endocytic events are limited to sites where these microtubules are absent: the flagellar pocket and the complex cytostome-cytopharynx.

3. Cytoskeletal elements associated with endocytic entry sites

3.1. Microtubules

The FP is an invagination close to the basal body and surrounds the site of flagellum exit. It is found at the anterior and posterior regions of the cell body of *T. cruzi* epimastigotes and the procyclic form of *T. brucei,* respectively. It is involved in cell polarity and cell division. The FP membrane is compositionally and functionally distinct from both the cell body and flagellar membranes, despite being continuous with them [21]. The absence of SPMT at the FP site allows budding of endocytic and exocytic vesicles. It is also essential for defense against the innate and acquired immune responses of the host [22, 23]. Trypanosomatids' FP is precisely positioned and it is closely related to cytoskeletal elements which are highly conserved throughout trypanosomatids. Flagellar pocket architecture and composition have been described in several studies; however, it was best analyzed in the African trypanosome, *T. brucei* [22, 24–29]. Electron microscopy tomography revealed a set of MtQ that are nucleated adjacent to the basal bodies [3] (**Figure 2A** and **C**). Their polarity is opposite (plus end at the anterior of the cell) that of the subpellicular microtubules. MtQ encircles the FP and forms a membrane dislocation, as a deep and longitudinal channel through which nutrients achieve the lumen of the flagellar pocket [30] (**Figure 2A**). These microtubules are different from the SPMT because they do not depolymerize in media with high salt concentrations [31]. The same MtQ, with a helicoidal pattern located close to the FP, was observed in a 3D reconstruction of *T. cruzi* epimastigotes [4] (**Figure 2B** and **C**).

Several studies with *T. brucei* have shown that FP is divided into several domains, such as the neck of the FP. In the neck, MtQ joins SPMT and is associated with FAZ; thus, MtQ defines the flagellum and flagellar pocket axes [32]. The flagellar collar is a horseshoe-shaped, electron-dense cytoskeletal structure located in the flagellar neck; the first protein characterized at this domain was BILBO1. Using RNA interference (RNAi)-mediated ablation of this protein in the procyclic form of *T. brucei,* it was shown that BILBO1 is essential for flagellar pocket biogenesis and cell survival [29].

T. brucei also possesses a cytoskeletal feature of unknown function, named bilobe. TbMORN1 was the first protein described as an exclusive component of the bilobe [33]. The bilobe is located near the Golgi apparatus; it is thought to function in Golgi apparatus replication and as an adaptor during cytokinesis [34]. Recently, TbMORN1 depletion experiments suggested that this protein allows access of endocytic tracers into the flagellar pocket [35].

Although *T. cruzi* epimastigotes and amastigotes uptake nutrients by endocytosis via the flagellar pocket they primarily obtain nutrients from the cytostome-cytopharynx complex [36–38]. The latter was also found in a free-living kinetoplastid, *Bodo sp.* [39] as well as in

Trypanosoma vespertilionis and *Trypanosoma dionisii* [15]. The cytostome is an opening in the anterior region of the cell surface that is followed by a membrane invagination called cytopharynx. Recently, using serial electron microscopy tomography and FIB-SEM to reconstruct the entire length of the cytostome-cytopharynx present in *T. cruzi*, it was shown that this invagination is supported by seven microtubules. A triplet of these microtubules was located underneath the cytostome aperture membrane. A quartet of microtubules originated underneath the flagellar pocket membrane in a staggered formation and followed the preoral ridge before reaching the cytopharynx invagination [4] (**Figure 3A**). This quartet was different from the MtQ described in *T. brucei*. These two sets of microtubules assisted the cytopharynx in forming a "gutter" and creating a microtubule-free side where vesicles can bud or fuse.

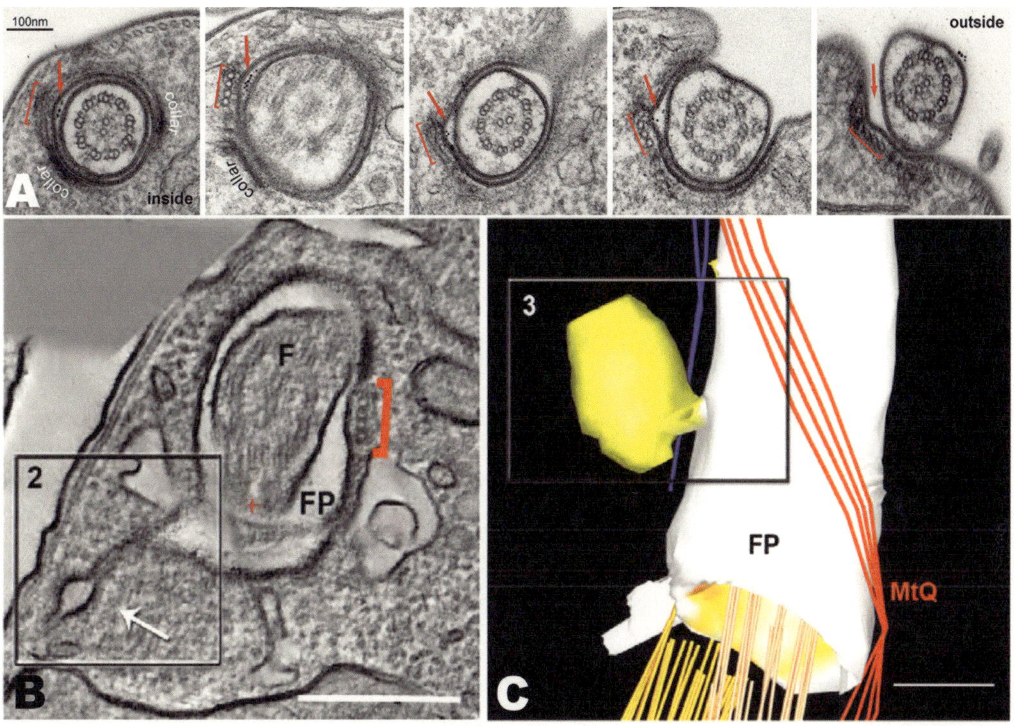

Figure 2. Microtubule quartet (MtQ) is associated with the FP of *Trypanosoma brucei* and *T. cruzi*. (A) Electron micrographs showing the MtQ (brackets) forming a channel that permits endocytic tracers (red arrows) to enter into the *T. brucei* FP [30]; (B–C) the MtQ is nucleated near the basal bodies and encircles the FP (brackets) *T. cruzi* epimastigotes [4]. Scale bars: 200 nm.

Electron microscopy tomography and dual beam scanning electron microscopy are used to observe epimastigotes of *T. cruzi* that were synchronized using hydroxyurea, a compound that induces G1/S cell cycle arrest, we observed that the cytostome-cytopharynx complex is completely disassembled during cell division. This complex is formed *de novo* from the flagellar pocket membrane during cytokinesis. This experiment elucidated the biogenesis of this structure [40] (**Figure 3B1–B3**).

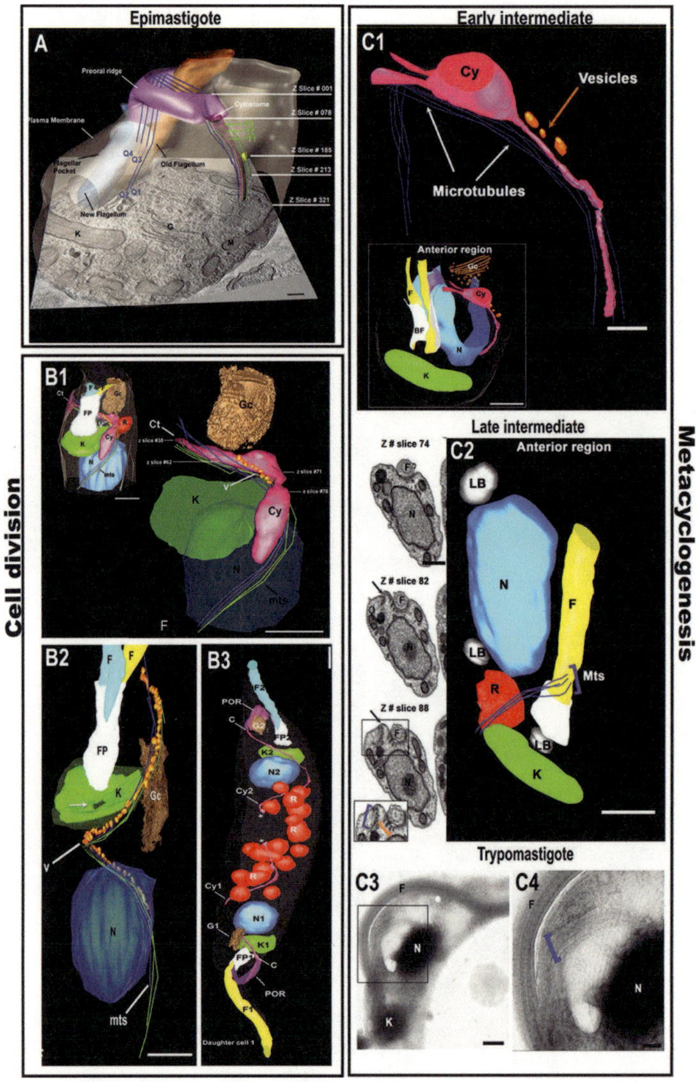

Figure 3. Ultrastructural changes of the cytostome-cytopharynx complex during the *T. cruzi* life cycle. (A-C1) Tomographic reconstruction of the cytostome-cytopharynx complex. (A) The cytostome-cytopharynx complex of epimastigotes is sustained by a microtubule triplet (green) that extends from underneath the cytostome membrane and a microtubule quartet (blue) that appears to start close to the flagellar pocket. It continues underneath the preoral ridge membrane [4]. (B) Cell division: (B1) in early G2, the cytopharynx was shorter. (B2) After this, the cytostome-cytopharynx complex becomes fragmented during the G2 phase. The microtubules maintained their helical arrangement. The vesicles were aligned with the microtubules, following the path of the cytopharynx. (B3) At the end of cytokinesis, both daughter cells possess a complete cytostome-cytopharynx complex (3D model of the cell) [40]; (C) The metacyclogenesis process. (C1) Some structures were omitted in order to highlight the structure of the cytopharynx (inset square). The reconstruction revealed the enlarged proximal portion of cytopharynx. Despite this alteration, this invagination still has the two microtubule sets and adjacent vesicles that were observed in the epimastigotes. Tomogram representing the intermediate forms; (C2) sequence of cell images showing different portions of the cytostome-cytopharynx complex present in a late intermediate form; the reconstruction shows the absence of the cytopharynx membrane; the microtubule triplet is still present; (C3–4) negative staining of a metacyclic trypomastigote shows the presence of the microtubule triplet. The cytostome-cytopharynx complex was not detected [41]. Flagellar pocket (FP-white), preoral ridge (POR-purple), cytostome-cytopharynx (Cy-pink), vesicles (V-orange), flagella (F-yellow and light blue), kinetoplast (K-green), nucleus (N-blue), and the Golgi complex (G-gold). Scale bars: 200 nm.

The same methodology was used to obtain 3D reconstructions of the ultrastructural changes present in the intermediate forms of *T. cruzi* during the metacyclogenesis process (differentiation of epimastigotes into metacyclic trypomastigotes). It was shown that the migration of the kinetoplast/flagellar pocket to the posterior region drags the cytostome aperture. Thus, the invagination shortens from the end to the beginning. The late intermediate forms did not have a cytopharynx membrane; only the accompanying microtubules remained in this form. In the trypomastigote form, shorter microtubules were still observed [41] (**Figure 3C1–C4**).

3.2. Microfilaments

The presence of actin has been confirmed in some trypanosomatids species. Nevertheless, little is known about microfilament function or localization [7, 42]. In trypanosomatids, it does not appear to polymerize into highly structured cytoskeletal microfilaments [43].

The trypanosomatids genome contains putative actin and actin-binding protein sequences [44]. However, to date, few studies have successful visualized microfilaments in trypanosomatids. Sahasrabuddhe et al. visualized actin filaments in *Leishmania spp.* promastigotes using immunogold and immunofluorescence approaches. They detected actin in the flagellar pocket, flagellum, nucleus and kinetoplast (mitochondrial DNA, termed kDNA). The presence of actin in the flagellar pocket strongly suggests that it participates in endocytosis. Interesting, in *Leishmania*, actin also colocalized with SPMT. The study speculates that actin may cooperate with SPMT to help maintain the shape of the cell. This study suggested that the actin present in *Leishmania* is a new isoform which may differ functionally and structurally from that of eukaryotes [45]. Immunofluorescence microscopy observations showed that actin of *T. brucei* colocalized with the endocytic pathway; however, in procyclic forms, actin was distributed throughout the entire cell. The endocytic pathway and cell division are arrested in the actin-depleted, bloodstream stage of *T. brucei* [46] (**Figure 4A–B**). These observations suggest different roles for actin during the life cycle of *T. brucei*.

Molecular methods confirmed the expression of actin in trypanosomes, although the role of actin in *T. cruzi* has not been fully elucidated. Immunoblotting of parasite extracts with polyclonal serum raised against a recombinant version of the *T. cruzi* actin protein confirmed the presence of actin variants; the pattern of expression was different for each stage of the parasite life cycle [43], as previously observed with *T. brucei* [46].

Using cytochalasin to depolymerize actin resulted in morphological changes to the cytoskeletal elements associated with the cytostome-cytopharynx of *T. cruzi* epimastigotes; cytochalasin treatment also impaired transferrin uptake [48]. Comparative genomic analyses suggested that the actin-myosin system might function at the cytostome-cytopharynx during endocytosis [44]. The first detailed characterization of actin (TcActin) and actin-binding proteins in *T. cruzi* was carried out by De Melo et al.. The authors showed that TcActin was distributed in patch-like cytoplasmic structures during all stages of the *T. cruzi* life cycle (**Figure 4C–F**) [47]. Recently, confocal microscopy analyses of several trypanosomatids (*T. cruzi, T. brucei, L. major, Angomonas deanei, Crithidia fasciculata, Herpetomonas samuelpessoai, Strigomonas culicis* and *Phytomonas serpens*) showed that these protozoa express actin, which is present diffusely throughout the cytoplasm. In *T. cruzi* epimastigotes, actin seems to be localized at the cytopharynx [49], which may explain the disruption of endocytosis when epimastigotes are treated with cytochalasin.

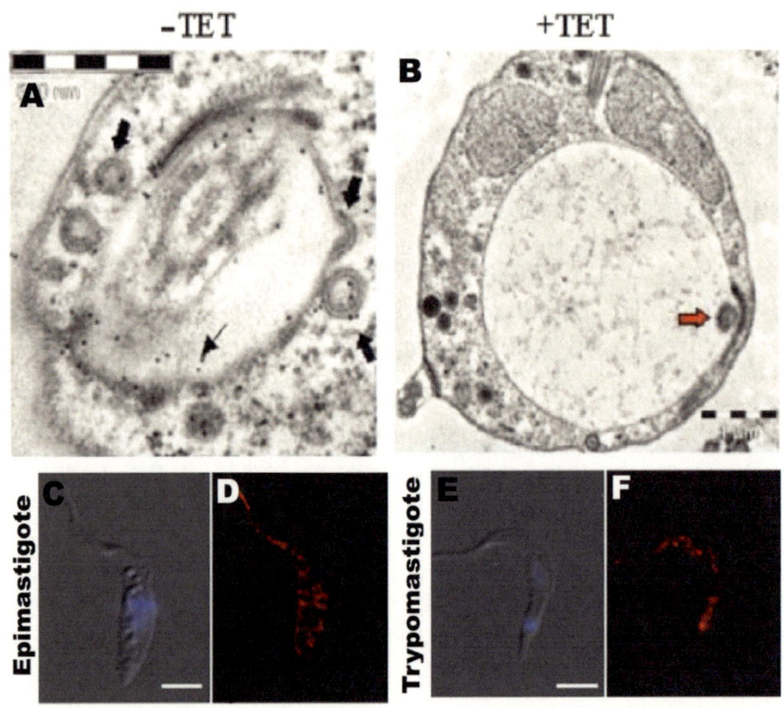

Figure 4. Effect of actin RNAi on the bloodstream form of *T. brucei* (A–B) and the subcellular distribution of actin in *T. cruzi* (C–H). (A) Electron micrograph of control bloodstream *T. brucei* (−TET). The flagellar pocket has budding vesicles. (B) Electron micrograph of RNAi-induced cells (+TET). These cells show loss of vesicles budding from the flagellar pocket membrane [46]; (C–F) Immunofluorescence confocal microscopy images of *T. cruzi* epimastigote and trypomastigote forms showing actin patches along the flagellum, and also spread all over the parasitic body [47].

4. Paraflagellar structure and flagellum attachment zone

The trypanosomatid flagellum consists of an evolutionarily conserved axoneme. It is also composed of peculiar elements such as the FAZ and the paraflagellar rod (PFR), a structure that is attached to the flagellum. FAZ is a specialized cytoskeletal region that links most of the flagellar membrane to the membrane of the cell body [5] (**Figure 5A–C**). The FAZ filament, the MtQ and the bilobe structure also help to link the flagellum to the cell surface [34, 50] (**Figure 2A**).

This complex has a left-handed, helical conformation around the cell body and within the microtubule array [51]. This region is considered a junctional complex and formed by lined apposed macular structures. The FAZ is an essential structure; disruption of FAZ assembly can lead to dramatic changes in morphogenesis.

The flagellum adhesion glycoprotein 1 (FLA1) was identified in *T. brucei*. FLA1 is critical for flagellar attachment and RNAi depletion of FLA1 results in flagellum detachment, cytokinesis defects and cell death [52]. Studies in *T. cruzi* indicate that the glycoprotein GP72 (homologous of FLA1) is an important molecule for flagellum attachment. When its gene is deleted, the flagellum is no more attached to the protozoan body [53].

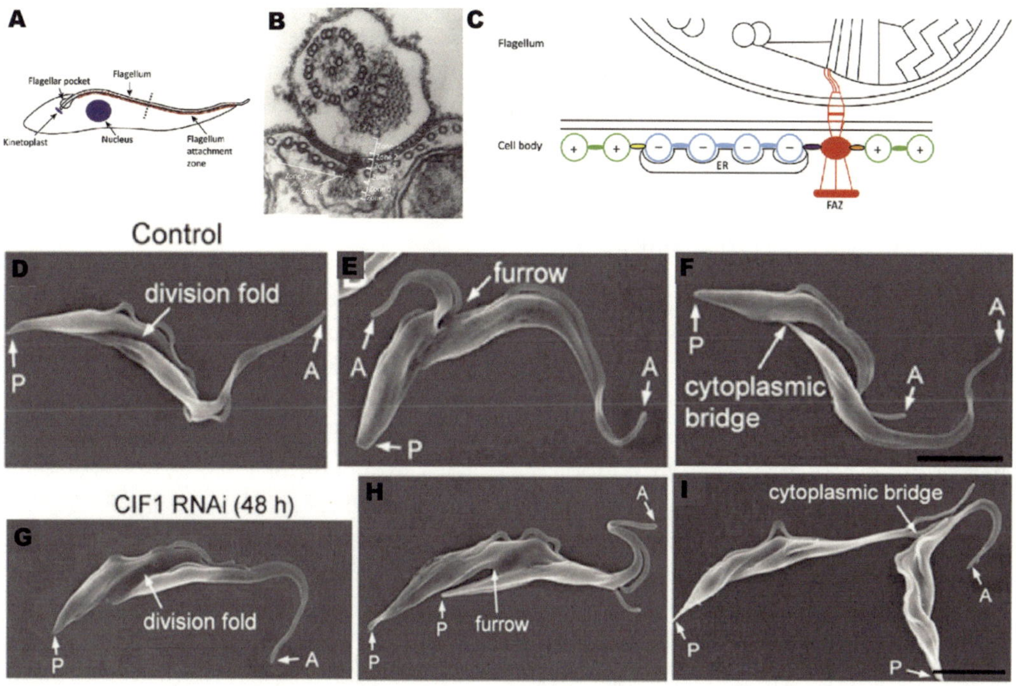

Figure 5. (A) Drawing of the main components of the *T. brucei* trypomastigote. The dashed diagonal line indicates the region seen in (B) and (C) [60]. (B) Electron micrograph showing a transversal section of the FAZ and its relative position to the microtubule quartet (MtQ-zone 7)) [60]. (C) Cartoon of the transversal section shown in (B) [60]. The connections among the paraflagellar rod, subpellicular microtubules (green), and FAZ (red) are emphasized. The drawing also shows the connection between the subpellicular microtubules, the microtubule quartet (MtQ-blue), and endoplasmic reticulum (ER). (D–I) Scanning electron micrographs of a cleavage furrow in control (D–F) and CIF2 RNAi treated cells (G–I); (D–F) control cells demonstrate a different direction for cytokinesis (arrows). It occurs from the posterior towards the anterior regions [58]. Scale bars: 5 μm.

The first molecular component of the *T. brucei* FAZ filament, FAZ1, was identified by screening an expression library with a monoclonal antibody. The knockdown of this protein resulted in disrupted FAZ and defects in cytokinesis. To date, FAZ1 is considered an essential protein needed to correctly assemble the FAZ in *T. brucei* [54]. Recently, immunoprecipitation assays detected a FLA-1 binding protein (FLA1BP). An RNAi approach targeting FLA1BP showed that this protein is involved in anchorage assembly of the new flagellum. Also, cell growth of these cultures was not disturbed [55]. RNAi knockdown of a coiled-coil-rich protein that contains a C2-domain (CC2D/a protein associated with the flagellum cytoskeleton) impairs the assembly of the FAZ filament; however, the formation of the four microtubules was not affected [56]. These results confirmed the participation of the FAZ filament in both flagellum attachment and cell morphogenesis.

Recently, depletion of ClpGM6, a calpain-like protein localized to the FAZ of *T. brucei* trypomastigotes, resulted in cells with a shorter FAZ. Consequently, these treated organisms were also missing portions of the basal body, the kinetoplast, the Golgi complex and the flagellar pocket. Cells produced long free flagellum, a characteristic phenotype of epimastigote-like cells [57]. Despite a normal growth rate, this phenotype highlights FAZ as an important structure in orchestrating basal body positions and determining the plane of cytokinesis. Recently, during *T. brucei* cytokinesis, a novel signaling pathway was described; it is composed of Polo-like kinase, CIF1, CIF2 and Aurora B kinase [58]. These signaling molecules behave as cytokinesis initiation regulators. *In vivo* studies show CIF2 interacting with CIF1. Furthermore, both of them colocalize at the new FAZ tip during early cell cycle stages [58, 59]. By inhibiting typical anterior-to-posterior cytokinesis, these studies characterized a backup cytokinesis pathway located at the posterior end of the cell [59] (**Figure 5D–I**).

In live parasites, flagellar beating produces a wave that gives the appearance of an "undulating membrane" on the sides of the cell body linked to flagellum as consequence of the FAZ arrangement. A detailed study using high-resolution microscopy to compare the swimming behavior of several trypanosome species that infect livestock showed that the waveforms are distinctive for each trypanosome species. This is due to variations in the microenvironment, such as differences in viscosity [61]. Inside the flagellar membrane of *T. brucei*, it was possible to observe a filamentous structure connecting the membrane associated with the FAZ to the proximal domain of the paraflagellar rod [31]. The paraflagellar rod is an extra-axoneme structure unique to kinetoplastids, euglenoids and dinoflagellates [62–64]. Transmission electron microscopy observations showed the paraflagellar rod as a lattice-like structure that is localized along the entire length of the axoneme once it exits from the flagellar pocket; it is linked to the FAZ. The PFR has three distinct portions: proximal, intermediate and distal. These portions are defined by their location relative to the axoneme (**Figure 6A**) [65].

Replicas of quick-frozen, freeze-fractured, deep-etched and rotary-replicated *T. cruzi* epimastigotes examined by transmission electron microscopy (TEM) provided detailed observations of PFR in straight and bent flagella. Based on these observations, an animated model for the PFR structure during flagellar beating was proposed [66] (**Figure 6B–G**). The PFR structure is absent in the amastigote forms of *Leishmania* and other trypanosomatids [7]. Studies of the biogenesis of *L. amazonensis* flagellum during amastigote-promastigote differentiation show different stages of differentiation (**Figure 6H–L**), initial stages of the process, the early intermediate forms presented an expansion of the flagellar tip and the duplication of flagellum was observed (**Figure 6H**). Interestingly, PFR formation was observed inside the flagellar pocket (**Figure 6J**). In later stages of differentiation, intermediate cells display a longer flagellum (even if shorter than in promastigotes) that contains a PFR (**Figure 6K–L**) [67]. Only after this stage typical flagellar beating was observed, suggesting that the presence of PFR is a prerequisite for flagellum motility [67].

The paraflagellar major proteins are PFR1 and PFR2; null mutant and RNAi ablation of PFR2 demonstrated that this protein is required for efficient flagellar beating in *L. mexicana* and *T. brucei* [68]. These parasites displayed an incomplete paraflagellar structure, showing that this protein is essential for the correct assembly of PFR. Paraflagellar rod components can be divided into four groups: (a) those involved in the formation of the lattice-like pattern; (b) those with a role in metabolism and in adenine nucleotide signaling; (c) those that participate in calcium signaling; and (d) those with unknown function [69]. Some studies showed the presence of calmodulin (CaM) within PFR; depletion of CaM resulted in catastrophic failure of the PFR architecture and disruption of the links between PFR and axoneme [70].

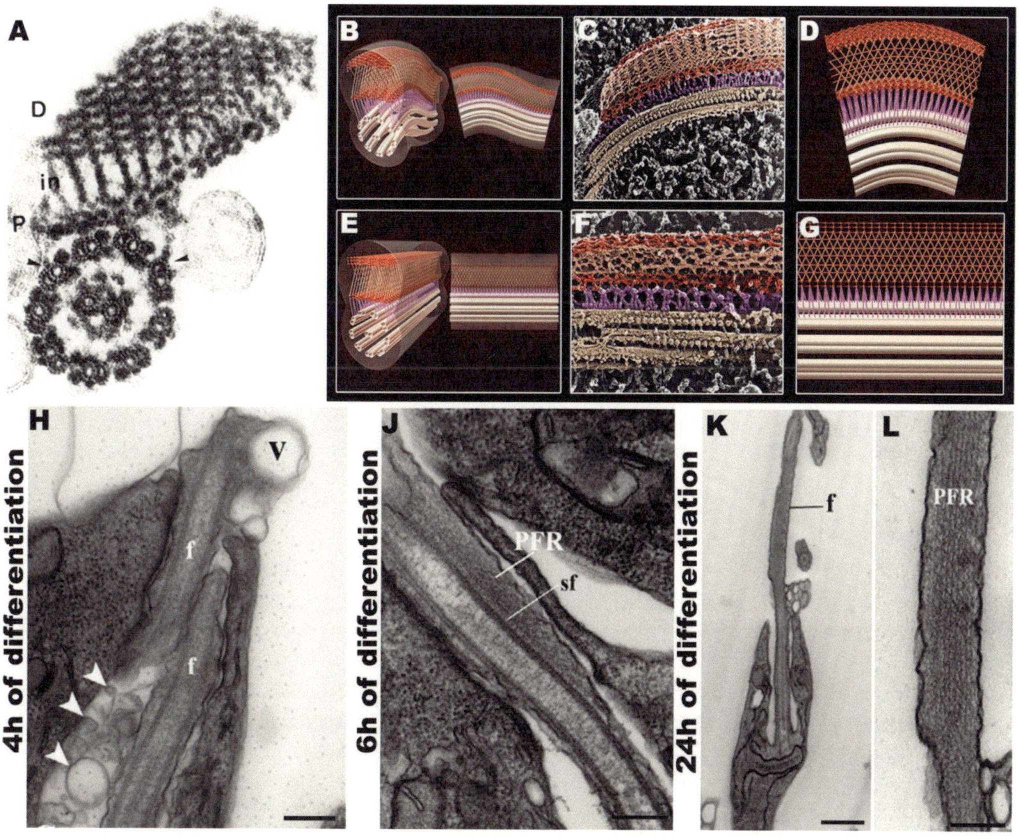

Figure 6. (A) Transmission electron microscopy images of a thin section showing the *T. cruzi* paraflagellar portions relative to the axoneme: proximal (p), intermediary (in) and distal (d) [65]. (B) One frame of an animation showing the flagellum in a bent state; (C) deep-etching replica image of the flagellum revealing the different portions of PFR in a bent flagellum; (D) one frame of an animation based on the deep-etching replica of a curved flagellum; (E) one frame of an animation showing a straight flagellum; (F) deep-etching replica image of the flagellum revealing the different portions of PFR in a straight flagellum; (G) an animation frame based on the deep-etching replica of a straight flagellum. (H–L) Steps of *Leishmania amazonensis* amastigote-promastigote differentiation; (B–G) [66]; (H) after 4 h of differentiation, the duplication of flagellum was observed with vesicles inside the flagellar pocket (arrowhead) and in flagellar tip (v). Scale bar: 200 nm; (J) at 6 h of differentiation, the PFR is observed inside the flagellar pocket. Scale bar: 200 nm; (K) at 24 h of differentiation, the flagellum presented axoneme and paraflagellar rod. Scale bar: 500 nm; (L) higher magnification of H showing the PFR [67].

The PFR in *T. cruzi* is composed of four proteins: PAR1–4 [71]. An important study indicated PAR4 as the target of *T. cruzi*-specific CD8+ T cell responses. Over expression of PAR4 improved PAR4-specific CD8+ T cell responses and provided significantly enhanced protection from infection; this chapter speculated that flagellar proteins can be used as antigens in potential vaccines against *T. cruzi* [72].

5. Conclusions

All the cytoskeletal structures and related proteins covered here are essential for the biology of trypanosomatids. Advances in high-resolution microscopies and molecular biology have provided more information regarding protein localization and function in these protozoa. This chapter provided an overview of unique and essentials cytoskeleton elements and proteins in trypanosomatids that may provide alternative targets in the future for chemotherapeutic drugs.

Author details

Juliana Cunha Vidal[1,2] and Wanderley de Souza[1,2*]

*Address all correspondence to: wsouza@biof.ufrj.br

1 Biophysics Institute of the Federal University of Rio de Janeiro, Rio de Janeiro, Brazil

2 National Center of Structural Biology and Bioimage CENABIO, Rio de Janeiro, Brazil

References

[1] Morriswood B. Form, fabric and function of a flagellum-associated cytoskeletal structure. Cells. 2015;4(4):726–47.

[2] Bastin P, Pullen TJ, Moreira-Leite FF, Gull K. Inside and outside of the trypanosome flagellum: a multifunctional organelle. Microbes and Infection/Institute Pasteur. 2000;2 (15):1865–74.

[3] Lacomble S, Vaughan S, Gadelha C, Morphew MK, Shaw MK, McIntosh JR, et al. Three-dimensional cellular architecture of the flagellar pocket and associated cytoskeleton in trypanosomes revealed by electron microscope tomography. Journal of Cell Science. 2009;122(Pt 8):1081–90.

[4] Alcantara CL, Vidal JC, de Souza W, Cunha ESNL. The three-dimensional structure of the cytostome-cytopharynx complex of *Trypanosoma cruzi* epimastigotes. Journal of Cell Science. 2014;127(Pt 10):2227–37.

[5] Vickerman K. On the surface coat and flagellar adhesion in trypanosomes. Journal of Cell Science. 1969;5(1):163–93.

[6] de Souza W, Souto-Padron T. The paraxial structure of the flagellum of trypanosomatidae. The Journal of Parasitology. 1980;66(2):229–36.

[7] Gull K. The cytoskeleton of trypanosomatid parasites. Annual Review of Microbiology. 1999;53:629–55.

[8] Sant'Anna C, Campanati L, Gadelha C, Lourenco D, Labati-Terra L, Bittencourt-Silvestre J, et al. Improvement on the visualization of cytoskeletal structures of protozoan parasites using high-resolution field emission scanning electron microscopy (FESEM). Histochemistry and Cell Biology. 2005;124(1):87–95.

[9] Souto-Padron T, de Souza W, Heuser JE. Quick-freeze, deep-etch rotary replication of *Trypanosoma cruzi* and Herpetomonas megaseliae. Journal of Cell Science. 1984;69:167–78.

[10] MacRae TH, Gull K. Purification and assembly in vitro of tubulin from Trypanosoma brucei brucei. The Biochemical Journal. 1990;265(1):87–93.

[11] Soares TC, de Souza W. Fixation of trypanosomatids for electron microscopy with the glutaraldehyde-tannic acid method. Zeitschrift fur Parasitenkunde. 1977;53(2):149–54.

[12] Robinson DR, Sherwin T, Ploubidou A, Byard EH, Gull K. Microtubule polarity and dynamics in the control of organelle positioning, segregation and cytokinesis in the trypanosome cell cycle. The Journal of Cell Biology. 1995;128(6):1163–72.

[13] de Souza W, Sant'Anna C, Cunha-e-Silva NL. Electron microscopy and cytochemistry analysis of the endocytic pathway of pathogenic protozoa. Progress in Histochemistry and Cytochemistry. 2009;44(2):67–124.

[14] Vedrenne C, Giroud C, Robinson DR, Besteiro S, Bosc C, Bringaud F, et al. Two related subpellicular cytoskeleton-associated proteins in Trypanosoma brucei stabilize microtubules. Molecular Biology of the Cell. 2002;13(3):1058–70.

[15] De Souza W. From the cell biology to the development of new chemotherapeutic approaches against trypanosomatids: dreams and reality. Kinetoplastid Biology and Disease. 2002;1(1):3.

[16] Pimenta PF, De Souza W. Fine structure and cytochemistry of the endoplasmic reticulum and its association with the plasma membrane of Leishmania mexicana amazonensis. Journal of Submicroscopic Cytology. 1985;17(3):413–9.

[17] Meyer H, De Souza W. Electron microscopic study of *Trypanosoma cruzi* periplast in tissue cultures. I. Number and arrangement of the peripheral microtubules in the various forms of the parasite's life cycle. The Journal of Protozoology. 1976;23(3):385–90.

[18] Sheriff O, Lim LF, He CY. Tracking the biogenesis and inheritance of subpellicular microtubule in Trypanosoma brucei with inducible YFP-alpha-tubulin. BioMed Research International. 2014;2014:893272.

[19] Robinson DR, Gull K. Basal body movements as a mechanism for mitochondrial genome segregation in the trypanosome cell cycle. Nature. 1991;352(6337):731–3.

[20] Wheeler RJ, Scheumann N, Wickstead B, Gull K, Vaughan S. Cytokinesis in Trypanosoma brucei differs between bloodstream and tsetse trypomastigote forms: implications for microtubule-based morphogenesis and mutant analysis. Molecular Microbiology. 2013; 90(6):1339–55.

[21] Borst P, Fairlamb AH. Surface receptors and transporters of Trypanosoma brucei. Annual Review of Microbiology. 1998;52:745–78.

[22] Landfear SM, Ignatushchenko M. The flagellum and flagellar pocket of trypanosomatids. Molecular and Biochemical Parasitology. 2001;115(1):1–17.

[23] Field MC, Carrington M. Intracellular membrane transport systems in *Trypanosoma brucei*. Traffic. 2004;5(12):905–13.

[24] Tardieux I, Webster P, Ravesloot J, Boron W, Lunn JA, Heuser JE, et al. Lysosome recruitment and fusion are early events required for trypanosome invasion of mammalian cells. Cell. 1992;71(7):1117–30.

[25] Webster P, Russell DG. The flagellar pocket of trypanosomatids. Parasitology Today. 1993;9(6):201–6.

[26] De Souza W, Bunn MM, Angluster J. Demonstration of concanavalin A receptors on *Leptomonas pessoai* cell membrane. The Journal of Protozoology. 1976;23(2):329–33.

[27] Coppens I, Opperdoes FR, Courtoy PJ, Baudhuin P. Receptor-mediated endocytosis in the bloodstream form of Trypanosoma brucei. The Journal of Protozoology. 1987;34(4): 465–73.

[28] Soares MJ, de Souza W. Endocytosis of gold-labeled proteins and LDL by *Trypanosoma cruzi*. Parasitology Research. 1991;77(6):461–8.

[29] Bonhivers M, Nowacki S, Landrein N, Robinson DR. Biogenesis of the trypanosome endo-exocytotic organelle is cytoskeleton mediated. PLoS Biology. 2008;6(5):e105.

[30] Gadelha C, Rothery S, Morphew M, McIntosh JR, Severs NJ, Gull K. Membrane domains and flagellar pocket boundaries are influenced by the cytoskeleton in African trypanosomes. Proceedings of the National Academy of Sciences of the United States of America. 2009;106(41):17425–30.

[31] Sherwin T, Gull K. The cell division cycle of *Trypanosoma brucei*: timing of event markers and cytoskeletal modulations. Philosophical Transactions of the Royal Society of London Series B, Biological Sciences. 1989;323(1218):573–88.

[32] Field MC, Carrington M. The trypanosome flagellar pocket. Nature Reviews Microbiology. 2009;7(11):775–86.

[33] Morriswood B, He CY, Sealey-Cardona M, Yelinek J, Pypaert M, Warren G. The bilobe structure of Trypanosoma brucei contains a MORN-repeat protein. Molecular and Biochemical Parasitology. 2009;167(2):95–103.

[34] Esson HJ, Morriswood B, Yavuz S, Vidilaseris K, Dong G, Warren G. Morphology of the trypanosome bilobe, a novel cytoskeletal structure. Eukaryotic Cell. 2012;11(6): 761–72.

[35] Morriswood B, Schmidt K. A MORN repeat protein facilitates protein entry into the flagellar pocket of *Trypanosoma brucei*. Eukaryotic Cell. 2015;14(11):1081–93.

[36] Porto-Carreiro I, Attias M, Miranda K, De Souza W, Cunha-e-Silva N. *Trypanosoma cruzi* epimastigote endocytic pathway: cargo enters the cytostome and passes through an early endosomal network before storage in reservosomes. European Journal of Cell Biology. 2000;79(11):858–69.

[37] Soares MJ. Endocytic portals in *Trypanosoma cruzi* epimastigote forms. Parasitology Research. 2006;99(4):321–2.

[38] Milder R, Deane MP. The cytostome of *Trypanosoma cruzi* and *T. conorhini*. The Journal of Protozoology. 1969;16(4):730–7.

[39] Attias M, Vommaro RC, de Souza W. Computer aided three-dimensional reconstruction of the free-living protozoan Bodo sp. (Kinetoplastida: Bodonidae). Cell Structure and Function. 1996;21(5):297–306.

[40] Alcantara CL, Vidal JC, de Souza W, Cunha ESNL. The cytostome-cytopharynx complex of *Trypanosoma cruzi* epimastigotes disassembles during cell division. Journal of Cell Science. 2016. DOI: 10.1242/jcs.187419

[41] Vidal, J.C., et al., Loss of the cytostome-cytopharynx and endocytic ability are late events in Trypanosoma cruzi metacyclogenesis. J Struct Biol, 2016. 196(3):319–328.

[42] De Souza W. Basic cell biology of *Trypanosoma cruzi*. Current Pharmaceutical Design. 2002;8(4):269–85.

[43] Cevallos AM, Segura-Kato YX, Merchant-Larios H, Manning-Cela R, Alberto Hernandez-Osorio L, Marquez-Duenas C, et al. *Trypanosoma cruzi*: multiple actin isovariants are observed along different developmental stages. Experimental Parasitology. 2011;127 (1):249–59.

[44] El-Sayed NM, Myler PJ, Bartholomeu DC, Nilsson D, Aggarwal G, Tran AN, et al. The genome sequence of *Trypanosoma cruzi*, etiologic agent of Chagas disease. Science. 2005;309(5733):409–15.

[45] Sahasrabuddhe AA, Bajpai VK, Gupta CM. A novel form of actin in Leishmania: molecular characterisation, subcellular localisation and association with subpellicular microtubules. Molecular and Biochemical Parasitology. 2004;134(1):105–14.

[46] Garcia-Salcedo JA, Perez-Morga D, Gijon P, Dilbeck V, Pays E, Nolan DP. A differential role for actin during the life cycle of Trypanosoma brucei. The EMBO Journal. 2004;23 (4):780–9.

[47] De Melo LD, Sant'Anna C, Reis SA, Lourenco D, De Souza W, Lopes UG, et al. Evolutionary conservation of actin-binding proteins in *Trypanosoma cruzi* and unusual subcellular localization of the actin homologue. Parasitology. 2008;135(8):955–65.

[48] Correa JR, Atella GC, Batista MM, Soares MJ. Transferrin uptake in *Trypanosoma cruzi* is impaired by interference on cytostome-associated cytoskeleton elements and stability of membrane cholesterol, but not by obstruction of clathrin-dependent endocytosis. Experimental Parasitology. 2008;119(1):58–66.

[49] Souza LC, Pinho RE, Lima CV, Fragoso SP, Soares MJ. Actin expression in trypanosomatids (Euglenozoa: Kinetoplastea). Memorias do Instituto Oswaldo Cruz. 2013;108(5):631–6.

[50] Gheiratmand L, Brasseur A, Zhou Q, He CY. Biochemical characterization of the bi-lobe reveals a continuous structural network linking the bi-lobe to other single-copied organelles in *Trypanosoma brucei*. The Journal of Biological Chemistry. 2013;288(5):3489–99.

[51] Portman N, Gull K. Proteomics and the Trypanosoma brucei cytoskeleton: advances and opportunities. Parasitology. 2012;139(9):1168–77.

[52] LaCount DJ, Barrett B, Donelson JE. Trypanosoma brucei FLA1 is required for flagellum attachment and cytokinesis. The Journal of Biological Chemistry. 2002;277(20):17580–8.

[53] Cooper R, de Jesus AR, Cross GA. Deletion of an immunodominant *Trypanosoma cruzi* surface glycoprotein disrupts flagellum-cell adhesion. The Journal of Cell Biology. 1993;122(1):149–56.

[54] Vaughan S, Kohl L, Ngai I, Wheeler RJ, Gull K. A repetitive protein essential for the flagellum attachment zone filament structure and function in *Trypanosoma brucei*. Protist. 2008;159(1):127–36.

[55] Sun SY, Wang C, Yuan YA, He CY. An intracellular membrane junction consisting of flagellum adhesion glycoproteins links flagellum biogenesis to cell morphogenesis in *Trypanosoma brucei*. Journal of Cell Science. 2013;126(Pt 2):520–31.

[56] Zhou Q, Liu B, Sun Y, He CY. A coiled-coil- and C2-domain-containing protein is required for FAZ assembly and cell morphology in *Trypanosoma brucei*. Journal of Cell Science. 2011;124(Pt 22):3848–58.

[57] Hayes P, Varga V, Olego-Fernandez S, Sunter J, Ginger ML, Gull K. Modulation of a cytoskeletal calpain-like protein induces major transitions in trypanosome morphology. The Journal of Cell Biology. 2014;206(3):377–84.

[58] Zhou Q, Gu J, Lun ZR, Ayala FJ, Li Z. Two distinct cytokinesis pathways drive trypanosome cell division initiation from opposite cell ends. Proceedings of the National Academy of Sciences of the United States of America. 2016;113(12):3287–92.

[59] Zhou Q, Hu H, Li Z. An EF-hand-containing protein in *Trypanosoma brucei* regulates cytokinesis initiation by maintaining the stability of the cytokinesis initiation factor CIF1.The Journal of Biological Chemistry, 2016; 291(28):14395–409.

[60] Sunter JD, Gull K. The flagellum attachment zone: 'the cellular ruler' of Trypanosome morphology. Trends in Parasitology. 2016;32(4):309–24.

[61] Bargul JL, Jung J, McOdimba FA, Omogo CO, Adung'a VO, Kruger T, et al. Species-specific adaptations of Trypanosome morphology and motility to the mammalian host. PLoS Pathogens. 2016;12(2):e1005448.

[62] Schlaeppi K, Deflorin J, Seebeck T. The major component of the paraflagellar rod of Trypanosoma brucei is a helical protein that is encoded by two identical, tandemly linked genes. The Journal of Cell Biology. 1989;109(4 Pt 1):1695–709.

[63] Hyams JS. The Euglena paraflagellar rod: structure, relationship to other flagellar components and preliminary biochemical characterization. Journal of Cell Science. 1982;55: 199–210.

[64] Cachon M, Cosson MP. Ciliary and flagellar apparatuses and their associated structures. Biology of the Cell/Under the Auspices of the European Cell Biology Organization. 1988;63(2):115.

[65] Farina M, Attias M, Souto-Padrón T, de Souza W. Further studies on the organization of the paraxial rod of Trypanosomatids. J. Protozool, 1986;33:552–557.

[66] Rocha GM, Teixeira DE, Miranda K, Weissmuller G, Bisch PM, de Souza W. Structural changes of the paraflagellar rod during flagellar beating in *Trypanosoma cruzi*. PloS One. 2010;5(6):e11407.

[67] Gadelha AP, Cunha-e-Silva NL, de Souza W. Assembly of the Leishmania amazonensis flagellum during cell differentiation. Journal of Structural Biology. 2013;184(2):280–92.

[68] Santrich C, Moore L, Sherwin T, Bastin P, Brokaw C, Gull K, et al. A motility function for the paraflagellar rod of Leishmania parasites revealed by PFR-2 gene knockouts. Molecular and Biochemical Parasitology. 1997;90(1):95–109.

[69] Portman N, Gull K. The paraflagellar rod of kinetoplastid parasites: from structure to components and function. International Journal for Parasitology. 2010;40(2):135–48.

[70] Ginger ML, Collingridge PW, Brown RW, Sproat R, Shaw MK, Gull K. Calmodulin is required for paraflagellar rod assembly and flagellum-cell body attachment in trypanosomes. Protist. 2013;164(4):528–40.

[71] Luhrs KA, Fouts DL, Manning JE. Immunization with recombinant paraflagellar rod protein induces protective immunity against *Trypanosoma cruzi* infection. Vaccine. 2003;21(21–22):3058–69.

[72] Kurup SP, Tarleton RL. The Trypanosoma cruzi flagellum is discarded via asymmetric cell division following invasion and provides early targets for protective CD8(+) T cells. Cell Host & Microbe. 2014;16(4):439–49.

Chapter 4

The Interplay between Cytoskeleton and Calcium Dynamics

Francisco Javier Martin-Romero,
Aida M. Lopez-Guerrero, Carlos Pascual-Caro and
Eulalia Pozo-Guisado

Additional information is available at the end of the chapter

http://dx.doi.org/10.5772/66862

Abstract

Cell motility is a complex cellular event that involves reorganization of cytoskeleton. This reorganization encompasses the transient polarization of the cell to facilitate the plasma membrane ruffling, a rearrangement of cortical actin cytoskeleton required for the development of cellular protrusions. It is known that extracellular Ca^{2+} influx is essential for cell migration and for the positive-feedback cycle that maintains leading-edge structures and ruffling activity. The aim of this review is to summarize our knowledge regarding the Ca^{2+}-dependent signaling pathways, Ca^{2+} transporters and sensors involved in cell migration. Also, we show here reported evidences that support for a crosstalk between Ca^{2+} transport and the reorganization of the cytoskeleton required for cell migration. In this regard, we will analyze the role of store-operated Ca^{2+} entry (SOCE) as a modulator of cytoskeleton and cell migration, but also the modulation of this Ca^{2+} entry pathway by microtubules and the actin cytoskeleton. As a main conclusion, this review will show that data reported in the last years support a role for SOCE in shaping cytoskeleton, but at the same time, SOCE is strongly dependent on cytoskeletal proteins, in an interesting interplay between cytoskeleton and Ca^{2+} dynamics.

Keywords: calcium, microtubules, actin, STIM1, ORAI1, cell migration, cortical cytoskeleton

1. Introduction

Calcium ions (Ca^{2+}) are essential intracellular transducers for cell signaling because of their role to bind Ca^{2+}-sensitive proteins that mediate key activities in signaling pathways. Upon cell stimulation through a variety of receptors and other types of physicochemical stimulations such as depolarization of plasma membrane, changes in osmolarity, physical distortion of cell surface, temperature, etc., a number of intracellular "second messengers" transmit the initial stimulation into the cell to trigger a proper response to these stimuli. The response is attained in some cases by transiently altering ion transport through plasma membrane as well as intracellular membranes, posttranslational modifications of proteins, changes in gene expression, or reshaping the cytoskeleton in a second-messenger-dependent manner.

To properly act as a second messenger the concentration of free cytosolic Ca^{2+} ($[Ca^{2+}]_i$) is substantially different across plasma membrane. While the physiological extracellular medium contains 1–3 mM Ca^{2+}, the level of the cytosolic milieu is within a narrow range, 80–120 nM [1, 2]. Considering this large chemical gradient of free Ca^{2+} concentration, cells can trigger a rapid and transient increase of $[Ca^{2+}]_i$ by increasing Ca^{2+} transport through plasma membrane. This transient Ca^{2+} increase triggers the action of a wide range of Ca^{2+}-dependent mediators that modify signaling pathways. Because Ca^{2+} is involved in numerous signaling pathways, differential features of Ca^{2+} increase are required to respond to diverse initial stimuli. This is achieved by the spatiotemporal control of the Ca^{2+} increase, i.e., the localization of the cytosolic Ca^{2+} spike within the cell, as well as the rate of the increase, the extent of the transient increase and the kinetics of the final decrease to basal levels [3].

This Ca^{2+} transport is therefore tightly regulated by specific channels, which are sensitive to hormones, cytokines, small molecules and other extracellular stimuli. The features of the transporters shape the characteristics of Ca^{2+} current (Ca^{2+} influx), modeling the kinetics of Ca^{2+} entry. Because a specific distribution of Ca^{2+} channels and transducers is required the cytoskeleton has been involved in the spatial regulation of Ca^{2+} entry. Interestingly, Ca^{2+} signaling strongly influences cytoskeleton dynamics, in an interesting interplay that is currently under study. In order to assess the crosstalk between Ca^{2+} signaling and cytoskeleton we will review here the recent literature regarding this topic, with special focus on the role of store-operated Ca^{2+} entry (SOCE) in cell migration, focal adhesion turnover and actin filaments reorganization. In addition, we will review the role of the microtubule cytoskeleton in the normal function of SOCE. A major regulator of SOCE is the protein STIM1, an endoplasmic reticulum resident protein that serves as intraluminal Ca^{2+} sensor and plasma membrane Ca^{2+} channel modulator. STIM1 is known to be a plus-end microtubule binding protein (+TIP). As a +TIP, STIM1 is transported throughout the cell while bound to microtubules, but it is released upon activation. The molecular basis of this regulation and the role of posttranslational modifications in STIM1 that are known to underlie this regulation, will be also described.

2. Ca^{2+} transporters and Ca^{2+} influx pathways

In eukaryotic cells plasma membrane (PM) contains different Ca^{2+} transport systems to control Ca^{2+} influx as well as Ca^{2+} extrusion. Because the temporal control of Ca^{2+} signaling is also strictly controlled, transporters are highly coordinated to let the Ca^{2+} spikes/waves last for a precise time, but this time ranges from microseconds (as in exocytosis) to hours, as observed during mammalian oocyte fertilization. To understand this control, we summarize here the most important Ca^{2+} transporters in eukaryotic cells. Ca^{2+} entry channels are divided into the following:

Voltage-operated channels (VOCs) are regulated by the net electric charge across the plasma membrane in the way that they open upon depolarization. This family consists of three different groups of channels, Ca_v1 (L-type channels), Ca_v2 (N-, P/Q and R-types) and Ca_v3 (T-type channels) [4].

Receptor-operated channels (ROCs), which are gated upon binding to agonists such as ATP, glutamate, or acetylcholine [5]. Within this group we should highlight the transient receptor potential (TRP) ion channel family, a large family of channels involved in sensory perception, smooth muscle contraction-relaxation cycles and cell proliferation [6]. Members of the TRPC (transient receptor potential canonical) family are also involved in the organization of heteromeric Ca^{2+} channels that respond to intracellular Ca^{2+} store depletion [7]. This is the case for TRPC1 that has been described to be part of complexes together with members of the ORAI family [8]. Other important members of this family are P2X receptors, which are Ca^{2+} channels gated by extracellular ATP [9] and glutamate receptors, all of them reviewed in Ref. [10].

Second-messenger-operated channels (SMOCs) are members of an important family of channels gated by intracellular second messengers, such as arachidonic acid-regulated Ca^{2+} (ARC) channel or TRPC6, a member of the TRPC which is sensitive to diacylglycerol (DAG) [11].

Store-operated channels (SOCs) are channels that are regulated by the filling state of intracellular Ca^{2+} stores, mainly the sarco(endo)plasmic reticulum (ER). Their activity is actually the result of an initial Ca^{2+} release from the ER, mainly through the activation of the phosphoinositide pathway. A wide range of stimuli triggers the breakdown of phosphatidylinositol 4,5-bisphosphate (PIP2) into diacylglycerol (DAG) and inositol 1,4,5-trisphosphate (IP3), which activates IP3 receptors (IP3R) at the ER, leading to the transient release of Ca^{2+}. Most of this Ca^{2+} is taken back to the ER by the sarco(endo)plasmic Ca^{2+} reticulum ATPase (SERCA), the Ca^{2+} pump in the ER membrane. However, some of this Ca^{2+} is extruded out of the cell by plasma membrane Ca^{2+} ATPases (PMCA), leading to the partial depletion of Ca^{2+} within the ER when the stimulation is repetitive. Thus, a system is required to replenish this Ca^{2+} and this action is mediated by SOCs that activate the Ca^{2+} influx pathway known as the store-operated Ca^{2+} entry (SOCE), described for the first time in the late 1980s by James W. Putney [12].

As stated above, there are some transporters considered as OFF systems, i.e., they are designed to decrease $[Ca^{2+}]_i$ to basal levels (to approximately 100 nM) and therefore to shut down Ca^{2+} signaling. These are mainly Ca^{2+} pumps, ATP-consuming transporters that pump Ca^{2+} ions against the Ca^{2+} gradient concentration [13]. PMCAs and SERCAs extrude Ca^{2+} from the cytosol to extracellular milieu and into the ER vesicles. In addition to these systems, the plasma membrane Na^+/Ca^{2+} exchanger contributes to Ca^{2+} extrusion in a Na^+-gradient-dependent manner and it is particularly important in cardiac cells and neuronal tissues [14]. Finally, a mitochondrial Ca^{2+} uniporter (MCU) and the secretory-pathway Ca^{2+}-ATPase (SPCA) are additional and widespread systems that restore cytosolic Ca^{2+} levels [15].

In summary, the combination of the activities of channels and pumps, either at the plasma membrane or at subcellular organelles and the particular expression profile of those transporters in different cells and tissues, makes possible for a cell-specific response to a wide range of stimuli.

3. Store-operated Ca^{2+} entry

As stated above, store-operated Ca^{2+} entry (SOCE) is a Ca^{2+} influx transport system that maintains the permanence of the Ca^{2+}-dependent signaling, because it preserves the intraluminal Ca^{2+} levels in the ER in the high micromolar range. For this extraordinary role, SOCE is a ubiquitous mechanism and is one of the most important pathways for Ca^{2+} entry in both excitable and nonexcitable cells [16]. However, the molecular nature of the members that control this Ca^{2+} influx pathway remained elusive until the description of two proteins, STIM1 and ORAI1, 10 years ago [17–22]. The mechanism that links luminal Ca^{2+} levels with plasma membrane Ca^{2+} entry is mediated by STIM1, a transmembrane protein located at the ER that acts as a Ca^{2+} sensor. STIM1 contains a Ca^{2+}-sensitive EF-hand domain at the intraluminal domain and activates plasma membrane Ca^{2+} channels (SOCs) upon Ca^{2+} depletion within the intracellular stores. This depletion triggers the oligomerization of STIM1 and the relocalization in ER-PM juxtapositions required for the binding and activation of plasma membrane SOCs.

One of the most important SOCs is ORAI1 (also known as CRACM1), although the ORAI family contains two additional members, ORAI2 and ORAI3. In addition to ORAI1, some of the transient receptor potential canonical (TRPC) channels can function in a STIM1-dependent mode. STIM1 directly activates TRPC1, TRPC4 and TRPC5 channels that can therefore act as SOCs [23]. TRPC1 also binds to ORAI1 and the TRPC1-ORAI1-STIM1 ternary complexes can be therefore considered SOCs [24]. STIM1 also activates TRPC3 and TRPC6, not by direct interaction, but mediating the heteromultimerization of TRPC3 with TRPC1 and TRPC6 with TRPC4 [23].

The mode of STIM1-dependent gating for ORAIs and TRPCs is also different. STIM1 gates ORAI1 by direct binding through the STIM1-ORAI1-activating region, or SOAR, also called CRAC-activating domain or CAD. In contrast, the C-terminal polybasic domain of STIM1 activates TRPC1 by an electrostatic gating mechanism that results in SOC channel activation [25, 26].

STIM1 can bind other types of Ca^{2+} channels, such as $Ca_v1.2$ channels, but this binding leads to the suppression of the activation of these channels [27, 28]. Interestingly, this inhibitory action is mediated by the domain of STIM1 that activates ORAI1 (SOAR or CAD) and this direct binding to $Ca_v1.2$ causes also the internalization of the channel, which explains the coordinated regulation of Ca^{2+} entry through VOCCS and SOCs. Thus, the combination of different Ca^{2+} channels provides a diversity of Ca^{2+} conductances in response to a wide variety of stimuli. However, the molecular mechanisms underlying the expected differential localization of STIM1 within cells are not clear. In this regard, the cytoskeleton seems to be a requirement for STIM1 function and binding to ORAI1 and therefore for the activation of SOCE. It was early observed that SOCE is sensitive to drugs that modify cytoskeletal components, such as cytochalasin D or nocodazole in NIH 3T3 cells [29], or vascular endothelial cells [30] and later confirmed in RBL-2H3 cells, bone marrow-derived mast cells [31], HEK293 cells [32] and platelets [33–35].

From the information given above, one can conclude that cytoskeleton is not only a critical modulator of SOCE, but also one can hypothesize that cytoskeleton might underlie the differential preference of STIM1 for different Ca^{2+} transporters.

4. Cytoskeletal components involved in the control of SOCE

After the molecular description of STIM1 and ORAI1 as the two major elements controlling SOCE, it was earlier found that the microtubule cytoskeleton was involved in the regulation of this Ca^{2+} entry pathway. The group of James W. Putney described that STIM1 tagged with fluorescent proteins colocalized with endogenous tubulin and that the treatment with nocodazole, which induces tubulin depolymerization, triggered the loss of this colocalization in HEK293 cells. Moreover, nocodazole had an inhibitory effect on SOCE that could be reverted by the overexpression of YFP (yellow fluorescent protein)-STIM1 [32], suggesting that the microtubule cytoskeleton has an important role in the activation of SOCE by facilitating the appropriate localization of STIM1 to activate SOCs. Similar findings were reported in COS-7 cells, where the treatment with nocodazole triggered the retraction of tubulin filaments from cell periphery leading to the progressive loss of SOCE, similarly to what is found in mitosis. On the contrary, placlitaxel, which stabilizes microtubules, enhanced SOCE [36], supporting for a role of microtubules in the normal function of SOCE.

It was later demonstrated that STIM1 directly binds to EB1, a microtubule plus-end binding (or tracking) protein (+TIP) and that STIM1 is transported through the surface of the ER network by this microtubule-dependent mechanism [37]. In addition, it was observed that the overexpression of ectopic STIM1 led to the stimulation of the ER extension, an effect that was explained by the direct attachment of the ER to the growing ends of microtubules, which was stimulated by the accumulation of STIM1. Therefore, this observation also revealed that STIM1 concentration could regulate ER extension and remodeling. In addition, Luis Vaca group reported that STIM1 binds to EB1 when traveling through the ER under resting conditions and that there is a dissociation of this STIM1-EB1 complex upon Ca^{2+} depletion of the ER, an event that facilitates the clustering of STIM1 in the periphery of the cell [38].

STIM1-EB1 binding was studied further and it was found that the sequence Thr-Arg-Ile-Pro (TRIP) in the C-terminus of STIM1 is responsible for the direct binding to EB1 [39]. This sequence belongs to a short polypeptide motif, S/TxIP, found in several +TIPs. For those +TIPs with an S/TxIP motif, a phospho-regulation of the binding with microtubules has been reported. The +TIPs, APC [40, 41], MCAK [42] and CLASP2 [43, 44] are phosphorylated in the vicinity of the S/TxIP sequence, negatively regulating their interaction with microtubules. In these +TIPs we find a high number of proline, serine and basic residues that give a positive charge to the surroundings of the EB1 binding domain. This is why the negative charge of the phosphorylation of residues in the surroundings of the S/TxIP motif explains the blocking of the binding to microtubule ends. In this regard, our group reported that, as it was reported for those +TIPs, the binding of STIM1 to the tip of microtubules growing ends is regulated by phosphorylation. Near the S/TxIP sequence, which in human STIM1 is found in residues 642–645 (Thr-Arg-Ile-Pro), our group have described three phosphorylatable residues, Ser575, Ser608 and Ser621, which are target for ERK1/2 activity *in vitro* and *in vivo* [44, 45]. Using overexpressed STIM1 mutated at ERK1/2 target residues, we reported that dephosphorylated STIM1 (i.e., with Ser-to-Ala substitution mutations) remained bound to EB1, whereas constitutive phosphorylated STIM1 (mimicked by Ser-to-Glu mutations et ERK1/2 target sites) detached from EB1 [46, 47]. By means of the generation of phospho-specific antibodies directed against the three individual residues, we reported that there is a dynamic phosphorylation of STIM1 during its activation. Thus, activation of SOCE by thapsigargin or 2,5-di-ter-butyl-1,4-benzohydroquinone, two SERCA inhibitors that trigger ER emptying, is accompanied by the increase of phosphorylation at the ERK1/2 target Ser residues (575, 608 and 621). Moreover, the washout of the inhibitor with a Ca^{2+}-containing medium, in order to let the refilling of the ER, carried out in parallel to STIM1 dephosphorylation [46]. In addition, Ser/Ala mutation inhibited SOCE and impaired the binding STIM1-ORAI1. In contrast, Ser/Glu mutations enhanced SOCE, whereas facilitated the clustering of STIM1 in response to store depletion, as well as the binding to ORAI1. Later, we reported that IGF-1 [48] and EGF [49] also trigger Ca^{2+} release from the ER and phosphorylation of STIM1, an effect that has been proven to be essential for the dissociation from EB1 and finally for the activation of SOCE (see **Figure 1**).

Taken together, those results support a mechanism that explains the reversible interaction of STIM1 and EB1 [46, 50]. This mechanism predicts that those stimuli that induce store depletion and ERK1/2 activation lead to phosphorylation of STIM1 at Ser575, Ser608 and Ser621, an event that promotes the dissociation of STIM1 from EB1. This dissociation enables STIM1 clustering and the binding to SOCs, in order to activate STIM1-dependent Ca^{2+} entry (SOCE). In addition, the activation of Ca^{2+} entry and the subsequent refilling of Ca^{2+} stores induce STIM1 dephosphorylation, promoting the association of STIM1 with EB1 and microtubules.

In addition to EB1, other members of the cytoskeleton have been involved in SOCE regulation. For instance, it has been reported that the microtubule-binding protein adenomatous polyposis coli (APC) is required for STIM1 puncta near ER-PM junctions, because reduced STIM1 was observed at these junctions in APC-depleted cells and this effect correlated well with the inhibition of SOCE [51]. The APC-binding domain was found in the C-terminus of STIM1 (residues 650–685), similarly to what it has been reported for STIM1-EB1. Thus, it can be assumed that upon depletion of the ER, STIM1 dissociates from EB1 and associates to APC to form puncta near ER-PM junctions and to activate ORAI1 and SOCE.

A downstream effector of SOCE is the transcription factor NFAT (nuclear factor of activated T-cells). Once Ca^{2+} entry becomes activated, the increase of $[Ca^{2+}]_i$ activates the Ca^{2+}-dependent phosphatase calcineurin (or protein phosphatase 2B), which dephosphorylates NFAT, promoting the nuclear translocation of NFAT that can be easily monitored using GFP-tagged NFAT. By means of this well-established feature of SOCE, Sharma et al. [52] designed a genome-wide RNA interference screen for NFAT activation in HeLa cells and they identified septin proteins as key regulators of SOCE. Knockdown of SEPT4 gene expression reduced SOCE

http://dx.doi.org/10.5772/66862

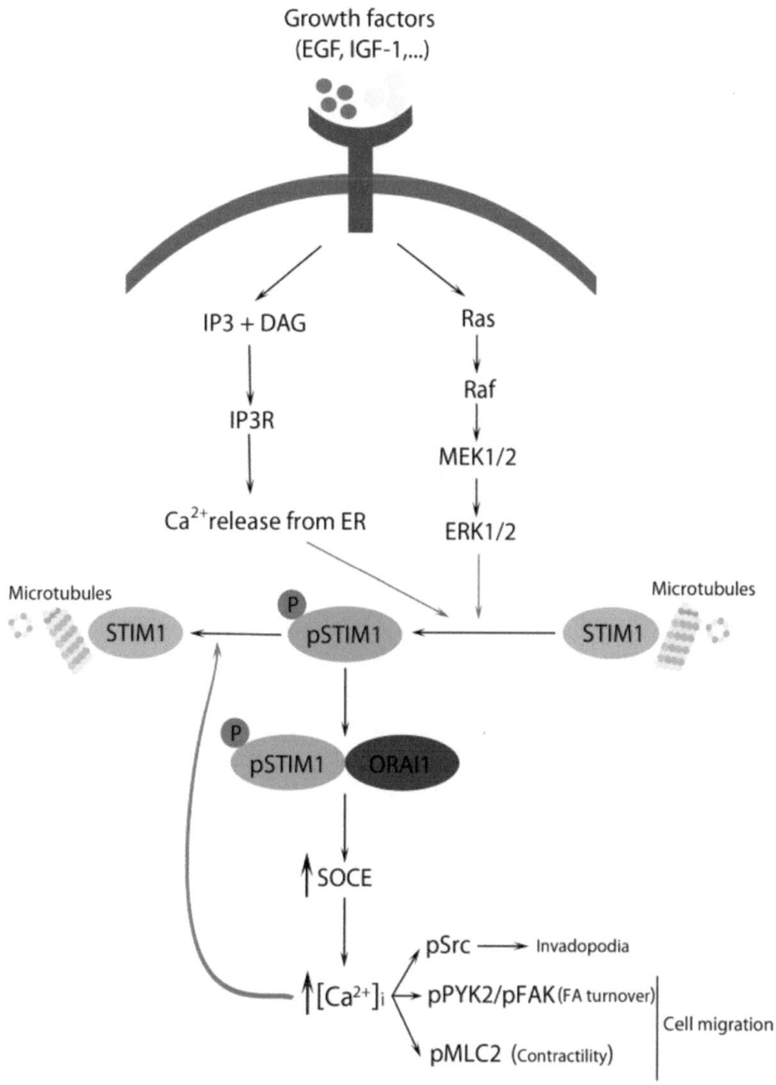

Figure 1. Involvement of SOCE in cell migration. The binding of ligands to some plasma membrane receptors activates the phosphoinositide pathway, releasing IP3 and activating Ca^{2+} release from the ER. In parallel, receptor tyrosine kinases trigger the activation of the MAPK pathway, via activation of Ras-Raf-MEK-ERK. Ca^{2+} mobilization from the ER and phosphorylation of STIM1 are two essential events for the activation of STIM1. Phospho-STIM1 releases from microtubules facilitating STIM1 clustering and translocation to PM-ER junctions to activate ORAI1. This activation increases $[Ca^{2+}]_i$ locally, an event required for the phosphorylation of Src, PYK2 and FAK kinases, as well as MLC2, which are well-known modulators of invadopodia formation, focal adhesion turnover and actomyosin contractility.

without affecting ER-Ca^{2+} release and without inhibiting SERCA or PMCAs. This knockdown also decreased Ca^{2+}-induced NFAT translocation by >95%. Septins were found to be required for proper organization of ORAI1 in the plasma membrane even before depletion of ER-Ca^{2+} stores. Septins also promoted the targeting of STIM1 to ER-PM junctions and the formation of stable ORAI1 clusters after store depletion. Because septins are considered scaffold proteins that recruit other proteins, preventing diffusion, septins should be considered promoters of the stable recruitment of STIM1-ORAI1 at ER-PM junctions. More recently it was found that Septin 7 inhibits the activation of Orai channels in Drosophila neurons (dOrai) and that the depletion of Septin 7 levels results in higher dOrai-mediated Ca^{2+} entry, independently of the filling state of the ER. In fact, overexpression of Septin 7 reduced Ca^{2+} entry, suggesting that, in Drosophila neurons, Septin 7 is a negative regulator of dOrai channel function [53]. These authors stated that hetero-hexamers septin filaments closely associate with the PM and near the ER in resting neurons and that the reduction of SEPT7 results in breaks in the linear septin filaments present in wild-type cells, leading to the formation of shorter septin filaments. ER-Ca^{2+} store depletion reorganizes these filaments, moving STIM1 to the peripheral ER, promoting the coupling of STIM/dOrai and the activation of Ca^{2+} entry through dOrai (SOCE). Shorter septin filaments due to SEPT7 knockdown leads to STIM1 recruitment to the peripheral ER in resting neurons and the activation of dOrai, resulting in a store-independent activation of dOrai [53].

Homer1 is another scaffolding protein that binds to TRPC channels through a PPPF sequence of the Ca^{2+} channel. However, this sequence is close to the STIM1 binding sequence, thus suggesting that Homer1 and STIM1 could be competitors for binding to TRPC. Yuan et al. [54] proposed that Homer couples TRPC channels to IP3 receptors (IP3R) to keep these channels closed, but dissociation of the TRPC-Homer-IP3R complexes gives STIM1 access to TRPC binding to gate these channels. A similar proposal was later reported for STIM1 and $Ca_v1.2$ channels in HEK293 cells [55], where the treatment with thapsigargin induces coimmunoprecipitation of Homer1 with STIM1 and the $Ca_v1.2$ $\alpha1$ subunit. Impairment of Homer1 function with the peptide PPKKFR or by siRNA specific for Homer1 reduced the association of STIM1 to $Ca_v1.2$ $\alpha1$, adding a new scaffolding protein to the list of regulators of Ca^{2+} entry.

In summary, the increasing number of members of the cytoskeleton involved in the specific interaction of STIM1 with different plasma membrane Ca^{2+} channels leads to the conclusion that the cytoskeleton strongly modulates SOCE, as well as the activation and inhibition of others Ca^{2+} transport systems.

5. How SOCE modulates cytoskeleton dynamics and cell migration

Because the localization and function of STIM1 and ORAI1 are strongly dependent on components of the actin and tubulin cytoskeleton, the question of how SOCE and Ca^{2+} dynamics influence cytoskeleton-dependent events, such as cell adhesion and migration, was rapidly considered by many groups.

Focal adhesions are complexes of macromolecules that serve as mechanical links between the extracellular substrate and the cytoskeleton. These molecular assemblies are highly dynamic. However, the molecular mechanism by which Ca^{2+} regulates focal adhesion turnover is still unclear. A few proteins that regulate focal adhesion assembly or disassembly are sensitive to changes in the intracellular free Ca^{2+} concentration. For instance, the Ca^{2+}-dependent protease calpain [56], which could cleave talin at focal adhesion sites, leads to an increase of focal adhesion disassembly rates. Indeed, disassembly of other focal adhesion components, like paxillin, vinculin and zyxin, has been shown to be dependent on the cleavage of talin by calpain, suggesting a role for the Ca^{2+}-dependent talin proteolysis in the regulation of focal adhesion turnover [57]. In this regard, it has been found that the reduction of STIM1 or ORAI1 gene expression, by RNA interference, dramatically affected the rate of focal adhesion turnover, slowing down cell migration *in vitro* and inhibiting metastasis of MDA-MB-231 cancer cells in nude mice [58]. The treatment of cells with SKF96365, a SOC inhibitor, also induced large focal adhesions *in vitro*, due to defective focal adhesion turnover. In addition, this blocking agent also inhibited tumor progression in mice [58].

Focal adhesion kinase (FAK) and proline-rich tyrosine kinase 2 beta (PTK2B or PYK2) are two well-known kinases involved in focal adhesion assembly [59]. Because focal adhesion targeting of PYK2 is required for the turnover and the Tyr402 autophosphorylation is required for PYK2 targeting, Chen et al. studied the correlation between EGF stimulation, PYK2 phosphorylation and STIM1 levels in SiHa cells [60]. EGF activates PYK2 phosphorylation at Tyr402, but this phosphorylation has been shown to be inhibited by the silencing of STIM1 [60]. In addition, STIM1 knockdown induced large focal adhesions, independently of the stimulation with EGF, further confirming the role of Ca^{2+} entry in facilitating the turnover of focal adhesions. Similarly, EGF triggered the phosphorylation of FAK at Tyr397, which is required for focal adhesion turnover and it was demonstrated that this phosphorylation is inhibited by STIM1 knockdown [60].

In this regard, it is known that EGF activates the phosphoinositide pathway, generating IP3 and activating Ca^{2+} release from the ER. EGF also activates de MAPK (mitogen-activated protein kinases) pathway. As a consequence STIM1 becomes phosphorylated at ERK1/2 target sites (Ser575, Ser608 and Ser621) upon EGF stimulation, leading to the dissociation from microtubules (i.e., EB1) and activating SOCE [49]. Casas-Rua et al. demonstrated that non-phosphorylatable mutants of STIM1, such as STIM1-S575A/S308A/S621A, blocked EGF signaling pathway, inhibiting cell migration in the endometrial adenocarcinoma Ishikawa cell line [49]. The impact of STIM1 was observed also at the genomic response level, because ectopic overexpression of STIM1-S575A/S608A/S621A blocked the epithelial-mesenchymal transition (EMT) of Ishikawa cells treated with EGF. The stimulation of cells with EGF induced a significant switch in E-cadherin localization from subplasma membrane region to a diffuse localization throughout the cytosol, as described for other epithelial cells [61]. EGF also triggered an increase of vimentin expression in well-defined cytoskeletal localization, as for other cells upon EGF stimulation [61, 62]. However, Ishikawa cells overexpressing STIM1-S575A/S608A/S621A-mCherry did not show significant increase in vimentin expression, nor E-cadherin relocalization [49], supporting for a role for phospho-STIM1 in the regulation of cell migration and cell transformation into a mesenchymal phenotype.

Cytosolic Ca^{2+} levels also regulate actomyosin, the macromolecular complex of actin and myosin that drives the mechanical forces for cell contractility during migration. Nonmuscle cell contractility is controlled by nonmuscle myosin II, through the phosphorylation of its regulatory light chains (MLC2) at Ser19 in a Ca^{2+}/calmodulin-dependent manner [63]. The contractile force that moves the cell body is then transmitted to focal adhesions by phospho-myosin II-based actomyosin contraction. This phosphorylation was abolished by the Ca^{2+} channel inhibitor SKF96365, or by knocking-down STIM1 expression [64], indicating that STIM1-dependent Ca^{2+} entry has a significant role in MLC2 activation and in the reorganization of the actin-myosin cytoskeleton in migrating cells.

In addition to focal adhesions, podosomes and invadopodia are also adhesion structures regulated by Ca^{2+} signaling. Sun et al. reported that melanoma invasion is promoted by STIM1- and ORAI1-mediated Ca^{2+} oscillations, which promote invadopodia formation and extracellular matrix (ECM) degradation [65]. These authors found that Ca^{2+} signaling mediated by STIM1 and ORAI1 is essential for invadopodia formation over a collagen matrix and that addition of SKF96365, or chelation of extracellular Ca^{2+} with EGTA, blocked invadopodia assembly. More interestingly, ectopic expression of STIM1 or STIM1+ORAI1 increased phosphorylation levels of Tyr416 in Src kinase, without affecting phospho-FAK. This effect was revealed for MCF-7 cells (human breast cancer cells), NMuMG (mouse mammary epithelial cell line) and WM793 melanoma cells and the activation of Ca^{2+} entry with thapsigargin or ionophore A23187 led to the same result, i.e., a rapid increase of pY416 Src levels. These data, together with the fact that STIM1 shRNA, chelation of extracellular Ca^{2+}, or the inhibition of SOCE with 2-APB, reduced pY416 Src levels, strongly suggesting a direct role of STIM1-ORAI1-mediated Ca^{2+} influx in the preservation of Src activity and invadopodia formation [65].

From the information given above it is now widely accepted that one of the major targets for SOCE is the cytoskeleton and that the dynamics of the focal adhesion assembly as well as the dynamics of actin and tubulin cytoskeleton are strongly influenced by the kinetics of Ca^{2+} entry through store-operated Ca^{2+} channels.

6. Localization and polarization of SOCE in migrating cells

From the given information arises the question about the spatial control of STIM1-ORAI1 localization. The polarization of Ca^{2+} entry pathways has been described in several cell types, especially in exocrine gland cells. In pancreatic acinar cells under stimulation of Ca^{2+} mobilizing receptors, the ER Ca^{2+} release is detected in the apical region of the cell and the $[Ca^{2+}]_i$ increase remains restricted at this region [66], although the signal propagates to basolateral regions at high concentration of agonists. In salivary gland acinar cells, $[Ca^{2+}]_i$ increase is also detected in the apical region and then propagates to the basal region [67]. Indeed, several studies have demonstrated that agonist-stimulated Ca^{2+} signaling in exocrine gland cells is highly polarized. TRPC1 is a key factor for SOCE in salivary gland acinar cells and pancreatic acinar cells. In addition, TRPC3 contributes to SOCE and contributes to the receptor-stimulated Ca^{2+} influx in exocrine pancreatic acinar cells [68]. In addition, SOCE is the major contributor to Ca^{2+} influx in salivary gland and pancreatic acinar cells, so it is expected an asymmetric distribution of STIM1, ORAI1 and TRPC1-3, some of the most widely studied members involved in the control of SOCE.

As expected, a polarized localization has been shown for all Ca^{2+} signaling proteins in exocrine acinar cells: IP3Rs, SERCAs and PMCA pumps, GPCRs, TRPC channels, ORAI1 channels and STIM1 [66, 69, 70], an asymmetric distribution required for the directed secretion carried out by these cells. However, little is known about the mechanisms involved in the targeting of Ca^{2+}-signaling complexes to these regions.

Because migrating cells are polarized cells, with polarized distribution of receptor tyrosine kinases, the study of Ca^{2+} entry in migrating cells is a task of particular importance. In this regard, Tsai et al. reported recently that STIM1 is enriched at the leading edge of migrating cells, measured with YFP-STIM1 and CFP-tagged ER marker [71]. The authors concluded that the polarization of STIM1 is microtubule-plus-end dependent, because STIM1-I644N/P645N, a mutant STIM1 that does not bind to EB1, failed to polarize in migrating cells. Because of the enrichment of receptor tyrosine kinases at the front of migrating cells, local Ca^{2+} pulses were observed with higher frequency at the leading edge, accompanied by a lower level of luminal ER Ca^{2+} and increased levels of PMCA activity at the front. In this report the authors also described that a DAG gradient is the result of the asymmetric activity of phospholipase C (PLC) at the leading edge of migrating cells, leading to the recruitment and activation of PKCbeta, a kinase involved in migration [72] by phosphorylating myosin [73], or other substrates of the cytoskeleton, such as GAP43, adducin, or fascin [74]. However, the precise mechanism that assembles de STIM1-ORAI1 and/or TRPC1 in a polarized distribution in cells is still unclear and it is expected that additional members of the cytoskeleton will solve this open question.

7. Conclusions

In conclusion, cytosolic-free Ca^{2+} concentration regulates the reorganization of cytoskeleton, focal adhesion turnover, invadopodia formation, actomyosin contractility and it is critical to trigger the development of lamellipodia as the leading structure during migration. But this dependence is reciprocal and Ca^{2+} influx through store-operated Ca^{2+} channels at the plasma membrane is fully dependent on the formation of endoplasmic reticulum-plasma membrane juxtapositions that are shaped by the reorganization of the cytoskeleton. In the last few years valuable knowledge has been gained regarding the activation of STIM1 by Ca^{2+} store depletion and how STIM1 relocalizes and activates ORAI1 at the PM-ER junctions. Here we have shown that both STIM1 and ORAI1 are involved in many aspects of cell migration and that gene silencing and specific inhibitors point out these two proteins in the regulation of Ca^{2+} influx pathways involved in supporting efficient cell adhesion and migration. However, a major lack of knowledge regarding the polarization of the signaling profile persists. Recent advances in genome editing will be valuable tools to knockout and knockin *STIM* and *ORAI* genes, as well as genes coding for cytoskeletal proteins involved in the reorganization of SOCE. With this coming era we will be able to monitor and study the behavior of tagged proteins at endogenous levels, as well as to study the loss of function of certain genes in any cell type, in an attempt to solve this open question in cell biology and signaling.

Author details

Francisco Javier Martin-Romero[1]*, Aida M. Lopez-Guerrero[1], Carlos Pascual-Caro[1] and Eulalia Pozo-Guisado[2]

*Address all correspondence to: fjmartin@unex.es

1 Department of Biochemistry and Molecular Biology, School of Life Sciences, University of Extremadura, Badajoz, Spain

2 Department of Cell Biology, School of Medicine, University of Extremadura, Badajoz, Spain

References

[1] Berridge, M.J., M.D. Bootman and H.L. Roderick, *Calcium signalling: dynamics, homeostasis and remodelling*. Nat Rev Mol Cell Biol, 2003. **4**(7): pp. 517–29.

[2] Berridge, M.J., P. Lipp and M.D. Bootman, *The versatility and universality of calcium signalling*. Nat Rev Mol Cell Biol, 2000. **1**(1): pp. 11–21.

[3] Berridge, M.J. (2014) *Cell Signalling Biology*. DOI: 10.1042/csb0001012

[4] Catterall, W.A. and N. Zheng, *Deciphering voltage-gated Na^+ and Ca^{2+} channels by studying prokaryotic ancestors*. Trends Biochem Sci, 2015. **40**(9): pp. 526–34.

[5] Frings, S., *Cyclic nucleotide-gated channels and calcium: an intimate relation*. Adv Second Messenger Phosphoprotein Res, 1997. **31**: pp. 75–82.

[6] Taberner, F.J., et al., *TRP channels interaction with lipids and its implications in disease*. Biochim Biophys Acta, 2015. **1848**(9): pp. 1818–27.

[7] Ambudkar, I.S. and H.L. Ong, *Organization and function of TRPC channelosomes*. Pflugers Arch, 2007. **455**(2): pp. 187–200.

[8] Ambudkar, I.S., et al., *TRPC1: the link between functionally distinct store-operated calcium channels*. Cell Calcium, 2007. **42**(2): pp. 213–23.

[9] Vial, C., J.A. Roberts and R.J. Evans, *Molecular properties of ATP-gated P2X receptor ion channels*. Trends Pharmacol Sci, 2004. **25**(9): pp. 487–93.

[10] Madden, D.R., *The structure and function of glutamate receptor ion channels*. Nat Rev Neurosci, 2002. **3**(2): pp. 91–101.

[11] Gudermann, T., et al., *Activation, subunit composition and physiological relevance of DAG-sensitive TRPC proteins*. Novartis Found Symp, 2004. **258**: pp. 103–18; discussion 118–22, 155–9, 263–6.

[12] Putney, J.W., Jr., *A model for receptor-regulated calcium entry*. Cell Calcium, 1986. **7**(1): pp. 1–12.

[13] MacLennan, D.H. and N.M. Green, *Structural biology. Pumping ions*. Nature, 2000. **405**(6787): pp. 633–4.

[14] DiPolo, R. and L. Beauge, *Sodium/calcium exchanger: influence of metabolic regulation on ion carrier interactions*. Physiol Rev, 2006. **86**(1): pp. 155–203.

[15] Babcock, D.F., et al., *Mitochondrial participation in the intracellular Ca^{2+} network*. J Cell Biol, 1997. **136**(4): pp. 833–44.

[16] Smyth, J.T., et al., *Activation and regulation of store-operated calcium entry*. J Cell Mol Med, 2010. **14**(10): pp. 2337–49.

[17] Liou, J., et al., *STIM is a Ca^{2+} sensor essential for Ca^{2+}-store-depletion-triggered Ca^{2+} influx*. Curr Biol, 2005. **15**(13): pp. 1235–41.

[18] Roos, J., et al., *STIM1, an essential and conserved component of store-operated Ca^{2+} channel function*. J Cell Biol, 2005. **169**(3): p. 435–45.

[19] Zhang, S.L., et al., *STIM1 is a Ca^{2+} sensor that activates CRAC channels and migrates from the Ca^{2+} store to the plasma membrane*. Nature, 2005. **437**(7060): pp. 902–5.

[20] Feske, S., et al., *A mutation in Orai1 causes immune deficiency by abrogating CRAC channel function*. Nature, 2006. **441**(7090): pp. 179–85.

[21] Luik, R.M., et al., *The elementary unit of store-operated Ca^{2+} entry: local activation of CRAC channels by STIM1 at ER-plasma membrane junctions*. J Cell Biol, 2006. **174**(6): pp. 815–25.

[22] Prakriya, M., et al., *Orai1 is an essential pore subunit of the CRAC channel*. Nature, 2006. **443**(7108): pp. 230–3.

[23] Yuan, J.P., et al., *STIM1 heteromultimerizes TRPC channels to determine their function as store-operated channels*. Nat Cell Biol, 2007. **9**(6): pp. 636–45.

[24] Ong, H.L., et al., *Dynamic assembly of TRPC1-STIM1-Orai1 ternary complex is involved in store-operated calcium influx. Evidence for similarities in store-operated and calcium release-activated calcium channel components*. J Biol Chem, 2007. **282**(12): pp. 9105–16.

[25] Yuan, J.P., et al., *SOAR and the polybasic STIM1 domains gate and regulate Orai channels*. Nat Cell Biol, 2009. **11**(3): pp. 337–43.

[26] Park, C.Y., et al., *STIM1 clusters and activates CRAC channels via direct binding of a cytosolic domain to Orai1*. Cell, 2009. **136**(5): pp. 876–90.

[27] Park, C.Y., A. Shcheglovitov and R. Dolmetsch, *The CRAC channel activator STIM1 binds and inhibits L-type voltage-gated calcium channels*. Science, 2010. **330**(6000): pp. 101–5.

[28] Wang, Y., et al., *The calcium store sensor, STIM1, reciprocally controls Orai and CaV1.2 channels*. Science, 2010. **330**(6000): pp. 105–9.

[29] Ribeiro, C.M., J. Reece and J.W. Putney, Jr., *Role of the cytoskeleton in calcium signaling in NIH 3T3 cells. An intact cytoskeleton is required for agonist-induced $[Ca^{2+}]i$ signaling, but not for capacitative calcium entry*. J Biol Chem, 1997. **272**(42): pp. 26555–561.

[30] Holda, J.R. and L.A. Blatter, *Capacitative calcium entry is inhibited in vascular endothelial cells by disruption of cytoskeletal microfilaments*. FEBS Lett, 1997. **403**(2): pp. 191–6.

[31] Oka, T., M. Hori and H. Ozaki, *Microtubule disruption suppresses allergic response through the inhibition of calcium influx in the mast cell degranulation pathway*. J Immunol, 2005. **174**(8): pp. 4584–89.

[32] Smyth, J.T., et al., *Role of the microtubule cytoskeleton in the function of the store-operated Ca^{2+} channel activator STIM1*. J Cell Sci, 2007. **120**(Pt 21): pp. 3762–71.

[33] Redondo, P.C., et al., *Dual role of tubulin-cytoskeleton in store-operated calcium entry in human platelets*. Cell Signal, 2007. **19**(10): pp. 2147–54.

[34] Rosado, J.A., S. Jenner and S.O. Sage, *A role for the actin cytoskeleton in the initiation and maintenance of store-mediated calcium entry in human platelets. Evidence for conformational coupling*. J Biol Chem, 2000. **275**(11): pp. 7527–33.

[35] Rosado, J.A. and S.O. Sage, *Farnesylcysteine analogues inhibit store-regulated Ca^{2+} entry in human platelets: evidence for involvement of small GTP-binding proteins and actin cytoskeleton*. Biochem J, 2000. **347 Pt 1**: pp. 183–92.

[36] Russa, A.D., et al., *Microtubule remodeling mediates the inhibition of store-operated calcium entry (SOCE) during mitosis in COS-7 cells*. Arch Histol Cytol, 2008. **71**(4): pp. 249–63.

[37] Grigoriev, I., et al., *STIM1 is a MT-plus-end-tracking protein involved in remodeling of the ER*. Curr Biol, 2008. **18**(3): pp. 177–82.

[38] Sampieri, A., et al., *Visualizing the store-operated channel complex assembly in real time: identification of SERCA2 as a new member*. Cell Calcium, 2009. **45**(5): pp. 439–46.

[39] Honnappa, S., et al., *An EB1-binding motif acts as a microtubule tip localization signal*. Cell, 2009. **138**(2): pp. 366–76.

[40] Zumbrunn, J., et al., *Binding of the adenomatous polyposis coli protein to microtubules increases microtubule stability and is regulated by GSK3 beta phosphorylation*. Curr Biol, 2001. **11**(1): pp. 44–9.

[41] Nathke, I.S., *The adenomatous polyposis coli protein: the Achilles heel of the gut epithelium*. Annu Rev Cell Dev Biol, 2004. **20**: pp. 337–66.

[42] Andrews, P.D., et al., *Aurora B regulates MCAK at the mitotic centromere*. Dev Cell, 2004. **6**(2): pp. 253–68.

[43] Kumar, P., et al., *GSK3beta phosphorylation modulates CLASP-microtubule association and lamella microtubule attachment*. J Cell Biol, 2009. **184**(6): pp. 895–908.

[44] Watanabe, T., et al., *Phosphorylation of CLASP2 by GSK-3beta regulates its interaction with IQGAP1, EB1 and microtubules*. J Cell Sci, 2009. **122**(Pt 16): pp. 2969–79.

[45] Pozo-Guisado, E., et al., *Phosphorylation of STIM1 at ERK1/2 target sites modulates store-operated calcium entry*. J Cell Sci, 2010. **123**(Pt 18): pp. 3084–93.

[46] Pozo-Guisado, E., et al., *Phosphorylation of STIM1 at ERK1/2 target sites regulates interaction with the microtubule plus-end binding protein EB1*. J Cell Sci, 2013. **126**(Pt 14): pp. 3170–80.

[47] Casas-Rua, V., et al., *Inhibition of STIM1 phosphorylation underlies resveratrol-induced inhibition of store-operated calcium entry*. Biochem Pharmacol, 2013. **86**(11): pp. 1555–63.

[48] Tomas-Martin, P., et al., *Phospho-STIM1 is a downstream effector that mediates the signaling triggered by IGF-1 in HEK293 cells*. Cell Signal, 2015. **27**(3): pp. 545–54.

[49] Casas-Rua, V., et al., *STIM1 phosphorylation triggered by epidermal growth factor mediates cell migration*. Biochim Biophys Acta, 2015. **1853**(1): pp. 233–43.

[50] Pozo-Guisado, E. and F.J. Martin-Romero, *The regulation of STIM1 by phosphorylation*. Commun Integr Biol, 2013. **6**(6): p. e26283.

[51] Asanov, A., et al., *A relay mechanism between EB1 and APC facilitate STIM1 puncta assembly at endoplasmic reticulum-plasma membrane junctions*. Cell Calcium, 2013. **54**(3): pp. 246–56.

[52] Sharma, S., et al., *An siRNA screen for NFAT activation identifies septins as coordinators of store-operated Ca^{2+} entry*. Nature, 2013. **499**(7457): pp. 238–42.

[53] Deb, B. K., T. Pathak and G. Hasan, *Store-independent modulation of Ca^{2+} entry through Orai by Septin 7*. Nat Commun, 2016. **7**: 11751 doi: 10.1038/ncomms11751

[54] Yuan, J.P., et al., *The closing and opening of TRPC channels by Homer1 and STIM1*. Acta Physiol (Oxf), 2012. **204**(2): pp. 238–47.

[55] Dionisio, N., et al., *Homer proteins mediate the interaction between STIM1 and Cav1.2 channels*. Biochim Biophys Acta, 2015. **1853**(5): pp. 1145–53.

[56] Huttenlocher, A., et al., *Regulation of cell migration by the calcium-dependent protease calpain*. J Biol Chem, 1997. **272**(52): pp. 32719–22.

[57] Franco, S.J., et al., *Calpain-mediated proteolysis of talin regulates adhesion dynamics*. Nat Cell Biol, 2004. **6**(10): pp. 977–83.

[58] Yang, S., J.J. Zhang and X.Y. Huang, *Orai1 and STIM1 are critical for breast tumor cell migration and metastasis*. Cancer Cell, 2009. **15**(2): pp. 124–34.

[59] Mitra, S.K., D.A. Hanson and D.D. Schlaepfer, *Focal adhesion kinase: in command and control of cell motility*. Nat Rev Mol Cell Biol, 2005. **6**(1): pp. 56–68.

[60] Chen, Y.F., et al., *Calcium store sensor stromal-interaction molecule 1-dependent signaling plays an important role in cervical cancer growth, migration and angiogenesis*. Proc Natl Acad Sci U S A, 2011. **108**(37): pp. 15225–30.

[61] Larue, L. and A. Bellacosa, *Epithelial-mesenchymal transition in development and cancer: role of phosphatidylinositol 3' kinase/AKT pathways*. Oncogene, 2005. **24**(50): pp. 7443–54.

[62] Lee, J.M., et al., *The epithelial-mesenchymal transition: new insights in signaling, development and disease*. J Cell Biol, 2006. **172**(7): pp. 973–81.

[63] Vicente-Manzanares, M., et al., *Non-muscle myosin II takes centre stage in cell adhesion and migration*. Nat Rev Mol Cell Biol, 2009. **10**(11): pp. 778–90.

[64] Chen, Y.T., et al., *The ER Ca^{2+} sensor STIM1 regulates actomyosin contractility of migratory cells*. J Cell Sci, 2013. **126**(Pt 5): pp. 1260–67.

[65] Sun, J., et al., *STIM1- and Orai1-mediated Ca^{2+} oscillation orchestrates invadopodium formation and melanoma invasion*. J Cell Biol, 2014. **207**(4): pp. 535–48.

[66] Yule, D.I., *Subtype-specific regulation of inositol 1,4,5-trisphosphate receptors: controlling calcium signals in time and space*. J Gen Physiol, 2001. **117**(5): pp. 431–4.

[67] Bruce, J.I., et al., *Modulation of $[Ca^{2+}]i$ signaling dynamics and metabolism by perinuclear mitochondria in mouse parotid acinar cells*. J Biol Chem, 2004. **279**(13): pp. 12909–17.

[68] Ambudkar, I.S., *Polarization of calcium signaling and fluid secretion in salivary gland cells*. Curr Med Chem, 2013. **19**(34): pp. 5774–81.

[69] Hong, J.H., et al., *Polarized but differential localization and recruitment of STIM1, Orai1 and TRPC channels in secretory cells*. Traffic, 2011. **12**(2): pp. 232–45.

[70] Bandyopadhyay, B.C., et al., *Apical localization of a functional TRPC3/TRPC6-Ca^{2+}-signaling complex in polarized epithelial cells. Role in apical Ca^{2+} influx*. J Biol Chem, 2005. **280**(13): pp. 12908–16.

[71] Tsai, F.C., et al., *A polarized Ca^{2+}, diacylglycerol and STIM1 signalling system regulates directed cell migration*. Nat Cell Biol, 2014. **16**(2): pp. 133–44.

[72] Rosse, C., et al., *PKC and the control of localized signal dynamics*. Nat Rev Mol Cell Biol, 2010. **11**(2): pp. 103–12.

[73] Ludowyke, R.I., et al., *Phosphorylation of nonmuscle myosin heavy chain IIA on Ser1917 is mediated by protein kinase C beta II and coincides with the onset of stimulated degranulation of RBL-2H3 mast cells*. J Immunol, 2006. **177**(3): pp. 1492–9.

[74] Larsson, C., *Protein kinase C and the regulation of the actin cytoskeleton*. Cell Signal, 2006. **18**(3): pp. 276–84.

Chapter 5

Role of Band 3 in the Erythrocyte Membrane Structural Changes Under Isotonic and Hypotonic Conditions

Ivana Pajic-Lijakovic and Milan Milivojevic

Additional information is available at the end of the chapter

http://dx.doi.org/10.5772/64964

Abstract

An attempt was made to discuss and connect various modeling approaches which have been proposed in the literature in order to shed further light on the erythrocyte membrane relaxation under isotonic and hypotonic conditions. Roles of the main membrane constituents: (1) the actin-spectrin cortex, (2) the lipid bilayer, and (3) the transmembrane protein band 3 and its course-consequence relations were considered to estimate the membrane relaxation phenomena. Cell response to loading conditions includes the successive sub-bioprocesses: (1) erythrocyte local or global deformation, (2) the cortex-bilayer coupling, and (3) the rearrangements of band 3. The results indicate that the membrane structural changes include: (1) the spectrin flexibility distribution and (2) the rate of its changes influenced by the number of band 3 molecules attached to spectrin filaments, and phosphorylation of the actin-spectrin junctions. Band 3 rearrangement also influences: (1) the effective bending modulus and (2) the band 3-bilayer interaction energy and on that base the bilayer bending state. The erythrocyte swelling under hypotonic conditions influences the bilayer integrity which leads to the hemolytic hole formation. The hemolytic hole represents the excited cluster of band 3 molecules.

Keywords: packing state changes of band 3 clusters, reversible hemolytic hole formation, the lipid bilayer bending state changes, the spectrin inter- and intrachain interactions, mathematical modeling

1. Introduction

Erythrocyte mechanics under isotonic and hypotonic conditions has been studied from engineering and biomedical stand points [1–14]. Rheological response of the erythrocyte membrane depends on the main membrane constituent rearrangement: (1) the actin cortex, (2) the lipid bilayer, and (3) the transmembrane protein band 3. The membrane fluctuations under isotonic condition induce alternating expansion and compression of the membrane parts in order to ensure surface and volume conservation. The membrane relaxation occurs within (millisecond order) affine regime and (second order) nonaffine regime. The affine regime corresponds to the spectrin interchain interactions while the nonaffine regime corresponds to the spectrin intrachain interactions. However, the membrane fluctuations under hypotonic condition induce volume increase by ensuring surface conservation. The membrane response under hypotonic condition includes the successive sub-processes: (1) erythrocyte swelling, (2) lifetime of the lipid structural integrity and the rearrangement of transmembrane protein band 3, and (3) the reversible hemolytic hole formation and hemoglobin (Hb) release to surrounding solution. Duration of the membrane relaxation depends on contributions of three sub-processes: (1) time for cell swelling $t_{sw} \in [5, 100\, s]$, (2) membrane lifetime $\tau_m \in (0,\ t_H)$ (where t_H is the hemolytic time), and (3) time for Hb release from already formed hemolytic hole during successive open-closed state changes $t_{release} \in [1{,}5\, s]$ [11, 13].

The band 3 rearrangement significantly influences the state of both, the lipid bilayer and the actin-spectrin cortex. Space distribution of band 3 molecules and their lateral diffusion influence the bending state of the lipid bilayer and its free energy [15–17]. Local changes of the bilayer bending state enhance anomalous sub-diffusion and eventually lead to hop-diffusion of lipids. These effects induce anomalous nature of energy dissipation during lipids structural ordering and could be quantified by the effective viscosity [18]. The bilayer structural changes have the feedback effects to band 3 protein-lipids positive hydrophobic mismatch effects [19–21]. De Meyer et al. [19] pointed to the cholesterol role in lipid mediated protein-protein interactions. These effects could lead to the protein tilt angle changes and the protein clustering. This clustering can be modulated by homotypic interactions of the protein transmembrane domains and electrostatic protein-lipid interactions [21]. Tilt angle changes influence packing state of band 3 clusters and their association-dissociation to spectrin. Band 3 molecules form various complexes with spectrin and influence its conformational changes. In-homogeneous distribution of band 3 molecules and their ability to clustering influence the rheological response of: (1) the bilayer, (2) the cortex, and (3) the nature of the bilayer-cortex mechanical coupling. Ehrig et al. [22] pointed that the lipid bilayer phase separation can be strongly affected by interaction with the actin cortex, which, depending on the temperature and membrane composition, can either lead to precipitation of highly dynamic membrane domains (rafts), or prevent large-scale phase separation.

For understanding the influence of band 3 rearrangement on complex nature of the membrane rheological response, it is necessary to consider three subpopulations of band 3 molecules under isotonic and hypotonic conditions. The first subpopulation (20–40%) as tetramers forms high affinity complexes with ankyrin (quantified by the dissociation constant ~5 nM) as reported by Tomishige et al. [23] and Kodippili et al. [24]. Band 3-ankyrin complexes are located near the center of spectrin tetramers. These complexes could survive the hypotonic conditions. Golan and Veatch [25] reported that 25% of band 3 population remains attached to the cortex under ionic strength 26 mM $NaPO_4$ solution at 37°C. The second subpopulation (~30%) as dimers forms lower affinity complexes with the adducin (the dissociation constant is ~100 nM) as reported by Franco and Low [26] and Kodippili et al. [24]. This subpopulation is located at the spectrin-actin junction complexes. The junctions of the network link 4–7 spectrin filaments [27]. The third one is freely diffusing subpopulation (~30%). The part of freely diffusing subpopulation increases under hypotonic conditions. Golen and Veatch [25] determined that mobile fraction of band 3 molecules under the external solution tonicity 46.0 mM $NaPO_4$ at 21°C was 11 ± 9%. Under tonicity of 5.2 mM $NaPO_4$ at 21°C, the mobile fraction of band 3 molecules was 72 ± 7%. Band 3 molecules are anion at pH = 7. The value of Stokes radius of band 3 dimmer is approximately 7.6 nm while for tetramer is 11 nm at room temperature and pH = 7.2 [28]. The total number of band 3 per single erythrocyte is ~1×10^6. Thermal fluctuations of the erythrocyte membrane could induce conformational changes of cytoplasmic domain of transmembrane protein band 3 [29]. Such conformational changes result in electrostatic interactions between highly anionic N-terminal domains of band 3 molecules and on that base intensified their short-range self-associative tendency [28]. Long-range self-associative tendency is induced by positive hydrophobic mismatch effects [21]. The band 3 clustering is pronounced under hypotonic conditions [30]. In this case, ~25% of band 3 molecules make aggregate of ~5000 mers [31]. All band 3 subpopulations through these complexes contribute to the spectrin conformational changes by reducing its mobility and influence the cortex stiffening.

The band 3 molecules within the freely diffusing subpopulation could form low affinity complexes to spectrin (the dissociation constant is ~1–10 μM) [25]. On that base, they influence spectrin conformations [32]. Gov [32] reported that the parts of the spectrin filament between two mid-point attachments behave as independent blobs. The main factor which influences the spectrin flexibility and conformations is $\frac{l}{L_p}$ (where l is the length of the filament part between two mid-point attachments and L_p is the spectrin persistence length $L_p = 15-25\ nm$ [33]). The spectrin filament parts are as follows: (1) flexible if $\frac{l}{L_p} \succ\succ 1$, (2) semiflexible if $\frac{l}{L_p} \approx 1$, and (3) rod-like if $\frac{l}{L_p} \prec\prec 1$. The average length of spectrin parts is equal to $\langle l \rangle = \frac{L_c}{\langle N_B \rangle}$ (where $L_c \approx 200\text{nm}$ is the spectrin contour length [27] and $\langle N_B \rangle \approx 4-10$ is the average number of attached band 3 molecules per single spectrin filament [34]). The length of the spectrin parts depends on the rearrangement of band 3 molecules which include their lateral diffusion and self-associative tendency [28]. Consequently, rheological behavior of the cortex is related to the spectrin flexibility distribution and the rate of its changes [35, 36]. Deeper insight into coarse-consequence relations between the band 3 rearrangement and the spectrin inter- and intrachain interactions as well as the bilayer bending offers a possibility for understanding the complex nature of the membrane relaxation phenomena.

2. Rearrangement of band 3 molecules-cluster packing state changes

Packing state of band 3 clusters and its changes during short-time lateral motion under isotonic and hypotonic conditions could be estimated by applying Edwards' statistics [30, 35, 37, 38]. Short-time motion of band 3 molecules includes: (1) Brownian diffusion within the mesh compartment of the spectrin-actin cortex and (2) hop diffusion between two compartments. Hop is observed at every 350 ms [23]. Band 3 molecules form clusters caused by positive hydrophobic mismatch effects during their lateral diffusion. This statistical approach is suitable for describing the cluster packing state changes under vibration field. Edwards introduced a new and very significant parameter for description of particle clusters named compactivity of the cluster part X representing the change of cluster part volume V_{cr} with entropy S_{cr} for given number of molecules N_p (i.e., considering it as a canonical ensemble):

$$X = \left(\frac{\partial V_{cr}}{\partial S_{cr}} \right)_{N_P} \tag{1}$$

The two limits of the compactivity have been introduced: (1) $X \to 0$ corresponds to the most compact particle rearrangement and (2) $X \to \infty$ corresponds to the least compact particle rearrangement. When $X \to 0$, the system picks out one particular configuration as being most likely, and when $X \to \infty$, the system picks out all configurations as being equally likely. Probability distribution of band 3 cluster states could be expressed as:

$$P = e^{(Y_r - W_r)/\lambda X} \tag{2}$$

where Y_r is effective volume of the cluster part (corresponding to the free energy F in classical statistical mechanics) and W_r is the "volume function" corresponding to the Hamiltonian and λ is constant adjusting dimensions. Then, we can write:

$$Y_r = V_{cr} + X \frac{\partial Y_r}{\partial X} = V_{cr} - X S_{cr} \tag{3}$$

Cluster is considered as canonical ensemble during short-time rearrangement. The partition function relating Y_r with the volume function of a cluster part $W_r(q)$ can be written in the form:

$$Z_p = e^{-\frac{Y_r}{\lambda X}} = \int\int ... \int \Omega_w e^{-\frac{W_r(q_1, q_2, ... q_n)}{\lambda X}} dq_1 dq_2 ... dq_n \tag{4}$$

where integration goes over all geometrical degrees of freedom (DOFs) q_i of all molecules in the cluster, $Z_p = Z_p(r,\ t_{eq})$ is the partition function. The key step is the identification of an exact volume function which makes it possible to pinpoint the configuration phase space and evaluate its dimensionality. We consider weakly interacting systems. To be precise, the interactions must be sufficient to lead to thermodynamical equilibrium, but weak enough that these interactions have negligible effects on the effective volume of individual molecules.

$$W_r\left(n_q\right) = N_{pr} w_p\left(q_1, q_2, \ldots, q_n\right) \tag{5}$$

where $w_p\left(q_1, q_2, \ldots, q_n\right)$ is the volume function of single molecule. Rearrangement of rigid molecules such as band 3 within the cluster could be described using low degrees of freedom (DOFs). For this case, the DOFs describe molecule's orientation in the cluster part located at r and its coordination number.

The band 3 cluster excitation under isotonic condition induces the molecule orientation changes presented in **Figure 1**. The excitation occurs during alternating expansion and compression of the membrane parts in order to ensure surface and volume conservation.

The corresponding volume function could be expressed as [30, 35]:

$$w_p\left(q_o, q_c\right) = v_0 + \Delta v_o q_o^2 \tag{6}$$

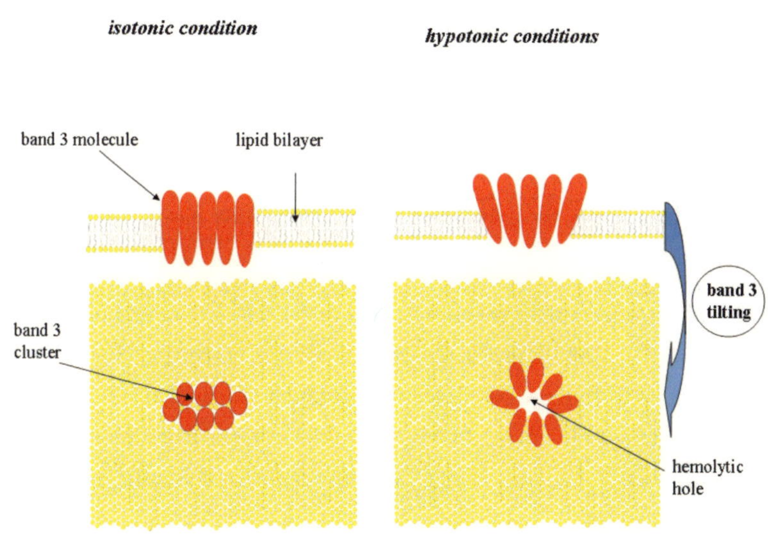

Figure 1. Schematic representation of the packing state changes for excited band 3 cluster under isotonic and hypotonic conditions.

where v_0 is the minimum specific volume of single molecule of band 3 $v_0 = \frac{4}{3}\langle r_s \rangle_{min}^{3} \pi$, $\langle r_s \rangle_{min}$ is the minimum of the molecule Stokes radius, Δv_o is the volume increment caused by the molecule orientation equal to $\Delta v_o = v_{max} - v_0$, v_{max} is the maximum specific volume of single molecule of band 3 $v_{max} = \frac{4}{3}\langle r_s \rangle_{max}^{3}\pi$, $\langle r_s \rangle_{max}$ is the maximum of the molecule Stokes radius, q_o accounts the molecule's orientation equal to $q_o^2 = \frac{1}{2}\sum_i q_{oi}^2$ while q_{oi} accounts various directions, Δv_o is the volume increment contributions caused by the DOFs q_o changes. For molecule loose packing, we can take $w_p = v_0 + \Delta v_o$ and a molecule has high values of the degrees of freedom in such system so q_i values are close to 1.

The band 3 cluster excitation under hypotonic conditions is more intensive than that obtained under isotonic condition due to changes the bilayer bending state during erythrocyte swelling. The excitation could induce changes the packing state from close packing to ring-like structure which represents the reversible hemolytic hole. These changes are influenced by the positive hydrophobic mismatch effects which could induce protein tilting [20]. Zade-Oppen [13] reported that the average hole opening time period is 270 ms, while the average hole closing time period is 260 ms. Seeman et al. [8] experimentally determined the diameter of the reversible osmotic holes in the range between 10 and 100 nm for human erythrocyte under hypotonic condition at pH = 7. Accordingly, the smaller hemolytic hole corresponds to ~4 band 3 molecules while the larger one corresponds to ~40 molecules [18, 30]. The result points out that cluster size is not the main factor for the hole formation. The main factor could be the hydrophobic mismatch effects between band 3 and the surrounding lipid bilayer as was shown in **Figure 1**. The corresponding volume function is as follows:

$$w_p(q_o, q_c) = v_0 + \Delta v_o q_o^2 + \Delta v_c q_c^2 \tag{7}$$

where q_c accounts the coordination number, Δv_c is the volume increment caused by changes the molecule coordination number equal to $\Delta v_c = \frac{R_H^2(t_R)\, \pi\, h_m}{N_r(t_R)}$, t_R is the relaxation time, h_m is the thickness of the lipid bilayer for already swollen erythrocyte, $R_H(t_R)$ is the radius of hemolytic hole, $N_r(t_R)$ is the number of molecules per cluster located at r. DOFs can have values in the range 0 to 1. When q_c=0 and q_{oi}=0 at $t_R \to t_{R\,eq}$, the volume function is $w_p = v_0$. When q_c=1 and q_{oi}=1 at t_R=0, the volume function is equal to $w_p = v_0 + \Delta v_o + \Delta v_c$. This excited cluster state represents the hemolytic hole.

Band 3 molecules change their states during migration. Consequently, for estimating the DOFs temporal changes it is necessary to consider the temporal changes of single molecule velocity.

Temporal changes of DOFs under vibration field could be described in the form of Langevin-type equations [30, 37]:

$$\frac{dq_i(t_R)}{dt_R} = -\frac{1}{\gamma_q}\frac{\partial w_p(q_i(t_R))}{\partial q_i(t_R)} + \phi_q(t_R) \tag{8}$$

where the stochastic random force $\phi_q(t_R)$ is formulated as white noise with correlation function $\langle \phi_q(t_R)\phi_q(t_R')\rangle = 2\lambda X \gamma_q \delta_{ij}\delta(t_R - t_R')$, and γ_q is the analog of frictional resistance. System structural changes could induce anomalous nature of energy dissipation derivative during molecules migration within the cluster. If the molecule migration causes damping effects as described by Tomishige et al. [23] (subdiffusion phenomenon), the derivative $\frac{dq_i(t_R)}{dt_R}$ could be replaced by the fractional derivative D_t^{γ} (where D_t^{γ} is the Caputo's fractional derivative, γ is the order of the fractional derivative such that $\gamma \prec 1$). Caputo's definition of the fractional derivative of some function $f(t)$ is given as follows [39]: $D_t^{\gamma}(f(t)) = \frac{1}{\Gamma(1-\gamma)}\frac{d}{dt}\int_0^t \frac{f(t')^{(1)}}{(t-t')^{\gamma}} dt'$ (where $\Gamma(1-\gamma)$ is the gamma function). Average equilibrium volume of single molecule is obtained as:

$$v_{p\,eq} = \frac{\iint \ldots \int w_{pr}(\bullet)\, e^{-\frac{w_{pr}(\bullet)}{\lambda X}}\, d[\bullet]}{Z_p} \tag{9}$$

where $Z_p(r, t_{Req})$ is the partition function, and $V_{cr}(r, t_{Req}) = N_p(r, t_{Req})\, v_{p\,eq}$ is the cluster volume.

3. Band 3 rearrangement influences the spectrin inter- and intrachain interactions and the bilayer bending

The spectrin interchain interactions depend on the number of band 3 molecules attached per single spectrin filament as was shown in **Figure 2**.

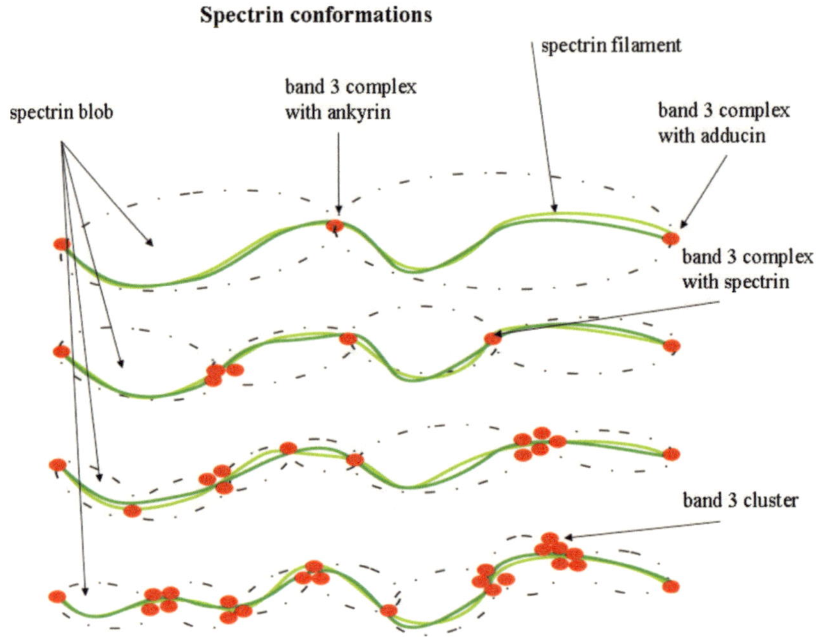

Figure 2. Spectrin conformational changes influenced by the number of attached band 3 molecules.

The spectrin filaments have been treated as flexible ($\frac{L_c}{L_p} \succ\succ 1$) [27] and semiflexible ($\frac{L_c}{L_p} \approx 1$) [40] (where L_c is the spectrin contour length equal to $L_c = \sum_{i=1}^{N_B - 1} l_i$, l_i is the length of i-th filament part between two mid-point attachments of band 3). The flexibility depends on the number of band 3 molecules attached per single spectrin filament. Spectrin filament without band 3-spectrin complexes behaves as flexible. Its conformations have been described as [27]:

$$F_s \approx N\mu_s \left(R - \left\langle r_g^2 \right\rangle^{1/2} \right) \tag{10}$$

Where $N \approx 3$ is the number of spectrin filaments per network units, μ_s is the surface shear modulus of the cortex equal to $\mu_s = \frac{k_B T}{\langle r_g^2 \rangle}$, k_B is Boltzmann constant and T is temperature, R is the end-to-end distance of spectrin filaments in the cortex, $\langle r_g^2 \rangle^{1/2}$ is the average filaments radius of gyration. If the band 3-spectrin low affinity complexes exist, the parts of the spectrin filament between the complexes behave as independent blobs [32]. The conformation changes within the blobs are the milliseconds order [32]. When $\frac{\langle l \rangle}{L_p} \succ\succ 1$ the filament parts are flexible, but if $\frac{\langle l \rangle}{L_p} \approx 1$ they become semiflexible, while if $\frac{\langle l \rangle}{L_p} \prec\prec 1$ they behave as rod-like polymers (where $\langle l \rangle$

is the average length of the filament part) [41]. Li et al. [40] treated the whole spectrin filaments as a semiflexible and proposed worm-like force for modeling of the spectrin conformations:

$$F_{wlc} = \frac{k_B T}{L_p}\left\{\frac{1}{4(1-x)^2} - \frac{1}{4} + x\right\} \tag{11}$$

where $x = \frac{R - \langle r_g^2 \rangle^{1/2}}{L_c}$ is the stretch ratio. The worm-like force corresponds to the condition $\frac{L_c}{L_p} \approx 1$ [41]. It is in accordance with the fact that the spectrin-band 3 complexes lead to decrease in the spectrin flexibility. It could be quantified by apparent increase in the spectrin persistence length $L_p \to L_{p\,eff}$ (where $L_{p\,eff}$ is the effective spectrin persistence length). The concept of effective persistence length has been introduced for describing the nature of interchain structural changes for worm-like chains such as proteins under stretching [42, 43]. On that base, the effective persistence length in our case could be expressed as [36]:

$$L_{p\,eff}(T, N_B) = L_p(T) + \Delta L_p(N_B) \tag{12}$$

Where $L_{p\,eff}(T, N_B)$ is the effective persistence length of spectrin filament, $L_p(T)$ is the spectrin persistence length for the filaments without mid-point attachments at the same temperature conditions and $\Delta L_p(N_B)$ is the contribution to the persistence length caused by the band 3 midpoint attachments. The collective phenomena among variously flexible spectrin filaments induce generation of the cortex in-homogeneities [36]. The in-homogeneities in the context of the cortex micro domains influence the cortex relaxation. The cortex relaxation modulus $G_C(t_R)$ could be expressed as [18, 36]:

$$G_C(t_R) = \int_{L_{p\min}}^{L_{p\max}} \rho_C(L_{p\,eff}, t_R)\; G_C(L_{p\,eff}, t_R)\; dL_{p\,eff} \tag{13}$$

where $G_C(L_{p\,eff}, t_R)$ is the cortex relaxation modulus within the domain and $\rho_C(L_{p\,eff}, t_R)$ is the spectrin flexibility distribution caused by the band 3 rearrangement.

The presence of the cortex micro domains is related to in-homogeneous distribution of: (1) band 3 molecules, (2) spectrin flexibility, and (3) the presence of the cortex defects as was shown in **Figure 3**. Cumulative effects as: (1) the spectrin intrachain interactions which lead to formation of the cortex micro domains, (2) longtime diffusion of band 3 molecules, and (3) longtime bending relaxation of the lipid bilayer are at the order of seconds [18].

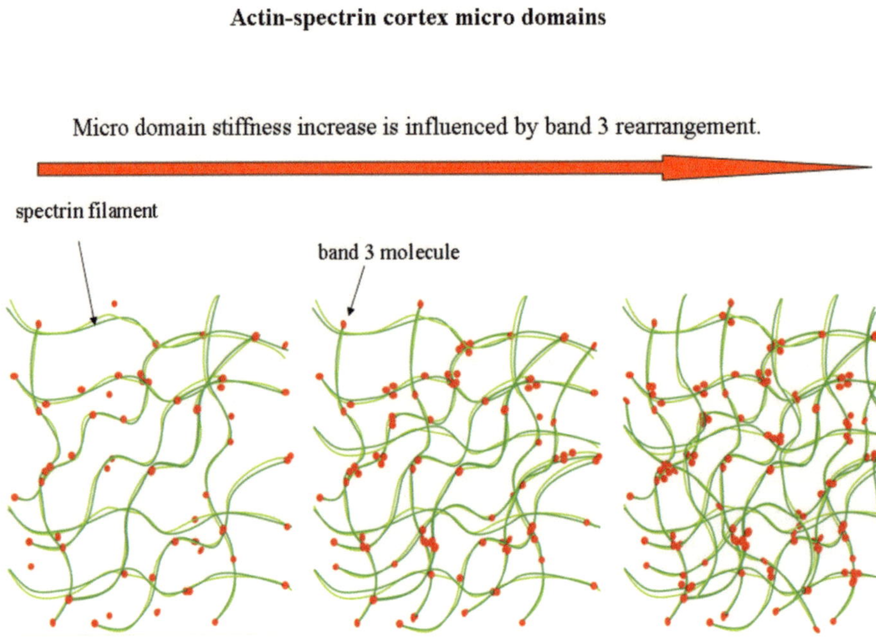

Figure 3. The cortex micro domains—schematic representation.

These local in-homogeneities of the cortex are caused by alternating expansion and compression of the membrane and the cortex-bilayer coupling. The bilayer bending is influenced by conformational changes of the two types of spectrin filaments [44]: (1) type 1—corresponds to the filaments grafted at one end or at both ends but not connected to the stretched cortex and (2) type 2—corresponds to the filaments grafted at both ends and on that base represents a part of the connected stretched cortex. Filaments within the type 1 induce a concave curvature of radius $R_{L\ 1}$, while the type 2 induce a concave curvature of radius $R_{L\ 2}$ such that $R_{L\ 1} = -R_{L\ 2}$. The bilayer-cortex coupling has been expressed by Helfrich-type bending-free energy functional [44]:

$$E(n_1, n_2) = \frac{1}{2}\kappa w \int \left(H - \bar{H}_1 n_1(r,s,t_R) - \bar{H}_2 n_2(r,s,t_R)\right)^2 d^2 s \tag{14}$$

Where s is the coordinate along the contour, κ is the bending modulus of the bilayer, w is the bilayer width, $n_1(r,\ s,\ t_R)$ and $n_2(r,\ s,\ t_R)$ are the relative densities of the types 1 and 2 of spectrin filaments, and the corresponding local mean curvature are $\bar{H}_1 = \frac{1}{R_{L\ 1}}$ and $\bar{H}_2 = \frac{1}{R_{L\ 2}}$. The overall curvature by spectrin filaments is expressed as $\bar{H}_1 n_1 + \bar{H}_2 n_2$. Band 3 molecules influence the lipid bilayer bending modulus. Shlomovitz and Gov [45] formulated the apparent bending modulus equal to:

$$\kappa_{app}(\varphi) = \kappa\left(1 - \varphi(r,t_R)\right) + \kappa'\varphi(r,t_R) \tag{15}$$

where $\kappa_{app}(\varphi)$ is the apparent bending modulus, κ is the bending modulus of the bilayer without band 3 molecules, κ' is the contribution of band 3 molecules to the bending modulus, $\varphi(r, t)$ is the local surface fraction of the band 3 molecules. Shlomovitz and Gov [45] expressed the influence of inclusions (band 3 molecules) on the lipid bilayer bending by formulating the free energy functional as:

$$E(\varphi) = \int \frac{1}{2}\kappa_{app}(\varphi)\left(H - \varphi\bar{H}\right)^2 d^2 r \tag{16}$$

Consequently, the model Eqs. (15) and (16) could be combined to describe the influence of: (1) spectrin conformationals and (2) band 3 molecules migration on the lipid bilayer bending state expressed in the form: $E(n_1, n_2, \varphi) = \int \frac{1}{2}\kappa_{app}(\varphi)\ w\int\left(H - (1-\varphi)\ \left(\bar{H}_1 n_1 + \bar{H}_2 n_2\right)\right)^2 d^2 s\ d^2 r$.
The collective phenomena related to the spectrin filaments migration is expressed by spatial-temporal changes of the conservative variable $n = n(r, s, t_R)$ [46] as:

$$\frac{1}{\dot{s}}\frac{\partial(\dot{s}\ n)}{\partial t_R} = \frac{D}{\dot{s}}\nabla_s^2(\dot{s}\ n) + \frac{1}{\dot{s}}\frac{\Lambda}{n_{sat}}\nabla_s\left(\dot{s}\ n\nabla_s\left(\frac{1}{\dot{s}}\frac{\delta E}{\delta\ n}\right)\right) \tag{17}$$

where $D = \frac{k_B T}{\Lambda}$ is the spectrin collective diffusion coefficient which accounted for the intrachain interactions, Λ is the filaments mobility parameter, n_{sat} is the maximal packing density of the filaments, and ∇_s is the derivative along the contour s. Spectrin filament mobility depends on the number of attached band 3 molecules. Consequently, the effective diffusivity could be formulated as $D_{eff} = D_{eff}(\varphi)$and introduced in eq. 17. Shlomovitz and Gov [45] modeled the collective migration of band 3 molecules as:

$$\frac{\partial\varphi}{\partial t_R} = D_B\nabla^2\varphi + \Lambda_B\nabla\left(\varphi\nabla\left(\frac{\delta E}{\delta\ \varphi}\right)\right) \tag{18}$$

where ∇ is the derivative along the space, Λ_B is the band 3 lateral mobility and D_B is the band 3 diffusion coefficient equal to $D_B = \frac{k_B T}{\Lambda_B}$. Lateral motion of the band 3 molecules induces the anomalous nature of energy dissipation which includes damping effects [23]. These damping effects are induced from band 3 association-dissociation to spectrin filaments. Pajic-Lijakovic [35, 36] proposed fractional Langevin equation for describing the lateral diffusion by applying the fractional derivatives [39]. Consequently, the time derivatives from Eqs. (17) to (18) could be replaced by the fractional derivative $D_t^{\alpha}(\bullet)$.

4. Conclusion

Rheological behavior of the cortex depends on the spectrin flexibility distribution and the rate of its changes [35, 36]. The spectrin flexibility primarily depends on the number of band 3 molecules attached per single spectrin filaments. Rearrangement of the band 3 molecules and their lateral diffusion also influence the bending modulus of the lipid bilayer and the band 3-bilayer interaction energy. Consequently, the band 3 rearrangement influences the cortex-bilayer coupling and on that base influences the membrane rheological behavior as a whole. The membrane structural changes induce anomalous nature of energy dissipation caused by these complex multi scale molecular dynamics.

Acknowledgements

This research was funded by grant III46001 from the Ministry of Science and Environmental Protection, Republic of Serbia.

Author details

Ivana Pajic-Lijakovic* and Milan Milivojevic

*Address all correspondence to: iva@tmf.bg.ac.rs

Faculty of Technology and Metallurgy, Belgrade University, Belgrade, Serbia

References

[1] Amin MS, Park YK, Lue N, Dasari RR, Badizadegan K, Feld MS, Popescu G. Microrheology of red blood cell membrane using dynamics scattering microscopy. Opt Express. 2007; 15(25):17001–17009.

[2] Popescu G, Park YK, Dasari RR, Badizadegan K, Feld MS. Coherence Properties of red blood cell membrane motion. Phys Rev E. 2007; 76:031902 1–5.

[3] Puig-de-Morales-Marinkovic M, Turner KT, Butler JP, Fredberg JJ, Suresh S. Viscoelasticity of the human red blood cell. Am J Physiol Cell Physiol. 2007; 293:C597–C605.

[4] Yoon YZ, Kotar J, Yoon G, Cicuta P. The nonlinear mechanical response of the red blood cell. Phys Biol. 2008; 5(3):036007 1–8.

[5] Yoon YZ, Kotar J, Brown AT, Cicuta P. Red blood cell dynamics: from spontaneous fluctuations to non-linear response. Soft Matter. 2011; 7:2042–2051.

[6] Park YK, Best CA, Badizadegan K, Dasari RR, Feld MS, Kuriabova T, Henle ML, Levine AJ, Gabriel Popescu G. Measurement of red blood cell mechanics during morphological changes. PNAS. 2010; 107(15):6731–6736.

[7] Danon D. Osmotic hemolysis by a gradual decrease in the ionic strength of the surrounding medium. J Cell Comp Physiol. 1961; 57:111–117.

[8] Seeman P, Cheng D, Iles GH. Structure of membrane holes in osmotic and saponian hemolysis. J Cell Biol. 1973; 56:519–527.

[9] Leiber MR, Steck TL. A description of the holes in human erythrocyte membrane ghosts. J Biol Chem. 1982; 257:11651–11659.

[10] Sato Y, Yamakose H, Suzuki Y. Mechanism of hypotonic hemolysis of human erythrocytes. Biol Pharm Bull. 1993; 16(5):506–512.

[11] Pribush A, Meyerstein D, Meyerstein N. Kinetic of erythrocyte swelling and membrane hole formation in hypotonic media. Bioch Biophys Acta. 2002; 1558:119–132.

[12] Evans EA, Hochmuth RM. Membrane viscoelasticity. Biophys J. 1976; 16:1–11.

[13] Zade-Oppen AMM. Repetitive cell 'jump' during hypotonic lysis of erythrocytes observed with simple flow chamber. J Microsc. 1998; 192:54–62.

[14] Markle DR, Evans EA, Hochmuth RM. Force relaxation and permanent deformation of erythrocyte membrane. Biophys J. 1983; 42:91–98.

[15] Gil T, Ipsen JH, Mouritsen OG, Sabra MC, Sperotto MM, Zuckermann MJ. Theoretical analysis of protein organization in lipid membranes. Biochem Biophys Acta. 1998; 1376:245–266.

[16] Evans AR, Turner MS, Sens P. 2003. Interactions between proteins bound to biomembranes. Phys Rev E. 2003; 67:041907 1–10.

[17] Sens P, Turner MS. Theoretical model for the formation of caveolae and similar membrane invaginations. Biophys J. 2004; 86:2049–2057.

[18] Pajic-Lijakovic I, Milivojevic M. Modeling analysis of the lipid bilayer-cytoskeleton coupling in erythrocyte membrane. Biomech Model Mechanobiol. 2014; 13(5):1097–1104.

[19] De Meyer FJM, Rodgers JM, Willems TF, Smit B. Molecular simulation of the effect of cholesterol on lipid-mediated protein-protein interactions. Biophys J. 2010; 99:3629–3638.

[20] Strandberg E, Esteban-Martin S, Ulrich AS, Salgado J. Hydrophobic mismatch of mobile transmembrane helices: merging theory and experiments. Biochim Biophys Acta. 2012; 1818:1242–1249.

[21] Milovanovic D, Honigmann A, Koike S, Gottfert F, Pahler G, Junius M, Muller S, Diederichsen U, Janshoff A, Grubmuller H, Risselada HJ, Eggeling C, Hell SW, van den Bogaart G, Jahn R. Hydrophobic mismatch sorts SNARE proteins into distinct membrane domains. Nature Commun. 2015; 6:5984 1–10.

[22] Ehrig J, Petrov EP, Schwille P. Near-critical fluctuations and cytoskeleton-assisted phase separation lead to sub diffusion in cell membranes. Biophys J. 2011; 100:80–89.

[23] Tomishige M, Sako Y, Kusumi A. Regulation mechanism of the lateral diffusion of band 3 in erythrocyte membranes by the membrane skeleton. J Cell Biol. 1998; 142(4):989–1000.

[24] Kodippili GC, Spector J, Hale J, Giger K, Hughes MR, McNagny KM, Birkenmeier C, Peters L, Ritchie K, Low PS. Analysis of the mobilities of band 3 populations asociatted with ankyrin protein and junctional complexes in intact murine erythrocytes. J Biol Chem. 2012; 287(6):4129–4138.

[25] Golan DE, Veatch W. Lateral mobility of band 3 in the human erythrocyte membrane studied by fluorescence photobleaching recovery: evidence for control by cytoskeletal interactions, PNAS. 1980; 77(5):2537–2541.

[26] Franco T, Low PS. Erythrocyte adducin: a structural regulator of the red blood cell membrane. Transfus Clin Biol. 2010; 17(3):87–94.

[27] Gov NS, Safran SA. Red blood cell membrane fluctuations and shape controlled by ATP-induced cytoskeletal defects. Biophys J. 2005; 88:1859–1874.

[28] Taylor AM, Boulter J, Harding SE, Colfen H, Watts A. Hydrodynamics properties of hyman erythrocyte band 3 solubilized in reduced triton X-100. Biophys J. 1999; 76:2043–2055.

[29] Salhany JM, Cordes KA, Sloan RL. Mechanism of band 3 dimmer dissociation during incubation of erythrocyte membranes at 37 degrees C. Biochem J. 2000; 345:33–41.

[30] Pajic-Lijakovic I, Ilic V, Bugarski B, Plavsic MB. The rearrangement of erythrocyte band 3 molecules and reversible osmotic holes formation under hypotonic conditions. Europ Biophys J Biophys Lett. 2010; 39(5):789–800.

[31] Blackman SM, Cobb CE, Beth AH, Piston DW. The orientation of eosin-5-maleimide on human erythrocyte band 3 measured by fluorescence polarization microscopy. Biophys J. 1996; 71:194–208.

[32] Gov NS. Less is more: removing membrane attachments stiffness the RBC cytoskeleton. New J Phys. 2007; 9:430 1–14.

[33] Boal D. Mechanics of the Cell. 2nd edition. New York: Cambridge University Press, 2012.

[34] Waugh RE, Agre P. Reductions of erythrocyte membrane viscoelastic coefficients reflect spectrin deficiencies in hereditary spherocytosis. J Clin Invest. 1988; 81:133–141.

[35] Pajic-Lijakovic I. Erythrocytes under osmotic stress–modeling considerations. Progr Biophys Mol Biol. 2015; 117(1):113–124.

[36] Pajic-Lijakovic I. Role of band 3 in erythrocyte membrane structural changes under thermal fluctuations-modeling considerations. J Bioen Biomem. 2015; 47(6):507–518.

[37] Edwards S, Grinev DV. Statistical mechanics of vibration-induced compaction of powders. Phys Rev E. 1988; 58(4):4758–4762.

[38] Edwards SF. The full canonical ensemble of a granular system. Phys A. 2005; 353:114–118.

[39] Podlubny I. Fractional Differential Equations, Mathematics in Science and Engineering. London: Academic Press, 1999, 198, 78.

[40] Li J, Dao M, Lim CT, Suresh S. Spectrin-level modeling of the cytoskeleton and optical tweezers stretching of the erythrocyte. Biophys J. 2005; 88:3707–3719.

[41] Pritchard RH, Huang YYS, Terentjev EM. Mechanics of biological networks: from the cell cytoskeleton to connective tissue. Soft Matter. 2014; 10:1864–1884.

[42] Bouchiat C, Wang MD, Allemand JF, Strick T, Block SM, Croquette V. Estimating the persistence length of a worm-like chain molecule from force-extension measurements. Biophys J. 1999; 76:409–413.

[43] Nir G, Lindner M, Dietrich HRC, Girshevitz O, Vorgias CE, Garini Y. HU protein induces incoherent DNA persistence length. Biophys J. 2011; 100:784–790.

[44] Kabaso D, Shlomovitz R, Auth T, Lew VL, Gov NS. Curling and local shape changes of red blood cell membranes driven by cytoskeletal reorganization. Biophys J. 2010; 99:808–816.

[45] Shlomovitz R, Gov NS. Curved inclusions surf membrane waves. Europhys Lett. 2008; 84:58008 1–6.

[46] Kabaso D, Shlomovitz R, Auth T, Lew VL, Gov NS. Cytoskeletal Reorganization of Red Blood Cell Shape: Curling of Free Edges and Malaria Merozoites. In: Iglic A, editor. Advances in Planar Lipid Bilayers and Liposomes, Volume 13; Amsterdam: Elsevier, 2011. p. 73–102.

Chapter 6

Dystrophin–Glycoprotein Complex in Blood Cells

Doris Cerecedo

Additional information is available at the end of the chapter

http://dx.doi.org/10.5772/66857

Abstract

The Dystrophin-Associated Protein Complex (DAPC), known as the Dystrophin–Glycoprotein Complex (DGC), comprises an array of glycoproteins that are essential for the normal function of striated muscle, in which they were first described, and for many other tissues, including blood. Understanding the role that these molecules play in muscle function has increased over the last decade, and some of the knowledge derived can be applied to other biological systems. However, there is no doubt that to date, some progress has been achieved in blood cells.

Multiple interactions have been described among the proteins comprising the DGC, it is now well established that the DGC possesses a crucial role for numerous signaling pathways, recruiting and regulating various signaling proteins into a macromolecular complex. The aim of this chapter is to summarize the current state of knowledge regarding DGC processing and assembly, mainly in muscle tissue and in blood cells, with a primary focus on the dystroglycan heterodimer and associated proteins, including ion channels and membrane lipids. In addition, and due to increasing evidence involving dystroglycan proteins in the pathophysiology of solid tissue cancer, Duchenne muscular dystrophy, and leukemia, current information on these topics will be included.

Keywords: DGC, dystroglycan, intermediate filaments, leukemia cells, adhered platelets

1. Introduction

Dystrophin-associated glycoprotein complex, known as the DGC, is a multimeric and multifaceted protein complex located in the plasma membrane and mediates interactions among the cytoskeleton, cell membrane, and extracellular matrix (ECM) of the muscle and nonmuscle tissues. Therefore, the DGC is involved in signaling pathways that regulate the structural organization of specialized membrane-contact zones, and on the basis of its different biochemical characteristics and localization, the DGC can be divided into the following three subcomplexes: the dystroglycan (Dg), the sarcoglycan, and the cytoplasmic, dystrophin-containing complex.

The dystroglycan subcomplex comprises α- and β-dystroglycan. α-Dystroglycan is the extracellular component that binds to α-laminin and to other components of the basal lamina (ECM), while β-dystroglycan is the transmembrane component. Both attach the intracellular cytoskeleton to the ECM, a task that is widespread in all human tissues and cells [1].

The sarcoglycan subcomplex is a multimember complex that, in association with dystroglycan, stabilizes interactions with the extracellular and transmembrane components of the DGC, as well as with dystrophin and its associated proteins. To date, six sarcoglycan genes have been identified that give rise to their respective proteins α-, β-, γ-, δ-, ε-, and ζ-sarcoglycan, which are organized in a tetrameric arrangement; however, it has been hypothesized that the six sarcoglycans can be arranged in an exameric structure [2].

The dystrophin subcomplex can form a mechanically strong bond with any costameric protein, forming a mechanically strong link between the sarcolemma and the costameric cytoskeleton through interaction with γ-actin filaments. Additionally, based on its structure, protein interactions, and the membrane defects associated with its absence or abnormality in dystrophic muscle, the dystrophin complex provides mechanical stabilization of the sarcolemmal membrane against the stresses imposed upon it during muscle contraction or stretch [3].

Therefore, the DGC appears to play both mechanical and nonmechanical roles in skeletal muscle and in nonmuscle cells, although neither the DGC structure nor the functions are completely understood at present.

This chapter focuses on recent insights into the specific roles of the DGC in different tissue cells, including blood cells, with special focus on dystroglycan biology and its feasible pathophysiologic implications in human leukemia cells and dystrophies.

2. Dystrophin–glycoprotein complex

Dystrophin is the protein that plays a central role in trans-sarcolemmal linkage between the basement membrane and the intracellular actin cytoskeleton, and is the product of the largest identified gene in the human genome [4].

The complexity of Duchenne muscular dystrophy (*DMD*) gene expression, which results in multiple transcripts and protein isoforms, has hampered understanding of the functions of individual dystrophin protein isoforms. The transcription of human DMD is controlled by the following three independent promoters, brain (B), muscle (M), and Purkinje (P) promoters, which indicate the tissue distribution of dystrophin expression, as well as four internal promoters (R for retinal, B for brain, S for Schwann cells, and G for general), which give rise to shorter transcripts encoding for the truncated COOH-terminal isoforms formed from the alternative splicing that generates dystrophin isoforms of 260 kDa (Dp260), 140 kDa (Dp140) [5], 116 kDa (Dp116) [6], and 71 kDa (Dp71) [7, 8]. When these COOH-terminal dystrophin submembrane cytoskeletal proteins interact with a large macromolecular protein complex, they constitute the dystrophin-associated protein complex (DAPC). The crucial structural role of this complex is based on its strategic localization, spanning the plasma membrane and linking with the ECM and the actin cytoskeleton. Since the original discovery of the dystrophin-glycoprotein complex (DGC) [9], a large number of studies have characterized the various components involved in dystrophin [10]. Dystrophin-associated proteins can be divided into sarcolemmal proteins (β-dystroglycan, α-sarcoglycan, β-sarcoglycan, γ-sarcoglycan, and δ-sarcoglycan, sarcospan), cytosolic proteins (dystrobrevins, syntrophins, neuronal nitric-oxide synthase [nNOS]), and extracellular proteins (α-dystroglycan and laminin) [11]. Several DGC components are also found in two or more isoforms, which are either generated by alternative splicing of a single gene or originate from distinct genes [12, 13].

The large, multi-subunit DGC is found in the sarcolemma of striated muscle fibers, and this is essential for maintaining the structural integrity of these fibers during contraction; therefore, the generally accepted role for the DGC is its acting as a molecular shock absorber and stabilizing the plasma membrane during muscle contraction. However, its role goes beyond that solely of a passive scaffold among the elements of the complex, anchoring these near sites-of-action or important partners, since genetic disruption of any of the DGC elements causes mislocalization, destabilization, and the loss-of-function of the cell [14].

As evidence of DGC signaling capacity, it has been reported that nNOS is associated with the DGC via α-dystrobrevin, and that there is a loss of nNOS from the sarcolemma in Duchenne muscular dystrophy (DMD) [15]. Additionally, the DGC promotes the mechanical activation of cardiac nNOS by acting as a mechanosensor in the regulation of AMP-activated protein kinase AMPK activity [16].

The complex also constitutes a scaffold for signaling molecules based on its association with several signaling proteins, including Grb2-Sos1 [17], MEK and ERK [18], heterotrimeric G protein subunits [19], archvillin [20], and nNOS [21].

3. Dystrophin-related proteins

Dystrophins share structural homology with a range of paralog proteins denominated the dystrophin-related proteins (DRP), such as utrophin, DRP2, dystrobrevin, and dystrotelin [22].

The utrophin gene possesses internal promoters and shorter protein products and is also modulated by alternative splicing [23]. Transcription of full-length utrophin (Up395) is driven by two independent promoters: Utrn-A and Utrn-B. The Utrn-A protein is the main isoform in adult skeletal muscles, in contrast with Utrn-B, which is found in the vascular muscle endothelium [24].

G-utrophin, or Up113, was the first short product identified as a structural homolog of Dp116, while Up140 and Up71 are homologous to the short dystrophins Dp140 and Dp71, respectively;

these short utrophins do not possess actin-binding sites in the N-terminal domain of the molecule [25]. Up71 is detected in nonmuscle tissues such as lung, kidney, thymus, liver, and brain, while Up140 is found in lung, muscle, kidney, thymus, liver, testes, and brain. Full-length utrophins are also detected in nonmuscle tissues, such ass those of the central nervous system (CNS), peripheral nerves, testes, kidney, spleen, liver, and lung, and in small arteries and veins [26, 27]. In 1995, utrophin was described as a component of the platelet cytoskeleton, participating in its reorganization [28], while in hematopoietic stem/progenitor cells, Up400 and Up140 comprised the main gene products [29]. In addition, Up71 has been described in platelets [30], as well as in neutrophils [31].

It has long been considered that utrophin and dystrophin share comparable functions during fetal development and adulthood, maintaining utrophin expression in adult dystrophic tissues, compensating for dystrophin loss, as has been observed in mdx skeletal and cardiac muscles [24, 32]. However, spontaneous upregulations also occur in nonmuscle tissues, such as in Dp71-deficient platelets [33] and, most importantly, in the brains of DMD mouse models [34]. However, their expression in distinct structures, as compared with dystrophin, may not reflect functional compensation [24].

4. Dual role of the Dp71 isoform

Dp71 (70–75 kDa) is the first product of the *DMD* gene detectable in pluripotent embryonic stem cells (ESC) during development. It decreases in differentiated ESC cultures and tumors [35] and is the major dystrophin expressed in nonmuscle cells, such as neural tissue [36], glia [37], spermatozoa [38], and astrocytoma cells [39]; in platelets, its participation has been suggested in cytoskeletal reorganization and/or signaling, and in thrombin-mediated platelet adhesion [28].

The variation in the molecular mass of Dp71 transcripts is consistent with the expression of Dp71 isoforms derived from transcripts alternatively spliced for exons 71 and/or 78 [40]. The splicing product of exon 78 produces the isoform known as Dp71d, which preserves the C-terminal, while Dp71f is the product of the absence of exon 78. Two other gene products resulting from an alternative splicing at exons 71–74 and/or 78 transcripts, $Dp71\Delta110^{a}$ and $Dp71\Delta110^{m}$, respectively, with a relative mass of 55 kDa, have been recently characterized [40, 41].

In 2005, $Dp71d/Dp71\Delta110^{m}$~DGC and Up400/Up71~DGC were described as participating with structural roles associated with the actin cytoskeleton in the formation of membrane scaffolds. They were probably involved in defining platelet shape, substrate adhesion, and granule migration, as well as possessing a signaling role, participating in signaling triggered by adhesion to glass and by interaction with agonists such as thrombin [30].

The presence of Dp71 and some DGC elements that form a nuclear complex at the plasma membrane and in the nucleus of muscle cells suggested their participation in nuclear structure and in the modulation of nuclear processes [42].

http://dx.doi.org/10.5772/66857

Figure 1. Schematic diagram of the dystrophin–glycoprotein complex (DGC) composed of Dp71 (left) and utrophin (right) in adhered platelets. Dystrophin is a linker between the cytoskeleton and the extracellular matrix (ECM). Dp71 and utrophin are associated with the dystroglycan complex and the dystrobrevin/syntrophin complex (α-Db/α-Syn). α-Dystroglycan (α-Dg) binds to ECM proteins and β-dystroglycan (β-Dg); β-Dg binds to the dystrophin, completing the link between the actin cytoskeleton and the ECM. Focal adhesion (magnified at the bottom of the figure) clusters the α- and β-integrin receptors and induces recruitment of focal adhesion proteins vinculin (Vin), talin (Tal), and α-actinin (α-Act), which connect directly with microfilaments and short dystrophins (Dp71) and indirectly with microtubules and intermediate filaments. The adhesion complex activates integrin-associated signaling cascades, including focal adhesion kinase (FAK). Dystroglycan plays a scaffold role, modulating the cytoplasmic protein kinases, and is in close association with integrin β1.

The neuronal cell line PC12 expresses at least two different Dp71 protein isoforms generated by the alternative splicing of exon 78 [43, 44]. The splicing isoform of Dp71 (Dp71d) contains ^{13}C-terminal amino acids encoded by exon 78, which are replaced by 31 new amino acids encoded by exon 79 in the Dp71f isoform upon removal of exon 78 [40]. Depletion of total Dp71 protein levels gives rise to impairment in nerve growth factor (NGF)-induced neurite outgrowth [45] and in the cell adhesion activity of PC12 cells [46], indicating that Dp71 is required for these neuronal functions. Dp71f assembles an adhesion complex comprising talin, α-actinin, paxillin, focal adhesion kinase (FAK), and actin, but not vinculin, contributing to cell stability [47].

During the platelet adhesion process, short dystrophins (Dp71d/Dp71Δ110^{m}) and utrophins (Up400/Up71) have demonstrated potential association with the integrin β-1 fraction and with focal adhesion system that includes α-actinin, vinculin, and talin. Apparently, in order to fulfill this hemosatic role, the coexistence of the DGC composed of short dystrophins or utrophins plays both a structural role in participation in stress-fiber assembly and in the centralization of cytoplasmic granules, and a regulatory role, incorporating FAK into the complex. The coexistence of dystrophin and utrophin complexes indicates structural and signaling mechanisms that are complementary to the actin network during the adhesion process [48] (**Figure 1**).

5. DGC components

The findings described in systematic proteomic studies indicate that dystrophin interacts closely with core members of the dystrophin-associated glycoprotein complex, such as dystroglycans, sarcoglycans, syntrophins, dystrobrevins, and sarcospan, but that it also forms indirect linkages with a large variety of other protein species, including tubulin, vimentin, desmin, annexin, and collagens [49].

5.1. Dystrobrevins

Dystrobrevins are proteins among dystrophin-related proteins that are encoded by two different genes, *α* and *β* and that possess significant homology to dystrophin. *α-Dystrobrevin* is expressed predominantly in muscle and brain, whereas *β-dystrobrevin* is expressed in nonmuscle tissues, which is abundant in brain, kidney, lung, and liver. Dystrobrevins have also been involved in intracellular signaling in muscle and nonmuscle tissues, either directly or through interaction with syntrophin, another element of the DGC. In humans, Sadoulet-Puccio et al. [50] found six isoforms of dystrobrevin (designated α-, β-, γ-, δ-, ε-, and ζ-dystrobrevin), which ranged in size from 22 to 80 kDa.

Human α-dystrobrevin and its few isoforms are expressed in the cytosol and the nucleus of the promyelocytic HL-60 cell line. A distinct distribution pattern of α-dystrobrevin, including colocalization with actin, was described in HL-60 promyelocytes, differentiated mature granulocytes, and in human neutrophils, supporting a signaling role [51]. In adhered platelets, it was suggested that actin filaments and microtubules contribute to α-granule and dense granule mobilization in adhered platelets, identifying α-dystrobrevins as part of the platelet transport machinery that is closely associated with the ubiquitous kinesin heavy chain (UKHC), this system is depicted in **Figure 2** [52].

5.2. Sarcoglycans

The sarcoglycan complex (SGC) is composed of α-, β-, γ-, and δ-sarcoglycan isoforms encoded by separate genes, and of sarcospan. Sarcoglycans are single transmembrane glycoproteins with the N-terminus oriented extracellularly for α-sarcoglycan and intracellularly for β-,

http://dx.doi.org/10.5772/66857

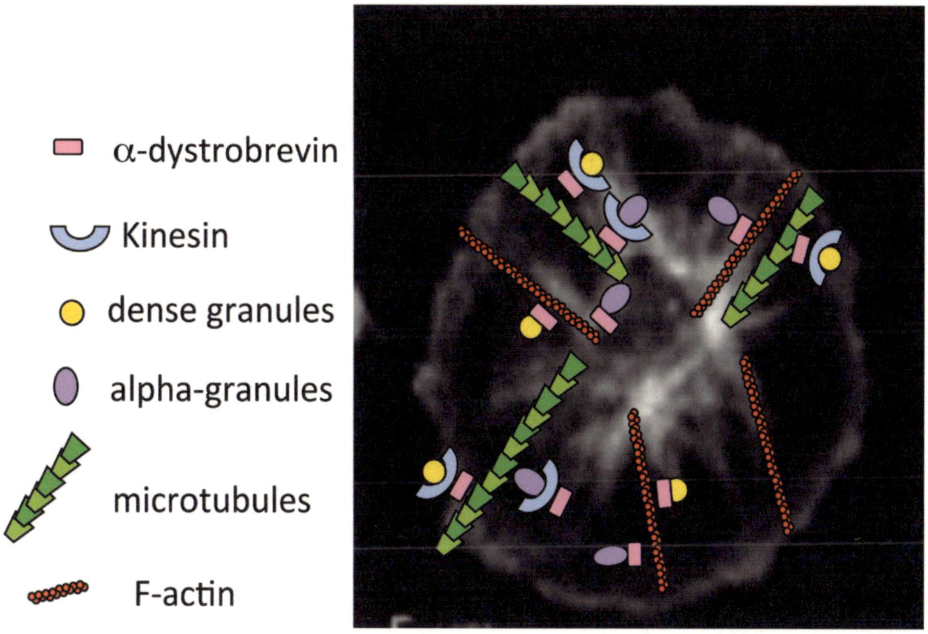

Figure 2. Platelet distribution of cytoskeleton elements. Schematic diagram of actin filaments, microtubules, and intermediate filaments in adhered platelets. Plectin is the protein that acts as a link among the three main components of the cytoskeleton.

γ-, and δ-sarcoglycans [53]. Contrariwise, sarcospan is composed of four transmembrane-spanning segments that are homologous to the tetraspanin family. The function of the SGC is not fully understood, but it appears to strengthen the interaction of β-dystroglycan with α-dystroglycan and dystrophin, as well as to play a role in intracellular signal transduction for sarcoglycan [54]. Sarcospan, a 25-kDa transmembrane protein, improves the cell-surface expression of the three major laminin-binding complexes, i.e., the dystrophin– and utrophin–glycoprotein complexes, as well as of an α7β1 integrin [55].

5.3. Syntrophins

Syntrophins are a multigene family of intracellular membrane-associated adaptor proteins and consist of five homologous isoforms: α1-syntrophin, β1-syntrophin, β2-syntrophin, γ1-syntrophin, and γ2-syntrophin; they possess a different cellular and subcellular localization, suggesting a distinct functional role [56]. In human platelets, a 54-kDa band corresponding to α-syntrophin is well expressed [30].

The pleckstrin homology (PH) and PDZ domains of syntrophins were shown to bind various proteins, including nitric oxide synthase (NOS), and have been implicated in the regulation of various plasma membrane ion channels, such as voltage-operated sodium channels and other nonvoltage gated channels, such as mechanosensitive Na^+ channels [57].

5.4. Dystrolglycans

The single dystroglycan gene encodes for a precursor protein that undergoes posttranslational proteolytic cleavage, which in turn produces two noncovalant DGC subunits: α- and β-dystroglycan. α-Dystroglycan is a dumbbell-shaped protein that binds to the laminin G domain in ECM components such as laminins, agrin, and perlecan. β-Dystroglycan (β-Dg) possesses a single transmembrane domain spanning the plasma membrane and an extracellular amino-terminal extracellular domain binding to the carboxy-terminal globular domain of α-Dg [58].

5.4.1. Dg involved in the signaling process

The β-Dg dual role (structural and signaling) has been demonstrated in various cell types and tissues. Examples of the former role are represented by the participation of β-Dg in cytoskeleton remodeling, where it is associated with actin [59, 60], while its signaling role is represented by its association with the extracellular signal-related kinase-mitogen-activated protein (ERK-MAP) kinase cascade [18], or with integrins modulates myoblast anchorage and migration [61]; this latter process is critically regulated by Src-mediated phosphorylation of β-Dg at tyrosine 890 [62].

Grb2–β-Dg interaction could facilitate the transduction of signals between the DGC and extracellular proteins and other signaling pathways [62]. However, when Dg is localized at the tips of dynamic filopodia, it directs local Cdc42 activation and recruits the guanine nucleotide exchange factor (GEF) Dbl to generate actin protrusions [60].

Dystroglycan is also a multifunctional adaptor or scaffold capable of interacting with components of the ERK-MAP kinase cascade, including MEK and ERK [18]. However, it has been established that integrin α6Aβ1 and dystroglycan play antagonistic roles in signaling to the Ras-Raf-MEK-ERK pathway in response to laminin [63].

5.4.2. Dg promoter of the adhesion process

Since 1995, dystroglycan-associated proteins, such as utrophin, have been considered residents of focal adhesions in nonmuscle cells [59, 64, 65] and, after direct interaction of the cytoplasmic tail of β-Dg with F-actin was described [66], Dg has been implicated in cell adhesion and spreading.

Dg was identified in podosomes at the early stages of myoblast spreading; these structures contain a regulatory complex comprising dystroglycan, Tks5, and Src [67]. Myoblast spreading occurred in relation to dystroglycan expression levels, which in turn altered the size and number of focal contacts, focal adhesions, and fibrillar adhesions. Dystroglycan-mediated cell adhesion and spreading took place through indirect interaction with vinculin by binding to the vinculin-binding protein vinexin [61], while an adhesome was made up of by laminin-Dg-myosin IIA, crucial for maintaining the shape of notochordal cells [65].

In addition to a specific role in the maintenance of muscle integrity, Dg plays a more ubiquitous role in cell adhesion, signaling, and polarity. During embryogenesis, the follicle-cell

http://dx.doi.org/10.5772/66857

epithelium (FCE) maintains the cell polarity promoted by the association between perlecan and Dg [68], while in astrocytes, end-feet in brain laminin induced a dramatic, polarized redistribution of cell-surface clusters or macrodomains, which colocalized extensively with β-Dg and AQP4 [69].

The cytoskeletal polymers—actin, microtubules, and intermediate filaments—are interlinked by coordinated protein interactions to form a complex three-dimensional (3D) cytoskeletal network; these components are depicted in **Figure 3**. Although these systems are composed of distinctly different proteins, they are in constant and intimate communication with each another and with intermediate filaments, and their associated proteins are important components in mediating this crosstalk [70].

In platelets, two members of type-III intermediate filament (IF) proteins, desmin and vimentin, maintain a close relationship with DGC components, such as β-dystroglycan [β-dg], α-syntrophin [α-syn], and α-dystrobrevin [α-db], and are codistributed at the granulomere zone, participating in α-granule distribution [71].

The epithelial sodium channel (ENaC) is associated with IF and with dystrophin-associated proteins (DAP) via α-syntrophin and β-dystroglycan. ENaC is apparently dispensable for

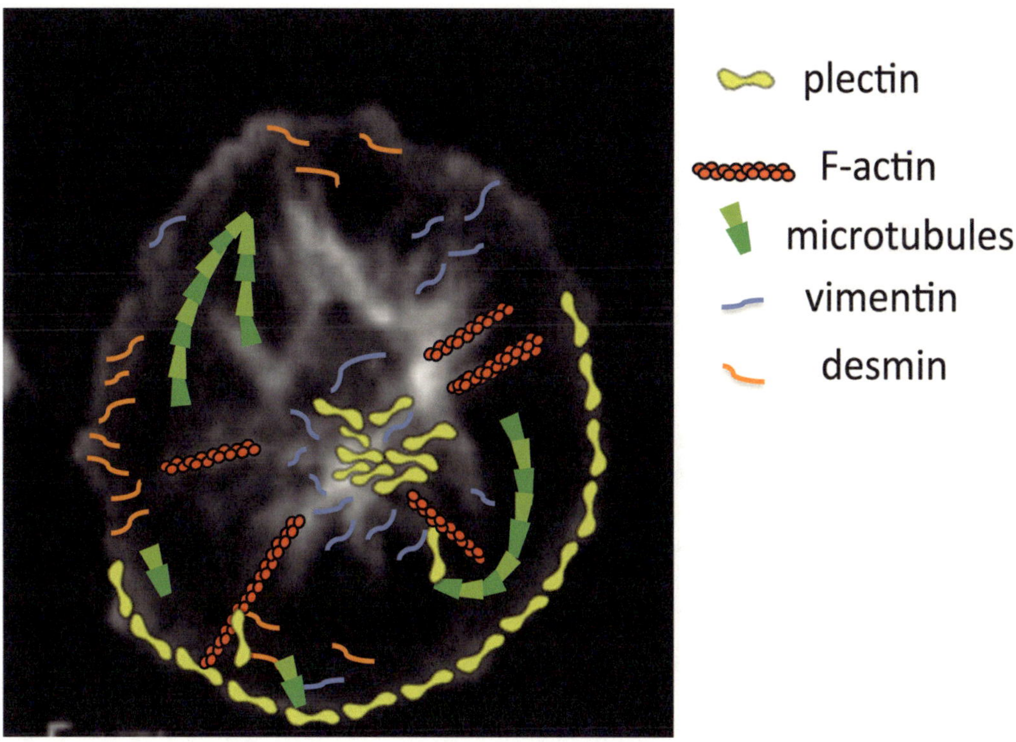

Figure 3. Schematic diagram of microtubules and actin filaments participating in the transport of alpha and dense granules in the platelet adhesion process, during which α-dystrobrevins are the regulatory and adaptor proteins for governing trafficking events.

migration and alpha- and dense-granule secretion, whereas Na^+ influx through this channel is fundamental for platelet collagen activation [72]. This channel is overexpressed in platelets from hypertensive subjects in relation with control subjects, and β-Dg is a scaffold for the organization of ENaC and associated proteins [73].

5.4.3. *Dg and its posttranscriptional modifications*

Posttranscriptional modifications in the Dg protein possess important implications in cellular functions. The transmembrane β-subunit, which interacts with α-Dg extracellularly and which also connects with several different cytolinker proteins intracellularly, is additionally subject to altered N-linked glycosylation [74]. Additional modifications to β-Dg, however, include phosphorylation on tyrosine [75, 76] and specific proteolytic cleavage events. Tyrosine phosphorylation of β-Dg serves as a molecular switch to regulate the binding of different cellular-binding partners [77], but it is also a signal of the internalization of Dg from the plasma membrane [78, 79] and may mediate some proteolytic events and nuclear translocation [80, 81].

β-Dg is subject to proteolysis at several key sites: matrix metalloproteinase (MMP)-mediated cleavage liberates the extracellular portion of β-Dg, MMP-9-mediated proteolytic cleavage of the β-Dg, and it has been implicated in dendritic outgrowth and arborization in primary hippocampal neurons [82]. The remaining 31 kDa transmembrane stub and cytoplasmic domain can be detected with antibodies at the carboxy terminus of the cytoplasmic domain. As yet unknown proteases generate smaller fragments corresponding to the cytoplasmic region of β-Dg [83, 84], most typically observed as a 26-kDa fragment, but occasionally as a 17-kDa fragment.

In hematopoietic stem/progenitor cells, a 50-kDa β-Dg is the main product, while in differentiated cells, such as neutrophils and platelets, the characteristic glycosylated 43 kDa band is present [29, 31]. A 65-kDa band was also observed in neutrophils; perhaps this molecular weight (MW) is due to a posttranscriptional modification such as SUMOylation.

Ezrin is able to interact with dystroglycan through a cluster of basic residues in the juxtamembrane region, and appears to be responsible for dystroglycan-mediated formation of filopodia [18]. Colocalization of endogenous dystroglycan with ezrin at the cleavage furrow and midbody during cytokinesis not only affords dystroglycan a role in organizing the contractile ring through direct or indirect associations with actin, but also can modulate the cell cycle by affecting extracellular signal-regulated kinase levels [85]. Recent experiments have demonstrated β-Dg trafficking from the cytoplasm to the nucleus by ezrin-mediated cytoskeleton reorganization, the latter dependent on IMPα2/β1 [86].

Due to the presence of a conventional nuclear localization sequence (NLS)-/Imp-dependent nuclear import pathway in the cytoplasmic juxtamembrane region of β-Dg [87], β-Dg and proteolytic fragments containing the nuclear localization signal can be targeted to the nucleus via an importin-dependent pathway [88], where it can exert effects on nuclear architecture [89].

5.4.4. Dg in the differentiation process

The expression has been described as the major components of DAPC visceral and subcutaneous rat adipose depots that are regulated during adipogenesis and by ECM components, suggesting an important role in adipocyte differentiation [90].

The human myeloid leukemia cell line HL-60 achieves increasing cessation after its exposure to all-trans-retinoic acid (ATRA) and dimethyl sulfoxide (DMSO) and becomes differentiated into granulocytes, evoking the biology of the disease *in vitro* [91, 92]. Recently, it was demonstrated that dystroglycans actively participate in the differentiation process, in that the expression levels of α-Dg (160 kDa), β-Dg (42kDa), and β-DgpY892 (42 kDa) were increased in differentiated compared with nondifferentiated cells. Additionally, low levels of β-Dg in differentiated HL-60 cells are accompanied by reducing actin-based protrusions, such as in filopodia and lamellipodia extrusion, avoiding motility or phagocytic capabilities, respectively [93]. Similar changes were also observed when HL-60 cells were transfected with a shRNA directed to dystroglycan; therefore, a direct consequence of the reduction in dystroglycan exerted a direct effect on actin cytoskeletal dynamics, either on its direct or indirect interaction with actin, but also interfering with actin regulatory pathways [66].

The Kasumi-1 cell line is a model system of acute myeloid leukemia (AML) with *t*(8;21) translocation and the corresponding functional consequences of the AML1–ETO fusion oncogene on myeloid differentiation [94]. *In vitro*, macrophages differentiated from myelomonocytic cell lines exhibited downregulation of adhesion molecules after tissue plasminogen activator (TPA) treatment [95]. The biochemical analysis of cytoplasmic or nuclear Kasumi-1 cell extracts revealed bands of 50, 38, and 30 kDa present in the nucleus of the cells, while the majority of 43 kDa β-Dg was found mainly in the cytoplasmic compartment, with the 38-kDa band also abundant in the cytoplasm of nondifferentiated Kasumi-1 and differentiated Kasumi-1 cells. The phosphorylated 31-kDa fragment of dystroglycan is the species that is most translocated to the nucleus of nondifferentiated cells, while the 50-kDa fragment comprised the most abundant species at the nucleus of differentiated cells. The diminished expression levels of Dg in differentiated Kasumi-1 cells compared with nondifferentiated cells could facilitate cell recruitment in solid tissues; apparently, the phosphorylated species may be ubiquitinated and processed by the proteasome. However, a direct consequence of a reduction in dystroglycan exerts an effect on actin cytoskeletal dynamics, but does not impair the differentiation process [96].

5.4.5. Dg in cell membrane organization

Several structures of the cell membrane play major roles in physiological functions through signaling and adhesion to neighbor cells and to ECM. Generic features, such as the cytoskeleton meshwork, rafts, and protein complexes, which are subjected to thermal motion, contribute to building membrane structures such as focal adhesions (FA) [97] and immune [98] and neuronal [99] synapses. The rapid and transient association of the partners of a given signaling pathway, localized in close proximity within narrow structures/domains, is a requirement for rapid and reliable signal transmission [100].

The existence of "rafts" supposes that membrane lipids and proteins associate with each other according to their affinities, due to their hydrophobicity and geometry [101]. Rafts were initially proposed as contributing to protein sorting along the synthesis pathway, and have also been associated with several membrane features, including signaling platforms and adhesion structures. Caveolae are cholesterol- and sphingolipid-enriched membrane invaginations [102], and caveolin-1 is the primary caveolae structural protein in several cells [103]. Therefore, caveolae and caveolin-1 play a key role in orchestrating the activation of pathways that underpin cell proliferation, migration, and contraction [104]. For example, direct interaction between caveolin-1 and β-Dg was demonstrated in contractile smooth muscle, where the distribution of caveolae is determined by their tethering to the actin cytoskeleton via caveolin-1 and the DGC [105].

In this regard, cholesterol demonstrated to be essential to modulate platelet cytoskeleton reorganization, while the association of caveolin-1 PY14 with intermediate filaments, as well as with focal adhesion proteins via vinculin, was a determinant in adhered platelets, where β-Dg participation was a key scaffold component for caveolin-1 and FAK [106].

In general, diseases of the DGC are incurable, in part because the majority of these give rise to great damage resulting from the loss of these proteins. However, there is increasing evidence that proteins in the DGC may play a significant role in the pathophysiology of more common diseases such as cancer, in which the DGC has been implicated.

Throughout these years of basic research, it has been observed that dystroglycan functional changes, either for posttranscriptional modifications or for deregulation of the protein, simultaneously affect both scaffolding and signaling roles. These changes modify cell adhesion and motility, MAPK signaling, or its translocation to the nuclei that, in the prostate, is associated with the ETV1 transcription factor, acting directly on cancer progression and the pathophysiology of the disease [81]. Therefore, a complete understanding of the role of DGC elements in the pathophysiology of a disease would allow the identification of strategies for the development of specific therapeutics.

Previous studies demonstrated that preventing tyrosine phosphorylation of β-Dg in mdx mouse alleviated the dystrophic phenotype in a genetic mouse model, ameliorating many of the main pathological symptoms associated with dystrophin deficiency [78]. The use of dasatinib was found to decrease β-Dg phosphorylation levels in tyrosine and to increase the relative levels of nonphosphorylated β-Dg in the sapje zebrafish, improving its physical condition [79].

6. Conclusion

Since 1980, the dystrophin–glycoprotein complex has been considered only as a group of multiproteins working together to ensure the function of muscle tissue; however, along these years and according to basic research, dystrophin has acquired prime status and has become the central component of a scaffold of proteins expressed in a variety of tissues including blood. Within the complex elements, dystroglycan has received the majority of our attention and has been identified as participating in the clustering of membrane receptors, integrins, and ion channels, modulating cellular signal integration, such as in the differentiation process.

Despite all of these advances, it remains difficult to dissect the specific function of a particular protein and, given the close association and interdependence of the different elements of the complex, it should be difficult to define the specific contribution of each of the complex's protein elements. However, the improvement and development of biochemical and molecular tolls will undoubtedly aid in elucidating novel therapies to counteract common diseases such as cancer.

Author details

Doris Cerecedo

Address all correspondence to: dcereced@prodigy.net.mx

Laboratorio de Hematobiología, Escuela Nacional de Medicina y Homeopatía (ENMH), Instituto Politécnico Nacional (IPN), Mexico City, Mexico

References

[1] Le Rumeur, E., S.J. Winder, and J.F. Hubert. Dystrophin: more than just the sum of its parts. Biochim Biophys Acta, 2010, **1804**(9): pp. 1713–1722.

[2] Ozawa, E., Y. Mizuno, Y. Hagiwara, T. Sasaoka, and M. Yoshida. Molecular and cell biology of the sarcoglycan complex. Muscle Nerve, 2005, **32**(5): pp. 563–576.

[3] Hanft, L.M., I.N. Rybakova, J.R. Patel, J.A. Rafael-Fortney, and J.M. Ervasti. Cytoplasmic gamma-actin contributes to a compensatory remodeling response in dystrophin-deficient muscle. Proc Natl Acad Sci U S A, 2006, **103**(14): pp. 5385–5390.

[4] Tennyson, C.N., H.J. Klamut, and R.G. Worton. The human dystrophin gene requires 16 hours to be transcribed and is cotranscriptionally spliced. Nat Genet, 1995, **9**(2): pp. 184–190.

[5] Lidov, H.G., S. Selig, and L.M. Kunkel. Dp140: a novel 140 kDa CNS transcript from the dystrophin locus. Hum Mol Genet, 1995, **4**(3): pp. 329–335.

[6] Byers, T.J., H.G. Lidov, and L.M. Kunkel. An alternative dystrophin transcript specific to peripheral nerve. Nat Genet, 1993, **4**(1): pp. 77–81.

[7] Hugnot, J.P., H. Gilgenkrantz, N. Vincent, P. Chafey, G.E. Morris, A.P. Monaco, Y. Berwald-Netter, A. Koulakoff, J.C. Kaplan, A. Kahn, et al. Distal transcript of the dystrophin gene initiated from an alternative first exon and encoding a 75-kDa protein widely distributed in nonmuscle tissues. Proc Natl Acad Sci U S A, 1992, **89**(16): pp. 7506–7510.

[8] Lederfein, D., Z. Levy, N. Augier, D. Mornet, G. Morris, O. Fuchs, D. Yaffe, and U. Nudel. A 71-kilodalton protein is a major product of the Duchenne muscular dystrophy gene in brain and other nonmuscle tissues. Proc Natl Acad Sci U S A, 1992, **89**(12): pp. 5346–5350.

[9] Ervasti, J.M., K. Ohlendieck, S.D. Kahl, M.G. Gaver, and K.P. Campbell. Deficiency of a glycoprotein component of the dystrophin complex in dystrophic muscle. Nature, 1990, **345**(6273): pp. 315–319.

[10] Gao, Q.Q. and E.M. McNally. The dystrophin complex: structure, function, and implications for therapy. Compr Physiol, 2015, **5**(3): pp. 1223–1239.

[11] Ohlendieck, K. Towards an understanding of the dystrophin-glycoprotein complex: linkage between the extracellular matrix and the membrane cytoskeleton in muscle fibers. Eur J Cell Biol, 1996, **69**(1): pp. 1–10.

[12] Blake, D.J., R. Nawrotzki, N.Y. Loh, D.C. Gorecki, and K.E. Davies. Beta-dystrobrevin, a member of the dystrophin-related protein family. Proc Natl Acad Sci U S A, 1998, **95**(1): pp. 241–246.

[13] Peters, M.F., M.E. Adams, and S.C. Froehner. Differential association of syntrophin pairs with the dystrophin complex. J Cell Biol, 1997, **138**(1): pp. 81–93.

[14] Constantin, B. Dystrophin complex functions as a scaffold for signalling proteins. Biochim Biophys Acta, 2014, **1838**(2): pp. 635–642.

[15] Bredt, D.S. Endogenous nitric oxide synthesis: biological functions and pathophysiology. Free Radic Res, 1999, **31**(6): pp. 577–596.

[16] Garbincius, J.F. and D.E. Michele. Dystrophin-glycoprotein complex regulates muscle nitric oxide production through mechanoregulation of AMPK signaling. Proc Natl Acad Sci U S A, 2015, **112**(44): pp. 13663–13668.

[17] Oak, S.A., Y.W. Zhou, and H.W. Jarrett. Skeletal muscle signaling pathway through the dystrophin glycoprotein complex and Rac1. J Biol Chem, 2003, **278**(41): pp. 39287–39295.

[18] Spence, H.J., A.S. Dhillon, M. James, and S.J. Winder. Dystroglycan, a scaffold for the ERK-MAP kinase cascade. EMBO Rep, 2004, **5**(5): pp. 484–489.

[19] Zhou, Y.W., S.A. Oak, S.E. Senogles, and H.W. Jarrett. Laminin-alpha1 globular domains 3 and 4 induce heterotrimeric G protein binding to alpha-syntrophin's PDZ domain and alter intracellular Ca^{2+} in muscle. Am J Physiol Cell Physiol, 2005, **288**(2): pp. C377–388.

[20] Spinazzola, J.M., T.C. Smith, M. Liu, E.J. Luna, and E.R. Barton. Gamma-sarcoglycan is required for the response of archvillin to mechanical stimulation in skeletal muscle. Hum Mol Genet, 2015, **24**(9): pp. 2470–2481.

[21] Brenman, J.E., D.S. Chao, S.H. Gee, A.W. McGee, S.E. Craven, D.R. Santillano, Z. Wu, F. Huang, H. Xia, M.F. Peters, S.C. Froehner, and D.S. Bredt. Interaction of nitric oxide synthase with the postsynaptic density protein PSD-95 and alpha1-syntrophin mediated by PDZ domains. Cell, 1996, **84**(5): pp. 757–767.

[22] Jin, H., S. Tan, J. Hermanowski, S. Bohm, S. Pacheco, J.M. McCauley, M.J. Greener, Y. Hinits, S.M. Hughes, P.T. Sharpe, and R.G. Roberts. The dystrotelin, dystrophin and dystrobrevin superfamily: new paralogues and old isoforms. BMC Genomics, 2007, **8**: pp. 19.

[23] Blake, D.J., J.M. Tinsley, K.E. Davies, A.E. Knight, S.J. Winder, and J. Kendrick-Jones. Coiled-coil regions in the carboxy-terminal domains of dystrophin and related proteins: potentials for protein-protein interactions. Trends Biochem Sci, 1995, **20**(4): pp. 133–135.

[24] Baby, S.M., S. Bogdanovich, G. Willmann, U. Basu, O. Lozynska, and T.S. Khurana. Differential expression of utrophin-A and -B promoters in the central nervous system (CNS) of normal and dystrophic mdx mice. Brain Pathol, 2010, **20**(2): pp. 323–342.

[25] Wilson, J., W. Putt, C. Jimenez, and Y.H. Edwards. Up71 and up140, two novel transcripts of utrophin that are homologues of short forms of dystrophin. Hum Mol Genet, 1999, **8**(7): pp. 1271–1278.

[26] Lumeng, C.N., S.F. Phelps, J.A. Rafael, G.A. Cox, T.L. Hutchinson, C.R. Begy, E. Adkins, R. Wiltshire, and J.S. Chamberlain. Characterization of dystrophin and utrophin diversity in the mouse. Hum Mol Genet, 1999, **8**(4): pp. 593–599.

[27] Rivier, F., A. Robert, J. Latouche, G. Hugon, and D. Mornet. Expression of a new M(r) 70-kDa dystrophin-related protein in the axon of peripheral nerves from Torpedo marmorata. Comp Biochem Physiol B Biochem Mol Biol, 1997, **116**(1): pp. 19–26.

[28] Earnest, J.P., G.F. Santos, S. Zuerbig, and J.E. Fox. Dystrophin-related protein in the platelet membrane skeleton. Integrin-induced change in detergent-insolubility and cleavage by calpain in aggregating platelets. J Biol Chem, 1995, **270**(45): pp. 27259–27265.

[29] Teniente-De Alba, C., I. Martinez-Vieyra, R. Vivanco-Calixto, I.J. Galvan, B. Cisneros, and D. Cerecedo. Distribution of dystrophin- and utrophin-associated protein complexes (DAPC/UAPC) in human hematopoietic stem/progenitor cells. Eur J Haematol, 2011, **87**(4): pp. 312–322.

[30] Cerecedo, D., D. Martinez-Rojas, O. Chavez, F. Martinez-Perez, F. Garcia-Sierra, A. Rendon, D. Mornet, and R. Mondragon. Platelet adhesion: structural and functional diversity of short dystrophin and utrophins in the formation of dystrophin-associated-protein complexes related to actin dynamics. Thromb Haemost, 2005, **94**(6): pp. 1203–1212.

[31] Cerecedo, D., B. Cisneros, P. Gomez, and I.J. Galvan. Distribution of dystrophin- and utrophin-associated protein complexes during activation of human neutrophils. Exp Hematol, 2010, **38**(8): pp. 618–628 e613.

[32] Weir, A.P., E.A. Burton, G. Harrod, and K.E. Davies. A- and B-utrophin have different expression patterns and are differentially up-regulated in mdx muscle. J Biol Chem, 2002, **277**(47): pp. 45285–45290.

[33] Cerecedo, D., R. Mondragon, A. Candelario, F. Garcia-Sierra, D. Mornet, A. Rendon, and D. Martinez-Rojas. Utrophins compensate for Dp71 absence in mdx3cv in adhered platelets. Blood Coagul Fibrinolysis, 2008, **19**(1): pp. 39–47.

[34] Vaillend, C., J.M. Billard, T. Claudepierre, A. Rendon, P. Dutar, and A. Ungerer. Spatial discrimination learning and CA1 hippocampal synaptic plasticity in mdx and mdx3cv mice lacking dystrophin gene products. Neuroscience, 1998, **86**(1): pp. 53–66.

[35] Rapaport, D., O. Fuchs, U. Nudel, and D. Yaffe. Expression of the Duchenne muscular dystrophy gene products in embryonic stem cells and their differentiated derivatives. J Biol Chem, 1992, **267**(30): pp. 21289–21292.

[36] Bar, S., E. Barnea, Z. Levy, S. Neuman, D. Yaffe, and U. Nudel. A novel product of the Duchenne muscular dystrophy gene which greatly differs from the known isoforms in its structure and tissue distribution. Biochem J, 1990, **272**(2): pp. 557–560.

[37] Claudepierre, T., D. Mornet, T. Pannicke, V. Forster, C. Dalloz, F. Bolanos, J. Sahel, A. Reichenbach, and A. Rendon. Expression of Dp71 in Muller glial cells: a comparison with utrophin- and dystrophin-associated proteins. Invest Ophthalmol Vis Sci, 2000, **41**(1): pp. 294–304.

[38] Hernandez-Gonzalez, E.O., D. Mornet, A. Rendon, and D. Martinez-Rojas. Absence of Dp71 in mdx3cv mouse spermatozoa alters flagellar morphology and the distribution of ion channels and nNOS. J Cell Sci, 2005, **118**(Pt 1): pp. 137–145.

[39] Garcia-Tovar, C.G., J. Luna, R. Mena, C.I. Soto-Zarate, R. Cortes, A. Perez, G. Leon-Avila, D. Mornet, A. Rendon, and J.M. Hernandez. Dystrophin isoform Dp7l is present in lamellipodia and focal complexes in human astrocytoma cells U-373 MG. Acta Histochem, 2002, **104**(3): pp. 245–254.

[40] Austin, R.C., P.L. Howard, V.N. D'Souza, H.J. Klamut, and P.N. Ray. Cloning and characterization of alternatively spliced isoforms of Dp71. Hum Mol Genet, 1995, **4**(9): pp. 1475–1483.

[41] Austin, R.C., J.E. Fox, G.H. Werstuck, A.R. Stafford, D.E. Bulman, G.Y. Dally, C.A. Ackerley, J.I. Weitz, and P.N. Ray. Identification of Dp71 isoforms in the platelet membrane cytoskeleton. Potential role in thrombin-mediated platelet adhesion. J Biol Chem, 2002, **277**(49): pp. 47106–47113.

[42] Gonzalez-Ramirez, R., S.L. Morales-Lazaro, V. Tapia-Ramirez, D. Mornet, and B. Cisneros. Nuclear and nuclear envelope localization of dystrophin Dp71 and dystrophin-associated proteins (DAPs) in the C2C12 muscle cells: DAPs nuclear localization is modulated during myogenesis. J Cell Biochem, 2008, **105**(3): pp. 735–745.

[43] Cisneros, B., A. Rendon, V. Genty, G. Aranda, F. Marquez, D. Mornet, and C. Montanez. Expression of dystrophin Dp71 during PC12 cell differentiation. Neurosci Lett, 1996, **213**(2): pp. 107–110.

[44] Marquez, F.G., B. Cisneros, F. Garcia, V. Ceja, F. Velazquez, F. Depardon, L. Cervantes, A. Rendon, D. Mornet, H. Rosas-vargas, M. Mustre, and C. Montanez. Differential

expression and subcellular distribution of dystrophin Dp71 isoforms during differentiation process. Neuroscience, 2003, **118**(4): pp. 957–966.

[45] Acosta, R., C. Montanez, L. Fuentes-Mera, E. Gonzalez, P. Gomez, L. Quintero-Mora, D. Mornet, L.M. Alvarez-Salas, and B. Cisneros. Dystrophin Dp71 is required for neurite outgrowth in PC12 cells. Exp Cell Res, 2004, **296**(2): pp. 265–275.

[46] Enriquez-Aragon, J.A., J. Cerna-Cortes, M. Bermudez de Leon, F. Garcia-Sierra, E. Gonzalez, D. Mornet, and B. Cisneros. Dystrophin Dp71 in PC12 cell adhesion. Neuroreport, 2005, **16**(3): pp. 235–238.

[47] Cerna, J., D. Cerecedo, A. Ortega, F. Garcia-Sierra, F. Centeno, E. Garrido, D. Mornet, and B. Cisneros. Dystrophin Dp71f associates with the beta1-integrin adhesion complex to modulate PC12 cell adhesion. J Mol Biol, 2006, **362**(5): pp. 954–965.

[48] Cerecedo, D., R. Mondragon, B. Cisneros, F. Martinez-Perez, D. Martinez-Rojas, and A. Rendon. Role of dystrophins and utrophins in platelet adhesion process. Br J Haematol, 2006, **134**(1): pp. 83–91.

[49] Murphy, S., P. Dowling, M. Zweyer, R.R. Mundegar, M. Henry, P. Meleady, D. Swandulla, and K. Ohlendieck. Proteomic analysis of dystrophin deficiency and associated changes in the aged mdx-4cv heart model of dystrophinopathy-related cardiomyopathy. J Proteomics, 2016, 145(8): pp 24–36.

[50] Sadoulet-Puccio, H.M., T.S. Khurana, J.B. Cohen, and L.M. Kunkel. Cloning and characterization of the human homologue of a dystrophin related phosphoprotein found at the Torpedo electric organ post-synaptic membrane. Hum Mol Genet, 1996, **5**(4): pp. 489–496.

[51] Kulyte, A., R. Navakauskiene, G. Treigyte, A. Gineitis, T. Bergman, and K.E. Magnusson. Characterization of human alpha-dystrobrevin isoforms in HL-60 human promyelocytic leukemia cells undergoing granulocytic differentiation. Mol Biol Cell, 2002, **13**(12): pp. 4195–4205.

[52] Cerecedo, D., B. Cisneros, R. Suarez-Sanchez, E. Hernandez-Gonzalez, and I. Galvan. beta-Dystroglycan modulates the interplay between actin and microtubules in human-adhered platelets. Br J Haematol, 2008, **141**(4): pp. 517–528.

[53] Barresi, R., S.A. Moore, C.A. Stolle, J.R. Mendell, and K.P. Campbell. Expression of gamma-sarcoglycan in smooth muscle and its interaction with the smooth muscle sarcoglycan-sarcospan complex. J Biol Chem, 2000, **275**(49): pp. 38554–38560.

[54] Vainzof, M., E.S. Moreira, G. Ferraz, M.R. Passos-Bueno, S.K. Marie, and M. Zatz. Further evidence for the organisation of the four sarcoglycans proteins within the dystrophin-glycoprotein complex. Eur J Hum Genet, 1999, **7**(2): pp. 251–254.

[55] Crosbie, R.H., L.E. Lim, S.A. Moore, M. Hirano, A.P. Hays, S.W. Maybaum, H. Collin, S.A. Dovico, C.A. Stolle, M. Fardeau, F.M. Tome, and K.P. Campbell. Molecular and genetic characterization of sarcospan: insights into sarcoglycan-sarcospan interactions. Hum Mol Genet, 2000, **9**(13): pp. 2019–2027.

[56] Adams, M.E., M.H. Butler, T.M. Dwyer, M.F. Peters, A.A. Murnane, and S.C. Froehner. Two forms of mouse syntrophin, a 58 kd dystrophin-associated protein, differ in primary structure and tissue distribution. Neuron, 1993, **11**(3): pp. 531–540.

[57] Hillier, B.J., K.S. Christopherson, K.E. Prehoda, D.S. Bredt, and W.A. Lim. Unexpected modes of PDZ domain scaffolding revealed by structure of nNOS-syntrophin complex. Science, 1999, **284**(5415): pp. 812–815.

[58] Winder, S.J. The complexities of dystroglycan. Trends Biochem Sci, 2001, **26**(2): pp. 118–124.

[59] Spence, H.J., Y.J. Chen, C.L. Batchelor, J.R. Higginson, H. Suila, O. Carpen, and S.J. Winder. Ezrin-dependent regulation of the actin cytoskeleton by beta-dystroglycan. Hum Mol Genet, 2004, **13**(15): pp. 1657–1668.

[60] Batchelor, C.L., J.R. Higginson, Y.J. Chen, C. Vanni, A. Eva, and S.J. Winder. Recruitment of Dbl by ezrin and dystroglycan drives membrane proximal Cdc42 activation and filopodia formation. Cell Cycle, 2007, **6**(3): pp. 353–363.

[61] Thompson, O., C.J. Moore, S.A. Hussain, I. Kleino, M. Peckham, E. Hohenester, K.R. Ayscough, K. Saksela, and S.J. Winder. Modulation of cell spreading and cell-substrate adhesion dynamics by dystroglycan. J Cell Sci, 2010, **123**(Pt 1): pp. 118–127.

[62] Ilsley, J.L., M. Sudol, and S.J. Winder. The interaction of dystrophin with beta-dystroglycan is regulated by tyrosine phosphorylation. Cell Signal, 2001, **13**(9): pp. 625–632.

[63] Ferletta, M., Y. Kikkawa, H. Yu, J.F. Talts, M. Durbeej, A. Sonnenberg, R. Timpl, K.P. Campbell, P. Ekblom, and E. Genersch. Opposing roles of integrin alpha6Abeta1 and dystroglycan in laminin-mediated extracellular signal-regulated kinase activation. Mol Biol Cell, 2003, **14**(5): pp. 2088–2103.

[64] Belkin, A.M. and N.R. Smalheiser. Localization of cranin (dystroglycan) at sites of cell-matrix and cell-cell contact: recruitment to focal adhesions is dependent upon extracellular ligands. Cell Adhes Commun, 1996, **4**(4–5): pp. 281–296.

[65] Belkin, A.M. and K. Burridge. Localization of utrophin and aciculin at sites of cell-matrix and cell-cell adhesion in cultured cells. Exp Cell Res, 1995, **221**(1): pp. 132–140.

[66] Chen, Y.J., H.J. Spence, J.M. Cameron, T. Jess, J.L. Ilsley, and S.J. Winder. Direct interaction of beta-dystroglycan with F-actin. Biochem J, 2003, **375**(Pt 2): pp. 329–337.

[67] Thompson, O., I. Kleino, L. Crimaldi, M. Gimona, K. Saksela, and S.J. Winder. Dystroglycan, Tks5 and Src mediated assembly of podosomes in myoblasts. PLoS One, 2008, **3**(11): pp. e3638.

[68] Schneider, M., A.A. Khalil, J. Poulton, C. Castillejo-Lopez, D. Egger-Adam, A. Wodarz, W.M. Deng, and S. Baumgartner. Perlecan and dystroglycan act at the basal side of the Drosophila follicular epithelium to maintain epithelial organization. Development, 2006, **133**(19): pp. 3805–3815.

[69] Noel, G., D.K. Tham, and H. Moukhles. Interdependence of laminin-mediated clustering of lipid rafts and the dystrophin complex in astrocytes. J Biol Chem, 2009, **284**(29): pp. 19694–19704.

[70] Chang, L. and R.D. Goldman. Intermediate filaments mediate cytoskeletal crosstalk. Nat Rev Mol Cell Biol, 2004, **5**(8): pp. 601–613.

[71] Cerecedo, D., I. Martinez-Vieyra, R. Mondragon, M. Mondragon, S. Gonzalez, and I.J. Galvan. Haemostatic role of intermediate filaments in adhered platelets: importance of the membranous system stability. J Cell Biochem, 2013, **114**(9): pp. 2050–2060.

[72] Cerecedo, D., I. Martinez-Vieyra, L. Alonso-Rangel, C. Benitez-Cardoza, and A. Ortega. Epithelial sodium channel modulates platelet collagen activation. Eur J Cell Biol, 2014, **93**(3): pp. 127–136.

[73] Cerecedo, D., I. Martinez-Vieyra, A. Sosa-Peinado, J. Cornejo-Garrido, C. Ordaz-Pichardo, and C. Benitez-Cardoza. Alterations in plasma membrane promote overexpression and increase of sodium influx through epithelial sodium channel in hypertensive platelets. Biochim Biophys Acta, 2016, **1858**(8): pp. 1891–1903.

[74] Moore, C.J. and S.J. Winder. Dystroglycan versatility in cell adhesion: a tale of multiple motifs. Cell Commun Signal, 2010, **8**: p. 3.

[75] James, M., A. Nuttall, J.L. Ilsley, K. Ottersbach, J.M. Tinsley, M. Sudol, and S.J. Winder. Adhesion-dependent tyrosine phosphorylation of (beta)-dystroglycan regulates its interaction with utrophin. J Cell Sci, 2000, **113 (Pt 10)**: pp. 1717–1726.

[76] Sotgia, F., H. Lee, M.T. Bedford, T. Petrucci, M. Sudol, and M.P. Lisanti. Tyrosine phosphorylation of beta-dystroglycan at its WW domain binding motif, PPxY, recruits SH_2 domain containing proteins. Biochemistry, 2001, **40**(48): pp. 14585–14592.

[77] Moore, C.J. and S.J. Winder. The inside and out of dystroglycan post-translational modification. Neuromuscul Disord, 2012, **22**(11): pp. 959–965.

[78] Miller, G., C.J. Moore, R. Terry, T. La Riviere, A. Mitchell, R. Piggott, T.N. Dear, D.J. Wells, and S.J. Winder. Preventing phosphorylation of dystroglycan ameliorates the dystrophic phenotype in mdx mouse. Hum Mol Genet, 2012, **21**(20): pp. 4508–4520.

[79] Lipscomb, L., R.W. Piggott, T. Emmerson, and S.J. Winder. Dasatinib as a treatment for Duchenne muscular dystrophy. Hum Mol Genet, 2016, **25**(2): pp. 266–274.

[80] Mitchell, A., G. Mathew, T. Jiang, F.C. Hamdy, S.S. Cross, C. Eaton, and S.J. Winder. Dystroglycan function is a novel determinant of tumor growth and behavior in prostate cancer. Prostate, 2013, **73**(4): pp. 398–408.

[81] Mathew, G., A. Mitchell, J.M. Down, L.A. Jacobs, F.C. Hamdy, C. Eaton, D.J. Rosario, S.S. Cross, and S.J. Winder. Nuclear targeting of dystroglycan promotes the expression of androgen regulated transcription factors in prostate cancer. Sci Rep, 2013, **3**: pp. 2792.

[82] Bijata, M., J. Wlodarczyk, and I. Figiel. Dystroglycan controls dendritic morphogenesis of hippocampal neurons in vitro. Front Cell Neurosci, 2015, **9**: pp. 199.

[83] Cross, S.S., J. Lippitt, A. Mitchell, F. Hollingsbury, S.P. Balasubramanian, M.W. Reed, C. Eaton, J.W. Catto, F. Hamdy, and S.J. Winder. Expression of beta-dystroglycan is reduced or absent in many human carcinomas. Histopathology, 2008, **53**(5): pp. 561–566.

[84] Singh, J., Y. Itahana, S. Knight-Krajewski, M. Kanagawa, K.P. Campbell, M.J. Bissell, and J. Muschler. Proteolytic enzymes and altered glycosylation modulate dystroglycan function in carcinoma cells. Cancer Res, 2004, **64**(17): pp. 6152–6159.

[85] Higginson, J.R., O. Thompson, and S.J. Winder. Targeting of dystroglycan to the cleavage furrow and midbody in cytokinesis. Int J Biochem Cell Biol, 2008, **40**(5): pp. 892–900.

[86] Vasquez-Limeta, A., K.M. Wagstaff, A. Ortega, D.H. Crouch, D.A. Jans, and B. Cisneros. Nuclear import of beta-dystroglycan is facilitated by ezrin-mediated cytoskeleton reorganization. PLoS One, 2014, **9**(3): p. e90629.

[87] Oppizzi, M.L., A. Akhavan, M. Singh, J.E. Fata, and J.L. Muschler. Nuclear translocation of beta-dystroglycan reveals a distinctive trafficking pattern of autoproteolyzed mucins. Traffic, 2008, **9**(12): pp. 2063–2072.

[88] Lara-Chacon, B., M.B. de Leon, D. Leocadio, P. Gomez, L. Fuentes-Mera, I. Martinez-Vieyra, A. Ortega, D.A. Jans, and B. Cisneros. Characterization of an Importin alpha/beta-recognized nuclear localization signal in beta-dystroglycan. J Cell Biochem, 2010, **110**(3): pp. 706–717.

[89] Martinez-Vieyra, I.A., A. Vasquez-Limeta, R. Gonzalez-Ramirez, S.L. Morales-Lazaro, M. Mondragon, R. Mondragon, A. Ortega, S.J. Winder, and B. Cisneros. A role for beta-dystroglycan in the organization and structure of the nucleus in myoblasts. Biochim Biophys Acta, 2013, **1833**(3): pp. 698–711.

[90] Romo-Yanez, J., C. Montanez, and L.A. Salazar-Olivo. Dystrophins and DAPs are expressed in adipose tissue and are regulated by adipogenesis and extracellular matrix. Biochem Biophys Res Commun, 2011, **404**(2): pp. 717–722.

[91] Collins, S.J. and M.T. Groudine. Chronic myelogenous leukemia: amplification of a rearranged c-abl oncogene in both chronic phase and blast crisis. Blood, 1987, **69**(3): pp. 893–898.

[92] Breitman, T.R., S.E. Selonick, and S.J. Collins. Induction of differentiation of the human promyelocytic leukemia cell line (HL-60) by retinoic acid. Proc Natl Acad Sci U S A, 1980, **77**(5): pp. 2936–2940.

[93] Martinez-Zarate, A.D., I. Martinez-Vieyra, L. Alonso-Rangel, B. Cisneros, S.J. Winder, and D. Cerecedo. Dystroglycan depletion inhibits the functions of differentiated HL-60 cells. Biochem Biophys Res Commun, 2014, **448**(3): pp. 274–280.

[94] Banker, D.E., J. Radich, A. Becker, K. Kerkof, T. Norwood, C. Willman, and F.R. Appelbaum. The t(8;21) translocation is not consistently associated with high Bcl-2 expression in de novo acute myeloid leukemias of adults. Clin Cancer Res, 1998, **4**(12): pp. 3051–3062.

[95] Prieto, J., A. Eklund, and M. Patarroyo. Regulated expression of integrins and other adhesion molecules during differentiation of monocytes into macrophages. Cell Immunol, 1994, **156**(1): pp. 191–211.

[96] Escarcega-Tame, M.A., I. Martinez-Vieyra, L. Alonso-Rangel, B. Cisneros, S.J. Winder, and D. Cerecedo. Dystroglycan depletion impairs actin-dependent functions of differentiated Kasumi-1 cells. PLoS One, 2015, **10**(12): p. e0144078.

[97] Rossier, O. and G. Giannone. The journey of integrins and partners in a complex interactions landscape studied by super-resolution microscopy and single protein tracking. Exp Cell Res, 2016, **343**(1): pp. 28–34.

[98] Rossy, J., S.V. Pageon, D.M. Davis, and K. Gaus. Super-resolution microscopy of the immunological synapse. Curr Opin Immunol, 2013, **25**(3): pp. 307–312.

[99] Maglione, M. and S.J. Sigrist. Seeing the forest tree by tree: super-resolution light microscopy meets the neurosciences. Nat Neurosci, 2013, **16**(7): pp. 790–797.

[100] Cebecauer, M., M. Spitaler, A. Serge, and A.I. Magee. Signalling complexes and clusters: functional advantages and methodological hurdles. J Cell Sci, 2010, **123**(Pt 3): pp. 309–320.

[101] Simons, K. and E. Ikonen. Functional rafts in cell membranes. Nature, 1997, **387**(6633): pp. 569–572.

[102] Cohen, A.W., R. Hnasko, W. Schubert, and M.P. Lisanti. Role of caveolae and caveolins in health and disease. Physiol Rev, 2004, **84**(4): pp. 1341–1379.

[103] Bauer, M. and L. Pelkmans. A new paradigm for membrane-organizing and -shaping scaffolds. FEBS Lett, 2006, **580**(23): pp. 5559–5564.

[104] Halayko, A.J. and G.L. Stelmack. The association of caveolae, actin, and the dystrophin-glycoprotein complex: a role in smooth muscle phenotype and function? Can J Physiol Pharmacol, 2005, **83**(10): pp. 877–891.

[105] Sharma, P., S. Ghavami, G.L. Stelmack, K.D. McNeill, M.M. Mutawe, T. Klonisch, H. Unruh, and A.J. Halayko. beta-Dystroglycan binds caveolin-1 in smooth muscle: a functional role in caveolae distribution and Ca^{2+} release. J Cell Sci, 2010, **123**(Pt 18): pp. 3061–3070.

[106] Cerecedo, D., I. Martinez-Vieyra, D. Maldonado-Garcia, E. Hernandez-Gonzalez, and S.J. Winder. Association of membrane/lipid rafts with the platelet cytoskeleton and the caveolin PY14: participation in the adhesion process. J Cell Biochem, 2015, **116**(11): pp. 2528–2540.

Chapter 7

Biophysics of Fish Sperm Flagellar Movement: Present Knowledge and Original Directions

Galina Prokopchuk and Jacky Cosson

Additional information is available at the end of the chapter

http://dx.doi.org/10.5772/66863

Abstract

A fish spermatozoon has a minimalist structure: head, mid-piece and flagellum with the active inner core, called "axoneme". The axoneme represents a cylindrical scaffold of microtubular doublets arranged around a pair of single microtubules and assorted along the entire length with the dynein-ATPase motors. The mechanisms of wave generation along the flagellum becomes possible due to sliding of microtubules relative to each other and their propagation is a result of a balance between mechanical constraints and intra-flagellar biochemical actors that generate force.

How fish sperm flagella mechanics adapt to external constraints, such as vicinity of surfaces or viscosity, during the very short period of motility? By use of high-speed video microscopy, stroboscopic system, modelisation and simulation approaches, we show that fish sperm flagella respond to physical and chemical signals from environment in a very brief period of time.

This review chapter presents a brief description of the biological and biochemical features that characterize fish spermatozoa. Then it describes the biophysical aspects of flagellar movement covering various topics involved in fish sperm motility and offering a compilation of the recent knowledge acquired on different physical properties, such as wave propagation, energetics, hydrodynamics, temperature, viscosity, axonemal microtubules dynamics, among other aspects.

Keywords: spermatozoon, flagellum mechanics, hydrodynamics, motility, wave propagation, fish

1. Introduction

Spermatozoa of most fish species are immotile in the genital tract due to the specific constitution of the surrounding seminal plasma [1]. Osmotic pressure, concentration of K^+ ions, as well as pH level and sucrose concentration are considered as the main factors of seminal fluid preventing the initiation of movement of fish spermatozoa [2]. During natural spawning, ejaculated sperm cells are diluted with fresh- or seawater according to the fish habitat and right away initiate their motility by responding to changes in osmolality of the external milieu (hypo or hyper, respectively). Motility may be also induced in a laboratory designed saline solution with a certain pH, ionic and osmotic composition. Activation by the surrounding medium is immediately followed by a swimming response at full speed [3], which requires fast energy consumption by spermatozoa, thus leading to brevity of the motile period [4, 5]. In case of marine fishes, the duration of sperm motility is generally lasting for longer period as compared to freshwater species [6]. The total duration of flagellar activity of fish sperm lasts from minutes to tens of minutes [5]. However, European eel (*Anguilla anguilla*) [7] and European conger (*Conger conger*) spermatozoa [6] can swim for at least 30 min with little change in their motility characteristics such as high beat frequency, which can range up to 95 Hz. Same case for tilapia sperm where motility can last more than one hour whatever the surrounding osmolality conditions [8]. Arrest of fish sperm movement occurs partly because of rapid exhaustion of ATP by the cell and inability of the mitochondria to restore the energy content fast enough during the motility period [9], as well as due to some morphological changes mostly affecting membrane integrity and producing a curling of the flagellum [5, 10, 11]. For comparison, spermatozoa of mammals and invertebrates (e.g., oyster or sea urchin sperm) can swim for several hours [4].

It is worth to emphasize that during the swimming period, flagellar characteristics of fish spermatozoa change in many respects: wave velocity, wave amplitude, wavelength, number of waves along the flagellum length and degree of curvature of the wave [5, 12]. Whatever the wave parameter is considered, each one shows a decrease during the limited period of flagellar motility that altogether leads to a gradual but drastic lowering of the forward velocity of spermatozoa. Thus, it is clear that the behavior of the flagella is basically determining the global movement ability of the sperm cells [5, 12].

A main difficulty for observation of fish spermatozoa and quantification of their swimming parameters is that they are "fast swimmers" but for short duration [4]. This partly explains why most knowledge acquired on sperm flagellar movement comes from studies on the classical model of sea urchin sperm motility and on the mammals for more structurally complex sperm cells [13–15]. Nevertheless, fish spermatozoa are specifically interesting objects, due to their particular motility activation mode and their short motility duration that enables observation of the complete swimming period during a short time laps. Initiation of movement, the motility period and the arrest of motility of fish spermatozoa allow to develop specific studies on general understanding of regulation and signaling of sperm motility in terms of flagellar beating and wave parameters, thus leading to a better acquaintance of the fine-tuning of the internal axonemal mechanics (see [16] for a comprehensive overview on biochemical aspects of fish sperm motility).

The main aim of the present review is to describe existing methods for evaluation of the flagellum characteristics of fish sperm and present an overview of the literature embracing current understanding of their behavior from a biophysical, especially hydrodynamical, point of view.

http://dx.doi.org/10.5772/66863

2. Structure of fish spermatozoa

Fish spermatozoa present the same basic structural features as most of the male germ cells of other animals though the presence of organelles in fish sperm is reduced to a minimum: a head, a mid-piece and a flagellum (**Figure 1**) [17, 18].

The head is the carrier of hereditary information, mostly the nucleus with paternal DNA material. In most fish species, the head of spermatozoa presents an almost spherical shape with diameter varying from 2 to 4 μm. However, in some cases, such as sturgeon, paddlefish and eel spermatozoa, the shape of head is elongated: up to 9 μm long and 2 μm wide [7, 19, 20]. Such variation in head shape certainly influences the swimming performance because of differences induced in the viscous friction against the aquatic milieu.

Mid-piece is a receptacle of the centrioles and the mitochondria (usually from 2 to 9 per each spermatozoon), the latter generating energy (ATP) for sperm motility [9]. In several fish families, the sperm mitochondria were found ring-shaped [17]. Even though the mitochondrial DNA is present in the sperm cells, the male mitochondrial genes are not transmitted to the progeny [21]. In a mature spermatozoon, the protein synthesis machinery is absent, and therefore, no gene expression can occur. However, the sperm epigenomic transmission of information from father to progeny is nowadays corroborated by experimental results in mammals [22], but little is known in fish species. The centriolar complex of mid-piece consists of the proximal and the distal centrioles, the latter forming the basal body of the flagellum. This complex anchors the flagellum to the head of the sperm cell and is normally located in close vicinity of the nucleus. Such mechanical anchoring is crucial for the process of wave development. It is worth mentioned still that the mid-piece of fish spermatozoa remains separated from the flagellum by the cytoplasmic canal.

The length of fish sperm flagella varies from 20 to 100 μm depending on species. Flagellar bending is generated by a highly organized cylindrical system of microtubules, called the axoneme, emanating from the basal body [23]. The basal body is a barrel-like structure made of nine triplet of microtubules strongly associated together, which reminds a cartwheel in the lumen of the proximal portion of the basal body [5, 13, 15]. As explained later, the anchoring of the flagellum to the basal body is essential for the wave generation mechanism. In turn, the canonical axoneme consists of nine pairs of peripheral microtubular doublets and one central pair of singlet microtubules. This structural arrangement is illustrated in **Figure 1**. Although the patterns adopted during flagellar movement are distinct from those of ciliary movement, and flagella are typically much longer than cilia, such basic "9 + 2" structure of the axoneme is highly conserved and almost identical among eukaryotic cilia and flagella from protozoans to human. In the axonemal structure of some species, there are some variations though, for example, in Anguilliformes and Elopiformes sperm flagella present a "9 + 0" pattern lacking central microtubules [7, 19, 24]. This specific 9 + 0 structure is probably responsible for the helical shape (3D) of flagellar waves in those species, a feature that contrasts with planar flagellar waves developed by the 9 + 2 canonical structure [7, 19].

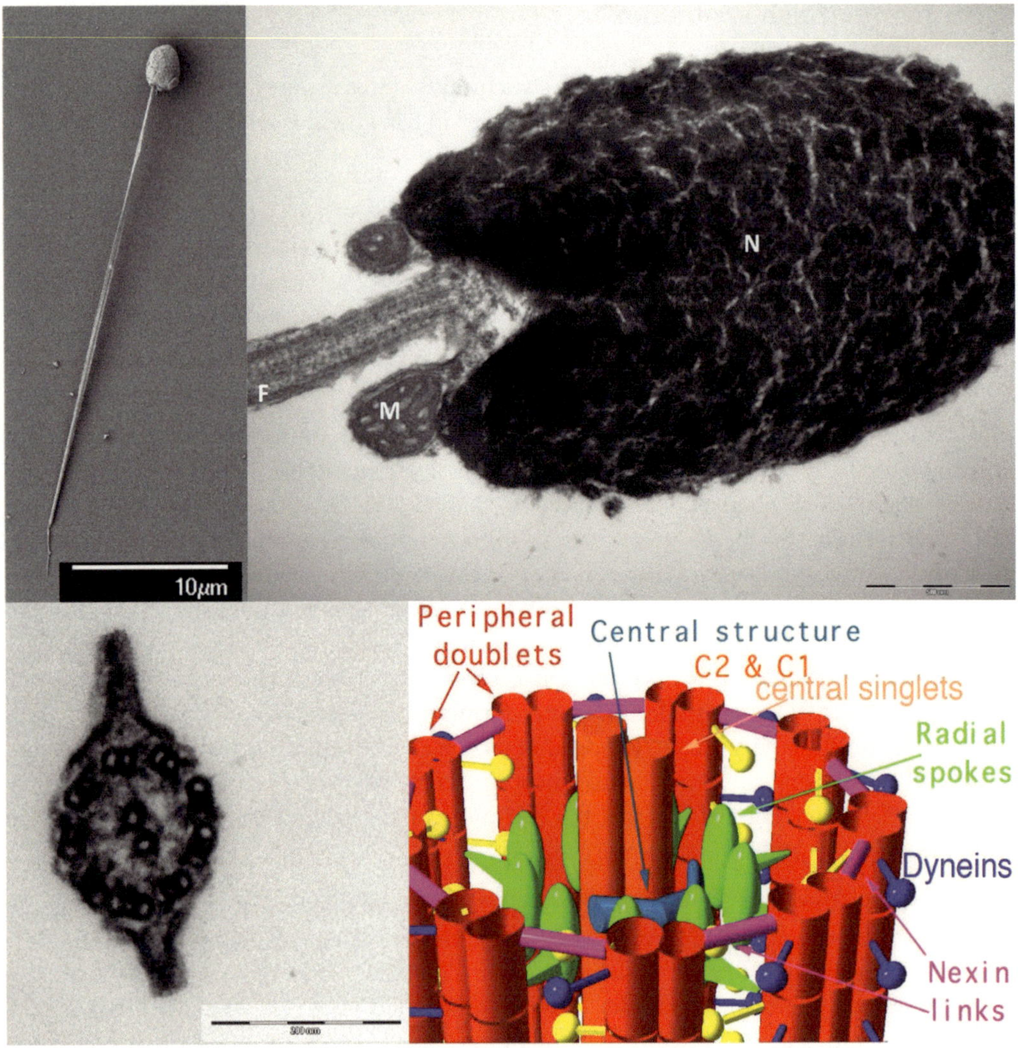

Figure 1. Morphology (top) and ultrastructure of the axoneme (bottom) of fish spermatozoon (Chinook salmon *Oncorhynchus tshawytscha*). Top: general view of spermatozoon with the ribbon-shaped flagellum and longitudinal section of the head region (N–nucleus), mid-piece with mitochondria (M) and flagellum (F) obtained by electron microscopy; scale bar = 500 nm. Bottom: A cross section of the axoneme at the distal part of the flagellum and a three-dimensional view of the arrangement of an axoneme. Microtubules are arranged according to the typical 9 + 2 structure with peripheral doublets (red), two central singlet microtubules (orange) and structures arranged around: the inner and outer dynein arms (blue and yellow), the radial spokes (green) and nexin links (pink).

The structural connections between the nine peripheral outer doublets and the sheath surrounding the central pair are named radial spokes. The central pair itself is enclosed in this sheath of proteins forming a series of projections that are well positioned to interact with each of the spoke heads and are among candidates to regulate the wave propagation [25]. Each of the outer doublets is connected to adjacent pairs of doublets by nexin links. The nexin protein has elastic properties that allow to resist the free sliding of the microtubules with respect to each other during movement and is homologous to the dynein regulatory protein [26]. The peripheral doublets are strung with rows of dynein arms along the entire length of microtubules. These dynein arms consist of macromolecular ATPase complex [27] used as basic motor actuating the whole axoneme and extend from an outer doublet toward an adjacent doublet at regularly spaced intervals along the entire length of each A microtubules [28]. Both the spokes and the dynein complex contain different calcium-binding proteins so as for flagella to be able to respond to regulation by free calcium concentration through altering their beating pattern [29, 30]. Altogether, axonemes are composed of at least 500 different protein components [15].

The bending process in an axoneme is caused by sliding between two adjacent doublets of outer microtubules forced to slide relatively to each other by the molecular motive force, generated by dynein motor activity initially described by Gibbons and Rowe [27]. According to Ref. [31], the inner arms that are both necessary and sufficient to generate flagellar bends determine the size and shape of the waveform and the outer dynein arms add power and increase beat frequency. Due to enzymatic hydrolysis of ATP, which induces force generation of the power stroke of individual dyneins, the dynein arms interact with tubulin of the B-tubule from the adjacent doublet, causing a process of active sliding in a cooperative way [32]. The presence of inter-doublet links between peripheral microtubules and intermittent sliding between some of them creates a tension that results in flagellum oscillations [33]. Since the relative sliding of the microtubules at the proximal end (near the head) of flagellum is restricted because of the strict structural link between axonemal doublets and the basal body (see **Figure 1**), and as each microtubule doublet maintains its approximate radial position due to protein arrangement in the core of the flagellum, the filament is thus forced to bend. There are also some passive sliding that occurs in other portions of the axoneme as a consequence of the active sliding of doublets pairs [28], as well as recovery sliding due to elasticity of nexin links. To prevent sliding disintegration, dynein arms probably also act as linkers between the doublet microtubules. Apparently, the dynein arms could alternatively act either as motors or as anchors, although the separate functions of rigor formation and that of force generation could be segregated into different dynein molecules [15].

The axoneme is fully encased by the cell membrane. Often, the plasma membrane also forms one or two fin-like ridges along the fish sperm flagellar tail, which are oriented along the horizontal axis defined by the central microtubules [34–37]. The ribbon shape instead of the usual cylindrical shape of the flagellum makes it brighter when observed by dark-field microscopy and allows to better visualization and recording of wave shapes [6]. This feature of the flagellar membrane has been documented among species belonging to many fish families: Poeciliidae, Jenysiidae, Pantodontidae and Embiotocidae [17, 38–40] and was shown to potentially contribute to improve the swimming efficiency [37].

Altogether, various and original characteristics of fish spermatozoa represent attractive biological objects generating model studies for specialists in fields of physics such as hydrodynamics and fluid mechanics. Flagellar movement can be explained by various functional models that account for the presumed mechanism on a theoretical basis and include features resulting from experiments. Such computer modeling approach aims to explain how, during the movement, several bends of opposite angular direction coexist along a flagellum and how these bends propagate along the flagellum. The precise nature of the spatial and temporal control mechanisms regulating the various flagellar and ciliary beating patterns is still not fully understood [41].

3. Physical aspects of flagellar movement

In terms of physical quantitative description, the analysis involves a viscous and incompressible fluid coupled to a single force-generating filament, the flagellum. In the past decades, several quantitative descriptions of the fluid dynamics of spermatozoa and ciliary propulsion have been attained successfully. The linear Stokes flow assumption has been used to investigate the hydrodynamic consequences of flagellar undulations taking into account the low value of the Reynolds number and the possibility to neglect inertial effects [42].

It has been hypothesized that flagellum is beating due to localized "contractions" propagated along the doublet microtubules [43]. According to the elaborated resistive force theory, active moments should balance both viscous and elastic moments present in the active filamentous flagellum. Bending waves could propagate along the flagellum if changes in length of contractile elements cause delayed changes in tension. Based on this theory, several researchers developed and explored more refined models for ciliary and flagellar motion. Using a series of photographs of cell position separated by very short periods (millisecond range) and flagellar motion parameters, Brokaw [44] was probably the first who suggested to compare computed cell trajectories and flagellar shape with experimental observations. Eventually, he proposed the model for the control of switching in which curvature controls the flagellar beat [45, 46]. The curvature control hypothesis maintains that when the flagellum bends to a sufficient curvature due to active forces, it triggers the inactivation of one set of dyneins and the activation of the set on the opposite side of the axoneme. The detachment of dynein in this case is regulated by doublet curvature [46, 47]. The degree of curvature is considered as a mechanical parameter of the axoneme that is in proportion of its resistance to bending. This model was expanded to include cross-bridge mechanics between microtubules [48]. The strain in a curved microtubule where the radius of curvature can reach up to 4 μm is very small (⇡ 1%), corresponding to strain in a tubulin dimer of angstrom range. Such a small strain is difficult to detect by an individual dynein microtubule-binding domain, except if dynein binds in a cooperative way. The degree of curvature of the axonemal microtubules could be controlled by a protein called "doublecortin." It was recently shown that this protein binds with higher affinity to curved microtubule lattices than to straight ones [49].

In an alternative model, the control of switching where dyneins behave as slipping links was proposed. These links detach when subjected to forces acting parallel to the long axis of the microtubule doublets and thus oppose sliding [50, 51]. Appearance of sliding forces on one side of the axoneme induces detachment of the dyneins on the other side (and vice versa) meaning that opposite sides are antagonistic.

A third model for regulating flagellar and ciliary beating is the geometric clutch theory developed by Lindemann [52]. This model treats the axoneme as dynamic elastic linkages exerting force between longitudinal arrays of doublet microtubules. The hypothesis is predicated that the transverse force that develops in the plane of bending of the axoneme changes the spacing between doublet microtubules and dynein bridges pull together adjacent doublets. The force generation of the attached dyneins creates sliding and bending. This active bending increases the transverse forces that pull the doublets apart and disengage the dyneins what allows the dyneins on the opposite side of the axoneme to attach [28].

Next model for cilia and flagella incorporates discrete representations of the dynein arms, the passive elastic structure of the axoneme including the doublets and nexin links [28, 53]. In this model, dynein activation is governed by a simple curvature control mechanism [46].

Recently, some authors developed a two-dimensional mathematical model of the axoneme that can incorporate any or all of these different feedback mechanisms above [54] in order to evaluate the validity of each model. This new model includes static curvature that is responsible of asymmetric beats. Results of these authors favor the curvature control mechanism as it gives best agreement with the bending waveforms of *Chlamydomonas flagella*, and predict that the motors respond to the time derivative of curvature rather than curvature itself.

The above paragraph presents two levels of investigation for physical description of flagellar beating: *either* at the organelle level where the flagellum is considered as an active filament (without considering its internal structure), *or* at the macro-molecular level where internal components of the flagellum interact between each other in a coordinated mode.

4. Evaluation of fish sperm motility parameters

Due to the short duration of fish sperm motility, special methods for recording the sperm motion [55], and especially obtaining high-resolution flagellar images [14], have been developed. An unclassified and non-exhaustive list of variables that are commonly used to describe the motility phase of fish sperm in details includes: velocity of head displacement, percentage of motile cells, duration of motility, linearity of track of sperm heads, shape of the flagellar waves and other criteria such as wave velocity or frequency [14].

Duration of motility. The duration of motility period is estimated as the time period elapsed from activation by transfer in a swimming medium to the full arrest of progressive motility for all spermatozoa [14].

It has been demonstrated that the duration of motility is temperature dependent and species specific [56–58]. In cyprinids, it was shown that extracellular and intracellular pH, as well as the ionic composition of the swimming media, influences the initiation and duration of sperm motility [59]. As stated above, motility duration of fish spermatozoa is frequently limited by flagellar damages appearing during the motility period, mostly in relation to osmotic stress imposed at motility initiation [60].

In both freshwater and marine fishes, two main and common flagellar damages were reported: cytoplasmic blebs emerge anywhere along flagellar length during the motility period which impairs the propagation of wave [61, 62] and curling structure at flagellar tip particularly close to the end of motility period, which shortens obviously the flagellar length and leads to decrease the efficiency of axonemal beating [10, 11, 63, 64]. Damages such as blebs and curling usually result from local membrane defects caused mainly by hypo-osmotic shock, and they are usually reversed when reestablishing the osmolality of the surrounding solution to correct values [10, 65].

Duration of motility is also closely related to energy stored in fish sperm cells [16, 66, 67], as fast motility needs large rate of energy consumption that cannot be compensated by mitochondrial ATP production [4, 9]. When intracellular ATP store becomes low, flagellar dynein ATPases start to function at low rates that causes the decrease of wave amplitudes and eventually slows down the progressive motion [68–71]. Due to the decrease of the ATP store during progress of the motility phase [9, 66], the proportion of motile cells in the sperm population also decreases as a function of time after activation, which also contributes to a decrease in fertilizing ability [72]. In addition, as a consequence of ATP hydrolyses, ADP is continuously accumulating and at the end of the motility period reaches its maximal value [72]. It was shown that the presence of ADP releases the inhibitory effects of high concentration of ATP in sea urchin sperm [73, 74]. In fish sperm, a low ATP/ADP ratio would oppositely favor dynein inhibition and contribute to the decrease of flagellar beat frequency [75]. For example, in trout sperm, ATP has a K_m value of 0.2 mM [76], while the K_i for ADP is about 0.27 mM [77], and at the end of the motility period, internal ADP concentration reaches 2.28 mM, while concentration of ATP is much lower. The importance of ATP as energetic compound for sperm motility [9] is related to another major energetic compound, the creatine phosphate [78–80].

Sperm velocity. Sperm velocity represents a global combination of several parameters such as head dimension (diameter of head), beat frequency, length of flagellum and physical parameters of wave propagation like wave length and amplitude [43], which contributes differentially to energetic exhaustion. Velocities of spermatozoa are greatest immediately after activation [4], for example, in halibut 150–180 μm/s [81, 82], in fugu 160 μm/s [83], in cod 65–100 μm/s [84] or 130 μm/s [64], in hake 130 μm/s [64, 85], in tuna 215–230 μm/s [86], in turbot 220 μm/s [70, 87] and in sea bass 120 μm/s (straight line velocity) [69, 88]. High initial velocity leads to shortened total duration of motility, because fish spermatozoa mostly rely on their preaccumulated energy store for operating their propulsive motors [4, 9]. The velocity characteristics may be modulated by sperm microenvironment and particularly pH and osmolality of the swimming medium [6]. For instance, by increasing the osmolality of the activation media, the number of wave and curvatures along the flagellum will increase and this can be accompanied by a decrease in sperm velocity.

The above paragraph presents an overview of the methods and results leading to quantitative description of the fish sperm movement characteristics.

5. Flagellum wave propagation

The numerous variables developed for description of sperm motility by a follow-up of head displacement as used in CASA (Computer-assisted sperm analysis) (see above) are not independent but rather redundant. Therefore, unrelated variables were designed to describe specifically the flagellar beating through their wave properties [89]. Initial studies developed on invertebrate's sperm flagella [90] were applied to flagella of marine fish spermatozoa [14, 91, 92] such as turbot [70], sea bass [69], cod and hake [64], as well as to freshwater species such as trout [78], salmon [93], carp [60], sturgeon [3] and pike [11]. Some of the flagellar wave parameters play a critical role for the displacement efficiency of the whole sperm cell and control the forward velocity of translation, for example, the amplitude and the length of each flagellar wave, the number of bends, the curvature of the bend pattern, the wave velocity and the flagellar beat frequency.

The traits of the motility behavior of sperm flagella of fish with external fertilization are quite similar in several respects [87, 94]. Normally, wave propagation occurs from head to flagellar tip leading to forward movement of the spermatozoon with head first [5, 15]. More recently, the appearance of first bends at motility activation was described in detail [3, 95]. In most cases, the first bend initiates from the region close to the head and propagates toward the flagellar tip [3, 95]. This bend is then followed by a next one with opposite direction of curvature so that several successive bends occupy the whole flagellar length, mostly during the earliest period of fish sperm motility [5, 15]. The bend initiation mechanism itself is still not fully explained [96]. Studies demonstrated that the axoneme of demembranated spermatozoa (after removal of the membrane of flagella by application of a mild detergent) could be reinitiated to produce waves, if energy in the form of ATP was provided to the system [27, 97]. In case of rainbow trout, chum salmon or sturgeon, flagellar axonemes need to be exposed to both cyclic AMP (cAMP) and ATP to become functionally motile [61, 76, 80, 98, 99]. However, this feature is not general as the presence of cAMP does not seem to be necessary for sperm motility initiation in many other fish species [100, 101].

In all fish species studied so far, the waves propagate the whole length of the sperm flagellum when observed right after activation. However, during the motility period of fish sperm, several types of modifications of the wave pattern appear, which are paralleled by a decrease in flagellar beat frequency [6, 94]. The second part of the motility period is identified by the restriction of the waves to the third or quarter length of the flagellum near the head, leading to inefficiency of translation of the wave, decrease of velocity, and is ending up by a full stop [11, 62, 72, 78, 87]. This has been interpreted in terms of an energy transfer deficiency from the mid-piece (ATP production in mitochondria) to the distal part of the flagellum where ATP is consumed [68]. Similar problems of energy availability in trout spermatozoa were related to insufficiency of the ATP/creatine-phosphate shuttle [78].

Most paradigms on wave generation and propagation along the axoneme of flagella state that there is a clear distinction between the dynein-dependent microtubule sliding actuated by the dynein oscillatory motor and the bending mechanism that should include regulator mechanism responsible for the wave propagation. During wave propagation, a bending/relaxing cycle propagates in register and in a frame-shifted manner with the clusters of dynein-ATPase motors operating along the axoneme [44]. The motor components and their actuating mechanism are nowadays well understood, but little is known about the elements responsible for the bending regulating [13].

The input of the above studies conducted at the intra-flagellar level shows how the coordination between all the flagellar participants is crucial for the optimization of the flagellum function.

6. Wave shape: analysis and quantification

As a general description, the wave shape of fish sperm flagella is of the arcsine type, that is, linear segments intercalated between two successive curvatures, similar to what occurs in tunicate sperm flagella or sea urchins [102].

Usually, it is assumed that flagellar waves are almost planar, that is, each sine wave is "flat" and the successive waves are coplanar. An exception to wave's flatness can be found in European eel spermatozoa, which possess a corkscrew wave shape [7, 19, 103]. However, this helical wave pattern has lower efficiency in terms of forward velocity of the spermatozoa even though flagella beats at high frequency, up to 95 Hz [7, 19]. It is also suggested that swimming in 2D partly prevents dispersion of spermatozoa far away from the egg. This hypothesis was recently tested in a simulation study [104] showing that the predicted physical advantage is related to the relative angle between sperm swimming plane and egg surface plane. In many cases and species, waves are not perfectly planar, but slightly deviate from a strict plane while successive waves are not coplanar. This feature was described for sperm flagella in several species. Such slight distortion from wave flatness would explain the ability of sperm cells to maintain swimming in the surface vicinity [105]. Actually, the majority of cells in a population of fish spermatozoa swim in the vicinity of glass surfaces [106]. Swimming in vicinity of surface is also a property that is observed in human sperm [107]. It is speculated that such an ability to swim in the vicinity of the egg surface probably represents a biological advantage for fertilization efficiency.

In addition, sperm cells may be observed rotating transiently during the motion. For example, in the case of paddlefish and sturgeon, due to the rotation of the whole sperm cell, each spermatozoon image appears alternatively with flagellar top view (waves in the plane of observation) or side view, with waves orthogonal to the observation plane [61, 93, 106]. Nevertheless, waves are not arranged according to a helical shape but rather as successive waves subscribed in different planes [105]. For sperm with quite symmetrical heads, nonplanar beating can occur with cell rolling during surface swimming, resulting in circular swimming trajectories in the direction of cell rolling, which is always the same within a species [108].

The wave shape of fish sperm flagella is affected by several factors, such as the energetic content (ATP), which controls wave amplitude [6, 72], the internal ionic concentration (ionic strength) that affects the constancy of the wave amplitude along flagellum as well as physical constraints imposed by the external milieu like viscosity and temperature [11, 12, 87, 109].

Energetic content. The rate of energy consumption (ATP hydrolysis) by sperm flagellum determines the flagellar beat frequency and therefore velocity of forward displacement of fish spermatozoon [9, 13, 110, 111]. As already widely discussed above, there is a progressive decrease in the flagellar beat frequency in individual fish spermatozoa during the motile phase [55, 112]. Therefore, in case of fish sperm flagella, the beat frequency measurement should be associated with the precise timing after activation of that measurement because of its fast decay as a function of time [12, 14].

Internal ionic concentration. The intracellular ionic concentration is indirectly governed by the external osmolality. It was shown that the change of extracellular osmolality, which perceived by spermatozoa when transferred from seminal fluid to external milieu, causes a rapid change in intracellular ionic concentration observed during the course of the motility phase [94, 109]. As a consequence, the flagellar axonemes become exposed to a more and more drastic intracellular environment for dynein motors leading to reduce the development of waves (dampening process described above) and eventually lead to a full arrest of motility [5, 69, 72]. In some species, dampening of flagellar waves during the motility period is accompanied by asymmetry of beating. The ability to develop either symmetric or asymmetric ways of beating results in imbalanced amplitude of the bends following each other. For instance, if both bends are of equal curvature, then the symmetrical movement of sperm flagella is developed, which leads the sperm cells to describe linear tracks [13]. In case of asymmetry, sperm movement becomes consequently circular and sperm cells describe circles of corresponding diameter [113]. In cases studied in detail, asymmetry of beating is related to Ca^{2+} regulation, probably through a Ca^{2+}-calmodulin-dependent phosphorylation of some axonemal proteins [113, 114]. This Ca^{2+}-induced asymmetry also occurs in freshwater species, such as trout [115] and carp [60], where due to increased Ca^{2+} concentration spermatozoa describe circular trajectories, which become tighter with time elapsed after activation. Nevertheless, it was shown that asymmetry of the flagellar waveform can appear in the absence of any cell signaling change or flagellar heterogeneity. In physics, such phenomenon occurs commonly in elastic filament dynamics (so-called buckling instability) when passive filaments are subjected to high tangential forces. This was demonstrated also for asymmetric flagellar bending [116].

Viscosity. Sperm migrating in high viscosity fluids commonly exhibits larger numbers of waves though of lower amplitudes [86, 107]. Thus, an increase in viscosity by addition of viscous compounds in the swimming medium actually leads to a lowering of propulsive velocity of the sperm [86]. Practically, viscous medium mimics the situation occurring for sperm cells in ovary fluids or jelly-like layers that surround eggs in some fish species [117]. In addition, variations in the constitutive morphology of individual spermatozoa within the species also influence their velocity in viscous media [118]. As already mentioned, the viscosity effects are enhanced in case of the ribbon-shaped flagella like those in fish spermatozoa possessing fins, the latter greatly increasing the surface of viscous interaction with the surrounding medium [37].

Temperature effects. Fish species are adapted to a large variety of temperatures (from several degrees below 0 to 40–50°C). Low temperature could represent an adverse factor for sperm to fertilize eggs because decreasing the temperature reduces the flagellar beat frequency

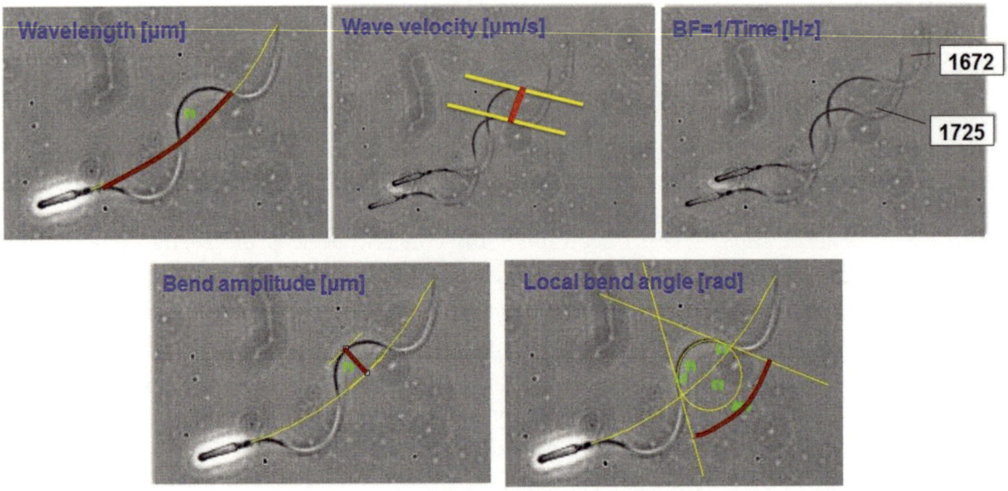

Figure 2. Series of images of sturgeon sperm illustrating the measurements of the flagellar wave parameters: wavelength, wave velocity, beat frequency, bend amplitude and local bend angle.

and consequently the sperm translational velocity [57]. This is probably compensated by the increase motility duration observed when temperature is decreased [57, 58, 119], the latter increasing statistically the chances of egg-sperm meeting. The relationship between beat frequency and temperature provides an opportunity to get access to important thermodynamic variables of the flagellar beating [120]. From a thermodynamic point of view, pressure represents an additional factor possibly acting on sperm performances, taking into account that sperm spawning occurs in deep water in some marine fish species [121]. Little is known about the pressure effects on flagellar behavior because of technical limitation for such studies [122].

In addition, measurement of flagellar parameters including beat frequency at different temperatures allows better understanding of energetic constraints involved in sperm movement [4, 56, 57]. The behavior of fish sperm flagella presents various and original interests as they are able to convert chemical energy into mechanical to generate movement [120]. Due to biophysical methods, such as hydrodynamic analyses of the beat patterns, it become possible to estimate the minimum intracellular consumption of chemical energy that needed for flagellar motion [16].

Some fish species reproduce at temperatures definitely lower than room temperature, and therefore, motility parameters should be measured at lower temperatures. As an option observation can be carried out in a temperature-controlled room. In addition, it is possible to control the temperature on the microscope itself, but this often leads to problem of condensation on the condenser and objective lenses. The temperature control of the glass slide can be simply set due to the contact with a cooling micro-Peltier plate. This also allows to measure local temperature with a micro-thermistor immersed in the observation drop [14].

Successive positions of flagellar waves can be observed in series of video frames obtained at several time intervals during the motility phase. This allows measurement of flagellar wave parameters (**Figure 2**).

The above paragraph shows that internal parameters (energy content as example) act in synergy with external parameters (viscosity as example) and complement each other in the signaling processes that allow fish spermatozoa to rapidly adapt to a large diversity of situations they are confronted to during their short-term motility.

7. Conclusions

The main aim of this review is to describe features of fish sperm motility, flagellar mechanics and different characteristic determining movement and show the interest of studying them from a biophysical point of view. To assess motility of fish sperm, a lot of variables, such as velocity of head displacement, percentage of motile cells, duration of motility, linearity of track of sperm heads, are commonly used to estimate the ability of a sperm population to achieve optimal fertilization. All these parameters describe the whole cell movement through the head displacement, but data that are more informative can be obtained from observation of flagellar behavior, as the flagellum is the actual source of movement generation. At present, most investigations on the mechanisms of flagellar beating and propulsion of spermatozoa were developed due to studies on mammalian and sea urchin flagella. Nevertheless, detailed records of fish sperm flagella from different species offer a unique opportunity to observe successive stages in the swimming period: activation step of motility itself, motility period and the gradual decrease leading to the end of the motility period. Due to the fact that fish spermatozoa swim at high speed and possess short period of motility, it has been historically difficult to observe their flagellar behavior. Therefore, most of knowledge about fish sperm motility was initially obtained from studies of the head movement using, for instance, CASA. However, additional methods to describe the details of wave flagellar movement, such as high magnification microscopy combined with stroboscopic illumination and high-speed video microscopy, were developed recently and become accessible. First attempts to obtain detailed description of fish flagellar behavior already reveal to be helpful for basic understanding of mechanistic and hydrodynamic aspects of their motile function and its adaptability. A further challenge will be to integrate the understanding of these basic mechanisms to the diversity of patterns exhibited during spermatozoa movement in different swimming fluids and under various signaling processes.

Acknowledgements

The study was funded by the Ministry of Education, Youth and Sports of the Czech Republic—projects "CENAKVA" (No. CZ.1.05/2.1.00/01.0024), "CENAKVA II" (No. LO1205 under the NPU I program), COST (No. LD14119) and COST Office (Food and Agriculture COST

Action FA1205: AQUAGAMETE), by the Grant Agency of the University of South Bohemia in Ceske Budejovice (No. 125/2016/Z) and by the Czech Science Foundation (GACR No. P502/15-12034S).

Author details

Galina Prokopchuk* and Jacky Cosson

*Address all correspondence to: progalinas@gmail.com

Faculty of Fisheries and Protection of Waters, South Bohemian Research Center of Aquaculture and Biodiversity of Hydrocenoses, Research Institute of Fish Culture and Hydrobiology, University of South Bohemia in Ceske Budejovice, Vodnany, Czech Republic

References

[1] Morisawa S, Morisawa M. Acquisition of potential for sperm motility in rainbow trout and chum salmon. J. Exp. Biol. 1986;**126:**89–96. PMID: 3806005

[2] Alavi SMH, Cosson J. Sperm motility in fishes. (II) Effects of ions and osmolality: a review. Cell Biol. Int. 2006;**30:**1–14. doi:10.1016/j.cellbi.2005.06.004

[3] Prokopchuk G, Dzyuba B, Bondarenko O, Rodina M, Cosson J. Motility initiation of sterlet sturgeon (*Acipenser ruthenus*) spermatozoa: describing the propagation of the first flagellar waves. Theriogenology. 2015;**84:**51–61. doi:10.1016/j.theriogenology.2015.02.011

[4] Cosson J. Frenetic activation of fish spermatozoa flagella entails short-term motility, portending their precocious decadence. J. Fish Biol. 2010;**76:**240–279. doi:10.1111/j.1095-8649.2009.02504.x

[5] Cosson J. The motility apparatus of fish spermatozoa. In: Alavi SMH, Cosson J, Coward K, Rafiee G, editors. Fish Spermatology. Oxford: Alpha Science International Ltd; 2008. p. 281–316.

[6] Cosson J, Groison AL, Suquet M, Fauvel C, Dreanno C, Billard R. Studying sperm motility in marine fish: an overview on the state of the art. J. Appl. Ichthyol. 2008;**24:**460–486. doi:10.1111/j.1439-0426.2008.01151.x

[7] Gibbons BH, Baccetti B, Gibbons IR. Live and reactivated motility in the 9 + 0 flagellum of Anguilla sperm. Cell Motil. 1985;**5:**333–350. doi:10.1002/cm.970050406

[8] Legendre M, Alavi SMH, Dzyuba B, Linhart O, Prokopchuk G, Cochet C, Dugué R, Cosson J. Adaptations of semen characteristics and sperm motility to harsh salinity: extreme situations encountered by the euryhaline tilapia *Sarotherodon melanotheron heudelotii* (Dumeril, 1859). Theriogenology. 2016. doi:10.1016/j.theriogenology.2016.04.066

[9] Cosson J. ATP: The sperm movement energizer. In: Kuester E, Traugott G, editors. Adenosine Triphosphate: Chemical Properties, Biosynthesis and Functions in Cells. New York: Nova Publisher Inc; 2013. p. 1–46.

[10] Perchec G, Cosson MP, Cosson J, Jeulin C, Billard R. Morphological and kinetic changes of carp (*Cyprinus carpio*) spermatozoa after initiation of motility in distilled water. Cell Motil. Cytoskeleton. 1996;**35:**113–120. doi:10.1002/(SICI)1097-0169(1996)35:2<113::AID-CM4>3.0.CO;2-B

[11] Alavi SM, Rodina M, Viveiros AT, Cosson J, Gela D, Boryshpolets S, Linhart O. Effects of osmolality on sperm morphology, motility and flagellar wave parameters in Northern pike (*Esox lucius* L.). Theriogenology. 2009;**72:**32–43. doi:10.1016/j.theriogenology.2009.01.015

[12] Cosson J, Prokopchuk G. Wave propagation in flagella. In: Rocha L, Gomes M, editors. Wave Propagation. Cheyenne: Academy Publish; 2014. p. 541–583.

[13] Gibbons IR. Cilia and flagella of eukaryotes. J. Cell Biol. 1981;**91:**107–124. doi:10.1083/jcb.91.3.107s

[14] Cosson J. Methods to analyse the movements of fish spermatozoa and their flagella. In: Alavi SMH, Cosson J, Coward K, Rafiee G, editors. Fish Spermatology. Oxford: Alpha Science International Ltd; 2008. p. 63–102.

[15] Cosson J, editor. Flagellar Mechanics and Sperm Guidance. Bentham Books Publisher; 2015. p. 424. doi:10.2174/97816810812811150101

[16] Dzyuba B, Bondarenko O, Fedorov P, Gazo I, Prokopchuk G, Cosson J. Energetics of fish spermatozoa: the proven and the possible. Aquaculture. 2016. doi:10.1016/j.aquaculture.2016.05.038

[17] Lahnsteiner F, Patzner RA. Sperm morphology and ultrastructure. In: Alavi SMH, Cosson J, Coward K, Rafiee G, editors. Fish Spermatology. Oxford: Alpha Science International Ltd; 2008. p. 1–61.

[18] Jamieson BGM, Leung LKP. Fish Evolution and Systematics: Evidence from Spermatozoa: with a Survey of Lophophorate, Echinoderm, and Protochordate Sperm and an Account of Gamete Cryopreservation. Cambridge: Cambridge University Press; 1991. 334 p. doi:10.1046/j.1420-9101.1992.5040721.x

[19] Gibbons BH, Gibbons IR, Baccetti B. Structure and motility of the 9 + 0 flagellum of eel spermatozoa. J. Submicr. Cytol. 1983;**15:**15–20. PMID: 6842644

[20] Linhartova Z, Rodina M, Nebesarova J, Cosson J, Psenicka M. Morphology and ultrastructure of beluga (*Huso huso*) spermatozoa and a comparison with related sturgeons. Anim. Reprod. Sci. 2013;**137:**220–229. doi:10.1016/j.anireprosci.2013.01.003

[21] DeLuca Steven Z, O'Farrell Patrick H. Barriers to male transmission of mitochondrial DNA in sperm development. Dev. Cell. 2012;**22:**660–668. doi:10.1016/j.devcel.2011.12.021

[22] Casas E, Vavouri T. Sperm epigenomics: challenges and opportunities. Front. Genet. 2014;**5**:330. doi:10.3389/fgene.2014.00330

[23] Inaba K. Molecular architecture of the sperm flagella: molecules for motility and signaling. Zool. Sci. 2003;**20**:1043–1056. doi:10.2108/zsj.20.1043

[24] Mattei C, Mattei X. Spermiogenesis and spermatozoa of the *Elopomorpha* (teleost fish). In: Afzelius B, editor. The Functional Anatomy of the Spermatozoon. Oxford: Pergamon Press; 1975. p. 211–221.

[25] Smith EF, Yang P. The radial spokes and central apparatus: mechano-chemical transducers that regulate flagellar motility. Cell Motil. Cytoskeleton. 2004;**57**:8–17. doi:10.1002/cm.10155

[26] Heuser T, Raytchev M, Krell J, Porter ME, Nicastro D. The dynein regulatory complex is the nexin link and a major regulatory node in cilia and flagella. J. Cell Biol. 2009;**187**:921–933. doi:10.1083/jcb.200908067

[27] Gibbons IR, Rowe AJ. Dynein: a protein with adenosine triphosphatase activity from cilia. Science. 1965;**149**:424–426. doi:10.1126/science.149.3682.424

[28] Dillon RH, Fauci LJ. An integrative model of internal axoneme mechanics and external fluid dynamics in ciliary beating. J. Theor. Biol. 2000;**207**:415–430. doi:10.1006/jtbi.2000.2182

[29] Gibbons BH, Gibbons IR. Calcium-induced quiescence in reactivated sea urchin sperm. J. Cell Biol. 1980;**84**:13–27. PMID: 7350165

[30] Eshel D, Brokaw CJ. New evidence for a "biased baseline" mechanism for calcium-regulated asymmetry of flagellar bending. Cell Motil. Cytoskeleton. 1987;**7**:160–168. doi:10.1002/cm.970070208

[31] Brokaw CJ. Control of flagellar bending: a new agenda based on dynein diversity. Cell Motil. Cytoskeleton. 1994;**28**:199–204. doi:10.1002/cm.970280303

[32] Sale WS, Satir P. Direction of active sliding of microtubules in Tetrahymena cilia. Proc. Natl. Acad. Sci. U.S.A. 1977;**74**:2045–2049. PMID: 266725

[33] Satir P. The present status of the sliding microtubule model of ciliary motion. In: Sleigh MA, editor. Cilia and Flagella. New York: Academic Press; 1974. p. 131–141.

[34] Nicander L. Comparative studies on the fine structure of vertebrate spermatozoa. In: Baccetti B, editor. Comparative Spermatology. Rome: Academic Press; 1970. p. 47–56.

[35] Matos E, Santos MNS, Azevedo C. Biflagellate spermatozoon structure of the hermaphrodite fish *Satanoperca jurupari* (Heckel, 1840) (Teleostei, Cichlidae) from the Amazon River. Braz. J. Biol. 2002;**62**:847–852. doi:10.1590/s1519-69842002000500014

[36] Suquet M, Dorange G, Omnes MH, Normant Y, Roux A, Fauvel C. Composition of the seminal Fluid and ultrastructure of the spermatozoon of turbot (*Scophthalmus maximus*). J. Fish Biol. 1993;**42**:509–516. doi:10.1111/j.1095-8649.1993.tb00355.x

[37] Gillies EA, Bondarenko V, Cosson J, Pacey AA. Fins improve the swimming performance of fish sperm: a hydrodynamic analysis of the Siberian sturgeon *Acipenser baerii*. Cytoskeleton. 2013;**70:**85–100. doi:10.1002/cm.21093

[38] Dadone L, Narbaitz R. Submicroscopic structure of spermatozoa of a cyprinodontiform teleost, *Jenynsia lineata*. Z. Zellforsch. Mikrosk Anat. 1967;**80:**214–219. doi:10.1007/BF00337457

[39] Stanley HP. An electron microscope study of spermiogenesis in the teleost fish *Oligocottus maculosus*. J. Ultrastruct. Res. 1969;**27:**230–243. doi:10.1016/S0022-5320(69)80014-6

[40] Lahnsteiner F, Berger B, Weismann T, Patzner RA. Sperm structure and motility of the freshwater teleost *Cottus gobio*. J. Fish Biol. 1997;**50:**564–574. doi:10.1111/j.1095-8649.1997.tb01950.x

[41] Brokaw CJ. Simulating the effects of fluid viscosity on the behaviour of sperm flagella. Math. Meth. Appl. Sci. 2001;**24:**1351–1365. doi:10.1002/mma.184

[42] Brennen C, Winet H. Fluid mechanics of propulsion by cilia and flagella. Ann. Rev. Fluid Mech. 1977;**9:**339–398. doi:10.1146/annurev.fl.09.010177.002011

[43] Gray J, Hancock GJ. The propulsion of sea-urchin spermatozoa. J. Exp. Biol. 1955;**32:**802–814.

[44] Brokaw CJ. Bending moments in free-swimming flagella. J. Exp. Biol. 1970;**53:**445–464. PMID: 5529723

[45] Brokaw CJ. Bend propagation by a sliding filament model for flagella. J. Exp. Biol. 1971;**55:**289–304. PMID: 5114025

[46] Brokaw CJ. Computer simulation of flagellar movement. Biophys. J. 1972;**12:**564–586. doi:10.1016/s0006-3495(72)86104-6

[47] Morita Y, Shingyoji C. Effects of imposed bending on microtubule sliding in sperm flagella. Curr. Biol. 2004;**14:**2113–2118. doi:10.1016/j.cub.2004.11.028

[48] Brokaw CJ, Rintala DR. Computer simulation of flagellar movement. III. Models incorporating cross-bridge kinetics. J. Mechanochem. Cell Motil. 1975;**3:**77–86. PMID: 1214108

[49] Bechstedt S, Lu K, Brouhard Gary J. Doublecortin recognizes the longitudinal curvature of the microtubule end and lattice. Curr. Biol. 2014;**24:**2366–2375. doi:10.1016/j.cub.2014.08.039

[50] Brokaw CJ. Molecular mechanism for oscillation in flagella and muscle. Proc. Natl. Acad. Sci. U.S.A. 1975;**72:**3102–3106. PMID: 1059095

[51] Camalet S, Jülicher F. Generic aspects of axonemal beating. New J. Phys. 2000;**2:**24–24. doi:10.1088/1367-2630/2/1/324

[52] Lindemann CB. Testing the geometric clutch hypothesis. Biol. Cell. 2004;**96:**681–690. doi:10.1016/j.biolcel.2004.08.001

[53] Dillon RH, Fauci LJ, Omoto C. Mathematical modeling of axoneme mechanics and fluid dynamics in ciliary and sperm motility. DCDIS Ser A Math Anal. 2003;**10:**745–757.

[54] Sartori P, Geyer VF, Scholich A, Jülicher F, Howard J. Dynamic curvature regulation accounts for the symmetric and asymmetric beats of *Chlamydomonas flagella*. Elife. 2016;5. doi:10.7554/eLife.13258

[55] Billard R, Cosson MP. Measurement of sperm motility in trout and carp. In: De Pauw N, Jaspers EJ, Ackefors H, Wilkins NP, editors. Aquaculture: A Biotechnology in Progress. Bredene: European Aquaculture Society; 1989. p. 499–503.

[56] Dadras H, Dzyuba V, Cosson J, Golpour A, Dzyuba B. The in vitro effect of temperature on motility and antioxidant response of common carp *Cyprinus carpio* spermatozoa. J. Therm. Biol. 2016;**59:**64–68. doi:10.1016/j.jtherbio.2016.05.003

[57] Dadras H, Dzyuba B, Cosson J, Golpour A, Siddique MAM, Linhart O. Effect of water temperature on the physiology of fish spermatozoon function: a brief review. Aquacult. Res. 2016. doi:10.1111/are.13049

[58] Billard R, Cosson MP. Sperm motility in rainbow trout *Parasalmo mykiss*: effect of pH and temperature. In: Zohar Y, Breton B, editors. Reproduction in Fish: Basic and Applied Aspects in Endocrinology and Genetics. Paris: INRA; 1988. p. 161–167.

[59] Marian T, Krasznai Z, Balkay L, Emri M, Tron L. Role of extracellular and intracellular pH in carp sperm motility and modifications by hyperosmosis of regulation of the Na^+/H^+ exchanger. Cytometry. 1997;**27:**374–382. doi:10.1002/(SICI)1097-0320(19970401)27:4<374::AID-CYTO9>3.0.CO;2-C

[60] Perchec Poupard G, Gatti JL, Cosson J, Jeulin C, Fierville F, Billard R. Effects of extracellular environment on the osmotic signal transduction involved in activation of motility of carp spermatozoa. J. Reprod. Fertil. 1997;**110:**315–327. doi:10.1530/jrf.0.1100315

[61] Cosson J, Linhart O, Mims SD, Shelton WL, Rodina M. Analysis of motility parameters from paddlefish and shovelnose sturgeon spermatozoa. J. Fish Biol. 2000;**56:**1348–1367. doi:10.1111/j.1095-8649.2000.tb02148.x

[62] Alavi SMH, Hatef A, Pšenicka M, Kašpar V, Boryshpolets S, Dzyuba B, Cosson J, Bondarenko V, Rodina M, Gela D, Linhart O. Sperm biology and control of reproduction in sturgeon: (II) sperm morphology, acrosome reaction, motility and cryopreservation. Rev. Fish Biol. Fish. 2012;**22:**861–886. doi:10.1007/s11160-012-9270-x

[63] Tsvetkova LI, Cosson J, Linhart O, Billard R. Motility and fertilizing capacity of fresh and frozen-thawed spermatozoa in sturgeons *Acipenser Baeri* and *A. ruthenus*. J. Appl. Ichthyol. 1996;**12:**107–112. doi:10.1111/j.1439-0426.1996.tb00071.x

[64] Cosson J, Groison AL, Suquet M, Fauvel C. Motility characteristics of spermatozoa in cod (*Gadus morhua*) and hake (*Merluccius merluccius*). Cybium. 2008;**32:**176–177.

[65] Linhart O, Alavi SMH, Rodina M, Gela D, Cosson J. After finishing of motility, common carp (*Cyprinus carpio*) sperm is able to re-initiate a second motility period and to fertilize eggs. Cybium. 2008;**32:**187–188.

[66] Billard R, Cosson MP. The energetics of fish sperm motility. In: Gagnon C, editor. Controls of Sperm Motility: Biological and Clinical Aspects. Boca Raton: CRC Press; 1990. p. 153–174.

[67] Ingermann R. Energy metabolism and respiration in fish spermatozoa. In: Alavi SMH, Cosson J, Coward K, Rafiee G, editors. Fish Spermatology. Oxford: Alpha Science International Ltd; 2008. p. 241–266.

[68] Perchec G, Jeulin C, Cosson J, Andre F, Billard R. Relationship between sperm ATP content and motility of carp spermatozoa. J. Cell Sci. 1995;**108:**747–753. PMID: 7769016

[69] Dreanno C, Cosson J, Suquet M, Cibert C, Fauvel C, Dorange G, Billard R. Effects of osmolality, morphology perturbations and intracellular nucleotide content during the movement of sea bass (*Dicentrarchus labrax*) spermatozoa. J. Reprod. Fertil. 1999;**116:**113–125. PMID: 10505062

[70] Dreanno C, Cosson J, Suquet M, Seguin F, Dorange G, Billard R. Nucleotide content, oxidative phosphorylation, morphology, and fertilizing capacity of turbot (*Psetta maxima*) spermatozoa during the motility period. Mol. Reprod. Dev. 1999;**53:**230–243. doi:10.1002/(SICI)1098-2795(199906)53:2<230::AID-MRD12>3.0.CO;2-H

[71] Billard R, Cosson J, Fierville F, Brun R, Rouault T, Williot P. Motility analysis and energetics of the Siberian sturgeon *Acipenser baerii* spermatozoa. J. Appl. Ichthyol. 1999;**15:**199–203. doi:10.1111/j.1439-0426.1999.tb00234.x

[72] Chauvaud L, Cosson J, Suquet M, Billard R. Sperm motility in turbot, *Scophthalmus marimus*: initiation of movement and changes with time of swimming characteristics. Environ. Biol. Fishes. 1995;**43:**341–349. doi:10.1007/bf00001167

[73] Kinoshita S, Miki-Noumura T, Omoto CK. Regulatory role of nucleotides in axonemal function. Cell Motil. Cytoskeleton. 1995;**32:**46–54. doi:10.1002/cm.970320106

[74] Omoto CK, Yagi T, Kurimoto E, Kamiya R. Ability of paralyzed flagella mutants of Chlamydomonas to move. Cell Motil. Cytoskeleton. 1996;**33:**88–94. doi:10.1002/(sici)1097-0169(1996)33:2<88::aid-cm2>3.0.co;2-e

[75] Omoto CK. Mechanochemical coupling in cilia. Int. Rev. Cytol. 1991;**131:**255–292. doi:10.1016/s0074-7696(08)62021-5

[76] Cosson MP, Cosson J, Andre F, Billard R. cAMP/ATP relationship in the activation of trout sperm motility: their interaction in membrane-deprived models and in live spermatozoa. Cell Motil. Cytoskeleton. 1995;**31:**159–176. doi:10.1002/cm.970310208

[77] Penningroth SM, Peterson DD. Evidence for functional differences between two flagellar dynein ATPases. Cell Motil. Cytoskeleton. 1986;**6:**586–594. doi:10.1002/cm.970060607

[78] Saudrais C, Fierville F, Loir M, Le Rumeur E, Cibert C, Cosson J. The use of phosphocreatine plus ADP as energy source for motility of membrane-deprived trout spermatozoa. Cell Motil. Cytoskeleton. 1998;**41:**91–106. doi:10.1002/(SICI)1097-0169(1998)41:2<91::AID-CM1>3.0.CO;2-I

[79] Fedorov P, Dzyuba B, Fedorova G, Grabic R, Cosson J, Rodina M. Quantification of adenosine triphosphate, adenosine diphosphate, and creatine phosphate in sterlet spermatozoa during maturation. J. Anim. Sci. 2015;**93:**5214. doi:10.2527/jas.2015-9144

[80] Dzyuba V, Dzyuba B, Cosson J, Rodina M. Enzyme activity in energy supply of spermatozoon motility in two taxonomically distant fish species (sterlet *Acipenser ruthenus*, Acipenseriformes and common carp *Cyprinus carpio*, Cypriniformes). Theriogenology. 2016;**85:**567–574. doi:10.1016/j.theriogenology.2015.09.040

[81] Billard R, Cosson J, Crim LW. Motility of fresh and aged halibut sperm. Aquat. Living Resour. 1993;**6:**67–75. doi:10.1051/alr:1993008

[82] Vermeirssen ELM, de Quero CM, Shields RJ, Norberg B, Kime DE, Scott AP. Fertility and motility of sperm from Atlantic halibut (*Hippoglossus hippoglossus*) in relation to dose and timing of gonadotropin-releasing hormone agonist implant. Aquaculture. 2004;**230:**547–567. doi:10.1016/s0044-8486(03)00414-9

[83] Takai H, Morisawa M. Change in intracellular K^+ concentration caused by external osmolality change regulates sperm motility of marine and freshwater teleosts. J. Cell Sci. 1995;**108:**1175–1181. PMID: 7622603

[84] Trippel EA, Morgan MJ. Sperm Longevity in Atlantic Cod (*Gadus morhua*). Copeia. 1994;**1994:**1025. doi:10.2307/1446727

[85] Groison A-L, Suquet M, Cosson J, Le Coz J-R, Jolivet A, Garren F. Biological characteristics of European hake (*Merluccius merluccius*) sperm. Cybium. 2008;**32:**178.

[86] Cosson J, Groison AL, Suquet M, Fauvel C, Dreanno C, Billard R. Marine fish spermatozoa: racing ephemeral swimmers. Reproduction. 2008;**136:**277–294. doi:10.1530/REP-07-0522

[87] Cosson J, Billard R, Cibert C, Dreanno C, Linhart O, Suquet M. Movements of fish sperm flagella studied by high speed videomicroscopy coupled to computer assisted image analysis. Pol Arch Hydrobiol. 1997;**44:**103–113.

[88] Abascal FJ, Cosson J, Fauvel C. Characterization of sperm motility in sea bass: the effect of heavy metals and physicochemical variables on sperm motility. J. Fish Biol. 2007;**70:**509–522. doi:10.1111/j.1095-8649.2007.01322.x

[89] Cosson J. Flagella parameters used as descriptors of fish spermatozoa motility. Anim. Reprod. Sci. 2016;**169:**128–129. doi:10.1016/j.anireprosci.2016.03.078

[90] Brokaw CJ. My favourite cell: the sea urchin spermatozoon. BioEssays. 1990;**12:**449–452. doi:10.1002/bies.950120910

[91] Cosson J. The ionic and osmotic factors controlling motility of fish spermatozoa. Aquacult. Int. 2004;**12:**69–85. doi:10.1023/B:Aqui.0000017189.44263.Bc

[92] Ishijima S. Comparative analysis of movement characteristics of lancelet and fish spermatozoa having different morphologies. Biol. Bull. 2012;**222:**214–221. PMID: 22815370

[93] Cosson J, Lahnsteiner F, Prokopchuk G, Valdebenito II. Initiation, prolongation, and reactivation of the sperm motility. In: Vladic T, Petersson E, editors. Evolutionary Biology of the Atlantic Salmon. Boca Raton: CRC Press; 2015. p. 63–107. doi:10.1201/b18721-7

[94] Cosson J, Dreanno C, Billard R, Suquet M, Cibert C. Regulation of axonemal wave parameters of fish spermatozoa by ionic factors. In: Gagnon C, editor. The Male Gamete: From Basic Science to Clinical Applications. Montréal: Cache River Press; 1999. p. 161–186.

[95] Dzyuba V, Cosson J. Motility of fish spermatozoa: from external signaling to flagella response. Reprod. Biol. 2014;**14:**165–175. doi:10.1016/j.repbio.2013.12.005

[96] Gibbons IR. Transient flagellar waveforms during intermittent swimming in sea urchin sperm. II. Analysis of tubule sliding. J. Muscle Res. Cell Motil. 1981;**2:**83–130. doi:10.1007/bf00712063

[97] Summers KE, Gibbons IR. Adenosine triphosphate-induced sliding of tubules in trypsin-treated flagella of sea-urchin sperm. Proc. Natl. Acad. Sci. U.S.A. 1971;**68:**3092–3096. PMID: 5289252

[98] Morisawa M, Okuno M. Cyclic AMP induces maturation of trout sperm axoneme to initiate motility. Nature. 1982;**295:**703–704. doi:10.1038/295703a0

[99] Morisawa M, Okuno M, Suzuki K, Morisawa S, Ishida K. Initiation of sperm motility in teleosts. J. Submicrosc. Cytol. 1983;**15:**61–65. PMID: 6302300

[100] Cosson M-P, Gagnon C. Protease inhibitor and substrates block motility and microtubule sliding of sea urchin and carp spermatozoa. Cell Motil. Cytoskeleton. 1988;**10:**518–527. doi:10.1002/cm.970100408

[101] He S, Jenkins-Keeran K, Woods LC. Activation of sperm motility in striped bass via a cAMP-independent pathway. Theriogenology. 2004;**61:**1487–1498. doi:10.1016/j.theriogenology.2003.08.015

[102] Brokaw CJ. Microtubule sliding in swimming sperm flagella: direct and indirect measurements on sea urchin and tunicate spermatozoa. J. Cell Biol. 1991;**114:**1201–1215. doi:10.1083/jcb.114.6.1201

[103] Woolley DM. Studies on the eel sperm flagellum. 2. The kinematics of normal motility. Cell Motil. Cytoskeleton. 1998;**39:**233–245. doi:10.1002/(sici)1097-0169(1998)39:3<233::aid-cm6>3.0.co;2-5

[104] Ishimoto K, Cosson J, Gaffney EA. A simulation study of sperm motility hydrodynamics near fish eggs and spheres. J. Theor. Biol. 2016;**389:**187–197. doi:10.1016/j.jtbi.2015.10.013

[105] Cosson J, Huitorel P, Gagnon C. How spermatozoa come to be confined to surfaces. Cell Motil. Cytoskeleton. 2003;**54:**56–63. doi:10.1002/cm.10085

[106] Boryshpolets S, Cosson J, Bondarenko V, Gillies E, Rodina M, Dzyuba B, Linhart O. Different swimming behaviors of sterlet (*Acipenser ruthenus*) spermatozoa close to solid and free surfaces. Theriogenology. 2013;**79:**81–86. doi:10.1016/j.theriogenology.2012.09.011

[107] Smith DJ, Gaffney EA, Blake JR, Kirkman-Brown JC. Human sperm accumulation near surfaces: a simulation study. J. Fluid Mech. 2009;**621:**289. doi:10.1017/s0022112008004953

[108] Woolley D. Motility of spermatozoa at surfaces. Reproduction. 2003;**126:**259–270. doi:10.1530/rep.0.1260259

[109] Inaba K, Dreanno C, Cosson J. Control of flatfish sperm motility by CO_2 and carbonic anhydrase. Cell Motil. Cytoskeleton. 2003;**55:**174–187. doi:10.1002/cm.10119

[110] Holwill ME. Kinetic studies of the flagellar movement of sea-urchin spermatozoa. J. Exp. Biol. 1969;**50:**203–222. PMID: 4237926

[111] Denehy MA. The propulsion of nonrotating ram and oyster spermatozoa. Biol. Reprod. 1975;**13:**17–29. PMID: 1222179

[112] Cosson MP, Cosson J, Billard R. Synchronous triggering of trout sperm is followed by an invariable set sequence of movement parameters whatever the incubation medium. Cell Motil. Cytoskeleton. 1991;**20:**55–68. doi:10.1002/cm.970200107

[113] Brokaw CJ. Calcium sensors in sea urchin sperm flagella. Cell Motil. Cytoskeleton. 1991;**18:**123–130. doi:10.1002/cm.970180207

[114] Morita M, Takemura A, Nakajima A, Okuno M. Microtubule sliding movement in tilapia sperm flagella axoneme is regulated by Ca^{2+}/calmodulin-dependent protein phosphorylation. Cell Motil. Cytoskeleton. 2006;**63:**459–470. doi:10.1002/cm.20137

[115] Cosson MP, Billard R, Letellier L. Rise of internal Ca^{2+} accompanies the initiation of trout sperm motility. Cell Motil. Cytoskeleton. 1989;**14:**424–434. doi:10.1002/cm.970140312

[116] Becker LE, Shelley MJ. Instability of elastic filaments in shear flow yields first-normal-stress differences. Phys. Rev. Lett. 2001;87. doi:10.1103/PhysRevLett.87.198301

[117] Yeates SE, Diamond SE, Einum S, Emerson BC, Holt WV, Gage MJG. Cryptic choice of conspecific sperm controlled by the impact of ovarian fluid on sperm swimming behavior. Evolution. 2013;**67:**3523–3536. doi:10.1111/evo.12208

[118] Tuset VM, Trippel EA, de Monserrat J. Sperm morphology and its influence on swimming speed in Atlantic cod. J. Appl. Ichthyol. 2008;**24:**398–405. doi:10.1111/j.1439-0426.2008.01125.x

[119] Alavi SMH, Cosson J. Sperm motility in fishes. I. Effects of temperature and pH: a review. Cell Biol. Int. 2005;**29**:101–110. doi:10.1016/j.cellbi.2004.11.021

[120] Cosson J . Fish spermatozoa motility: physical, and bio-energetic interactions with their surrounding media. In: Morisawa M, editor. Sperm Cell Research in the 21st Century: Historical Discoveries to New Horizons. Tokyo: Adthree Publishing Ltd; 2012. p. 152–156.

[121] Cosson J, Groison AL, Fauvel C, Suquet M. Description of hake (*Merlucius merlucius*) spermatozoa: flagellar wave characteristics and motility parameters in various situations. J Appl. Ichthyol. 2010;**26**:644–652. doi:10.1111/j.1439-0426.2010.01563.x

[122] Holwill ME. Hydrodynamic aspects of cilia and flagella. In: Sleigh MA, editor. Cilia and Flagella. London: Academic Press; 1974. p. 143–175.

Chapter 8

Cytoskeleton Rearrangements during the Execution Phase of Apoptosis

Jesús Porcuna Doncel, Patricia de la Cruz Ojeda, Manuel Oropesa-Ávila, Marina Villanueva Paz, Isabel De Lavera, Mario De La Mata, Mónica Álvarez Córdoba, Raquel Luzón Hidalgo, Juan Miguel Suarez Rivero, David Cotán and José Antonio Sánchez-Alcázar

Additional information is available at the end of the chapter

http://dx.doi.org/10.5772/66865

Abstract

Apoptosis is a regulated energy-dependent process for the elimination of unnecessary or damaged cells during embryonic development, tissue homeostasis and many pathological conditions. Apoptosis is characterized by specific morphological and biochemical features in which caspase activation has a pivotal role. During apoptosis, cells undergo characteristic morphological reorganizations in which the cytoskeleton participates actively. Traditionally, this cytoskeleton rearrangement has been assigned mainly to actinomyosin ring contraction, with microtubule and intermediate filaments both reported to be depolymerized at early stages of apoptosis. However, recent results have shown that microtubules are reformed during the execution phase of apoptosis forming an apoptotic microtubule network (AMN). Current hypothesis proposes that AMN is required to maintain plasma membrane integrity and cell morphology during the execution phase of apoptosis. AMN disruption provokes apoptotic cell collapse, secondary necrosis and the subsequent release of toxic molecules which can damage surrounding cells and promote inflammation. Therefore, AMN formation in physiological or pathological apoptosis is essential for tissue homeostasis.

Keywords: microtubules, actin, intermediate filaments, apoptosis, apoptotic microtubule network

1. Introduction

The term apoptosis refers to the process of programmed cell death characterized by a stereotypic sequence of cellular events including cell shrinkage, caspase activation and degradation of cell content, blebbing and maintenance of plasma membrane integrity and condensation and fragmentation of DNA, followed by ordered removal by phagocytes [1]. Apoptosis was first described in 1972 as a vital biological phenomenon, with both physiological and pathological implications [2]. Apoptosis regulates cell number in tissues serving as a quality control mechanism in order to eliminate damaged and senescent cells, and it has been proven to be essential during development of multicellular organisms. In the proliferative environment of embryonic development, apoptosis helps to shape organs, drives morphogenesis and deletes structures that won't be required any longer; for instance, the formation of the four-chamber architecture of the heart is a consequence of this process [3].

Traditionally, the process of apoptosis occurs in three distinct phases: induction, execution and clearance. The first one comprises all the intrinsic or extrinsic environmental changes that lead to the activation of the apoptotic cascade. Meanwhile, the execution phase is distinguished by the activation of a caspase-dependent proteolytic cascade [4]. Caspases are aspartic acid-specific proteases responsible for cellular components degradation. Some of them, such as caspase-8 and -9 act as initiators of the apoptotic signalling pathway, while other caspases such as caspases-3, -6 and -7 operate as executor caspases which actively participate in the degradation of intracellular proteins [5]. Eventually, the dying cell is engulfed by professional phagocytes or by neighbouring cells. This process of apoptotic cell clearance is essential for tissue turnover and homeostasis [6].

The fate of apoptotic cells in multicellular animals is their prompt elimination by professional phagocytes. However, cells that perform apoptosis in vitro cultures progress to secondary necrosis, which implies the loss of membrane integrity and the release of cellular content into the culturing medium [7]. In vivo, apoptotic cells can also undergo secondary necrosis when they are not properly eliminated due to massive cell death or impaired phagocytosis [8].

Efficient apoptotic cell removal is driven by the interaction with phagocytes through the expression of "eat-me" signals, the release of "find-me" signals, engulfment of the dying cell and its eventual digestion in phagocyte phagolysosomes. This interaction prevents undesired immune reactions by contributing to the development of an immunomodulatory environment [9]. On the other hand, secondary necrosis is thought to be pro-inflammatory and immunogenic, as it causes the release of endogenous damage-associated molecular patterns (DAMPs) [8]. Among DAMPs are proteolytically processed autoantigens, nucleosomes, proteases, calcium-binding protein (calgranulin), high-mobility group box 1 (HMGB-1) and urate crystals.

2. Cytoskeleton rearrangements during the execution phase of apoptosis

Cell contraction, plasma membrane blebbing, chromatin condensation and DNA fragmentation are typical hallmarks of the execution phase of apoptosis, which lasts approximately 1 h [10]. In order to achieve such dramatic morphologic changes, apoptotic cells make profound cytoskeleton reorganizations. On the other hand, caspase-mediated cytoskeleton proteins digestion ensures the proper dismantlement of the dying cell [11].

http://dx.doi.org/10.5772/66865

The eukaryotic cytoskeleton is mainly composed of actin filaments, microtubules and intermediate filaments. These three constituents act coordinately to increase tensile strength, allow cell motility, maintain plasma integrity, participate in cell division, contribute to cell morphology and provide a network for cellular transport [12]. It is widely accepted that actin cytoskeleton plays a central role in cell remodelling during the execution phase of apoptosis [13], while microtubules and intermediate filaments are disorganized at the onset of this phase [10]. However, recent work have demonstrated that microtubules are reorganized at later stages of apoptosis, giving rise to the formation of the apoptotic microtubule network (AMN) that

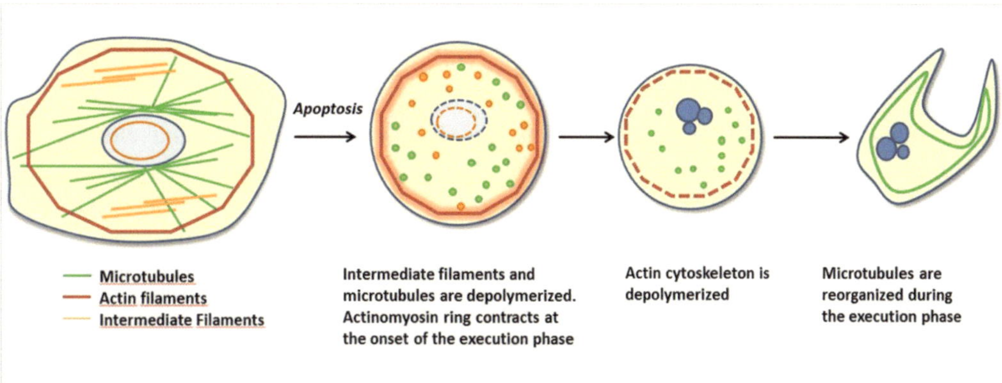

Figure 1. Cytoskeleton rearrangements during the execution phase of apoptosis.

contributes to the maintenance of the plasma membrane integrity [14, 15]. All these events are summarized in **Figure 1**.

Intermediate filaments are ubiquitous cytoskeletal components of 10 nm of diameter which provide mechanical strength and allow tissue growing among other functions. According to amino acid sequence identity, intermediate filaments can be classified into six types. Acidic keratins belong to type I, whereas basic keratins belong to type II intermediate filaments. They are typical components of hair and epithelium. Type III includes vimentin, desmin, glial fibrillary acidic protein and perinephrin. Nuclear lamins A and B, which support the cell nucleus, are represented by type V. Finally, type VI refers to nestin [16]. Intermediate filaments connect with other cytoskeletal components via cytolinker proteins of the plakin family, including desmoplakin, periplakin and plectin [17]. At the onset of the execution phase of apoptosis, type I keratins are targeted by caspases-3, -7 and -6 at their linker domain, whereas type II keratins are resistant to caspase-mediated proteolysis. Similarly, type III intermediate filaments, such as desmin or vimentin, are cleave by caspases. Once cleaved, intermediate filament subunits accumulate in the cytoplasm forming large aggregates. All these events contribute to cytoskeleton reorganizations [18]. Furthermore, caspase digestion of K18 (type I) has been proved to be indispensable for membrane integrity maintenance during apoptosis, as interference with keratin caspase cleavage shunts hepatocytes towards necrosis [19]. Keratins are not only caspase substrates but they also seem to regulate the induction phase of apoptosis.

Thus, it has been shown that deficiencies in keratins 8 and 18 favour tumour necrosis factor (TNF)/cycloheximide-induced cell death. Keratin 18 sequesters the adaptor molecule TNF receptor-associated death domain (TRADD) and thus prevents its interaction with the TNF receptor, regulating negatively apoptosis. In contrast, keratin type II K8 offers protection against Fas-mediated apoptosis.

With respect to the digestion of nuclear lamins, nuclear breakdown begins with the activation of caspases. Lamins appear to be specifically targeted by caspases 3 and 6 that become activated both via the intrinsic and extrinsic pathway of apoptosis [20]. However, little is known about the exact mechanism by which the cell nucleus dismantles. It has been established that active caspases are able to proteolyze lamins A and B, leading to lamina cleavage and chromatin condensation and fragmentation. These alterations contribute to the collapse of the nucleoskeleton [18].

Microtubules are depolymerized at the same time that intermediate filaments although the exact molecular mechanism involved is still unknown. Several hypotheses have been postulated to explain this process, which are not mutually exclusive [21]. Microtubules are polar protofilaments made up of α and β tubulin, which are involved in cell migration, growth, transport or mitosis [22]. Their dynamics is governed by microtubule-associated proteins (MAPs), Ran-GTP and proteins that bind to tubulin [22–24]. The cyclin-dependent kinase 1 Cdk1, associated with cyclin B is a key regulatory kinase which controls the entry in mitosis and regulates microtubule dynamics. In fact, it induces the depolymerization of interphase microtubules [25]. Cdk1 regulates some microtubule effectors such us MAP4, which reduces its ability to stabilize microtubules after phosphorylation [26]. In addition, Cdk1 is able to phosphorylate β-tubulin, thus inhibiting its incorporation to growing protofilaments. As Cdk1 and other Cdks activities have been observed during apoptosis, it has been suggested that they may act essential regulators in the apoptotic cytoskeleton reorganizations during apoptosis [27]. On the other hand, some authors have shown that microtubules depolymerization is associated with activation of the PP2A-like phosphatase, dephosphorylation of the microtubule regulator τ protein and deacetylation of tubulin [28]. This last mechanism can coexist with the hypothesis of Cdk1 regulation because PP2A downregulates the Cdk1 activator Cdc25 phosphatase [29].

In contrast to the apparent passivity of intermediate filaments and microtubules, the actin cytoskeleton is highly dynamic, and its remodelling turns out to be essential in the first stages of apoptosis. Actin filaments function in the generation and maintenance of cell morphology and polarity, endocytosis, intracellular trafficking, contractility, motility, cell division and apoptosis [30]. Actin is a 42-KDa globular protein (G-actin) which polymerizes to form actin filaments (F-actin). They adapt to the cell environment through actin-binding proteins (ABPs) which regulate actin cytoskeleton dynamics. A family of RhoGTPases are in charge of controlling ABPs in such a way that actin is organized into highly ordered structures such as stress fibres, lamellipodia and filopodia in non-apoptotic cells [13].

Once adherent cells have initiated apoptosis, they partially detach from the substrate by losing their focal adhesion sites. This is achieved by caspase-mediated cleavage of the focal adhesion kinase pp125, among other proteins [31]. Next, actin is reorganized into an actin-myosin II cortical ring with contractile force. Actinomyosin contraction is activated via the Rho/Rho-kinase (ROCK) signalling pathway which ends up with the phosphorylation of myosin light chain II (MLC-II) [32]. Rho-kinases are effectors of Rho-GTPase proteins, being RhoA and RhoC the most well characterized ROCK regulators [33]. Active GTP-bound Rho proteins activate ROCK by binding to their C-terminal portion of the coiled coil. Then, it induces actinomyosin contraction through two distinct mechanisms. First, it can increase the phosphorylation state of the MLCs by inhibiting the MLC phosphate or by directly phosphorylating MLC-II [32]. Alternatively, ROCK I but not ROCK II can be cleaved by caspase-3 at a conserved DETD1113/G sequence and its carboxy-terminal inhibitory domain is consequently removed [34]. The contractile force generated depends on other ROCK targets, such as the LIM kinase (a serine/threonine kinase containing LIM and PDZ domains), which phosphorylates and inactivates the actin stabilizer cofilin [35]. Cortical ring contraction results in the formation of membrane protrusions known as blebs. Their formation depends on a pressure gradient between the extracellular medium and the intracellular medium, taking place in areas of external negative pressure or in places where the plasma membrane is weakened as a consequence of caspase cleavage [11]. Thus, blebbing appears to be also dependent on the activation of ROCK by active caspases because it can be blocked by Y27632, a ROCK selective inhibitor. However, C3 toxin-induced inhibition of Rho is unable to block the formation of blebs [36].

Actinomyosin ring contraction is a fast process and coincides with the beginning of the execution phase. After that, cell content including organelles is packaged into apoptotic bodies and the actin cytoskeleton is dismantled by caspases [32]. Rho GTPases effectors contribute to this new cytoskeleton reorganization [11]. The Rho effector protein kinase C-related kinase (PRK1) is cleaved by caspase-3 generating a constitutively active kinase fragment that is able to induce actin structures disassembly [37, 38]. Similarly, the Rac effector p21-activated kinase (PAK2) may promote stress fibres dissolution after caspase cleavage [39]. Likewise, caspase-3 induces gelsolin fragmentation, contributing to the collapse of actin filaments in a calcium-independent manner [40].

At this time of the execution phase, apoptotic cells lack the main structured elements of cytoskeleton. It is then, when microtubules reorganize to give rise to the AMN [21]. Its organization, maintenance and properties will be reviewed in the next sections.

3. Influence of apoptotic cells on tissue remodelling

In tissues, apoptotic cells interact with their neighbouring partners and eventually must be eliminated. For example, in epithelia apoptotic cells are removed from the tissue by cell extrusion. Dying cell ejection is usually done apically, although it has been described that cells also extrude basally during Droshophila development. This coordinated process is necessary for tissue homeostasis and developmental morphogenesis [41]. In 2001, Rosenblatt et al. proposed a model to describe the sequence of events during epithelia extrusion [42]. They demonstrated that an actinomyosin ring is formed both in the apoptotic and in the neighbouring non-apoptotic cells. Apoptotic cells were proposed to send an undetermined early signal to the adjacent epithelial cells to induce the formation of an actinomyosin cable ring as well as the activation of small RhoGTPases. Recently, spinsohine-1-phosphate receptor 2 pathway has been proposed as the mediator between apoptotic and non-apoptotic cells [43]. Contraction of the cable ring extrudes the now late apoptotic cell out of the epithelium. In other words, during apoptosis within epithelia, dying cells are able to not only rearrange their cytoskeleton but also induce actinomyosin reorganizations within adjacent cells. Importantly, apoptotic cell membranes do not permeabilize until cell extrusion is completed [44].

The mechanical force produced during apoptosis is used not only to extrude dying cells from tissues but also to change the morphology of neighbouring cells to fill the space originally occupied by the dying cell [45]. This finding suggests that apoptotic forces might be harnessed throughout cell death-related morphogenesis. Mechanical forces arising from the apoptotic process had been originally proposed as an "apoptotic force theory" [46] that would be important during animal development including elimination of interdigital webs, dorsal closure or leg folding [46, 47]. Therefore, apoptosis should not be seen as a passive carving process. Instead, it is a generator of mechanical forces and an active player during tissue remodelling that helps to the correct morphogenesis of embryos and control of tissue dynamics [47].

4. Apoptotic microtubule network

The reorganization of microtubules during the execution phase of apoptosis has been examined in a variety of cell lines such as H460, A431, HeLa cells, primary human fibroblasts and pig LLCPK-1α cells, under several apoptosis inducers such as camptothecin (CPT), anisomycin, staurosporine, serum withdrawal, UV irradiation and TNF-related apoptosis-inducing ligand (TRAIL). In addition, apoptotic microtubules have also been observed in cell fragments and apoptotic bodies [48, 49]. These findings suggest that AMN may play an important role during the apoptotic process.

Commonly, AMN is arranged beneath plasma membrane, adopting a cortical structure that gives a "cocoon-like" structure which confines most of the intracellular content of apoptotic cells. Furthermore, apoptotic microtubules may extend from the body of apoptotic cells as long and thin spikes, suggesting a key structural role in maintaining apoptotic cell morphology and surface extensions (**Figure 2**). Apoptotic microtubules organization beneath plasma membrane also suggests that AMN may function as a kind of support to preserve plasma membrane integrity and/or as a barrier for confining the degradation reactions inside apoptotic cells.

Apart from plasma membrane protection, another function of apoptotic microtubules has been associated with the process of apoptotic body formation by helping to sustain the peripheral localization of chromatin within surface blebs and by facilitating cell fragmentation [48].

To exclude the possibility that AMN could be an artefact of the process of fixation in the immunofluorescence protocol, apoptotic microtubules formation has been also monitored in vivo in pig epithelial (LLCPK-1α) expressing GFP-tubulin (Green fluorescent protein-tubulin) and A431 cells expressing YFP-tubulin (Yellow fluorescent protein-tubulin) by live imaging [48,

http://dx.doi.org/10.5772/66865

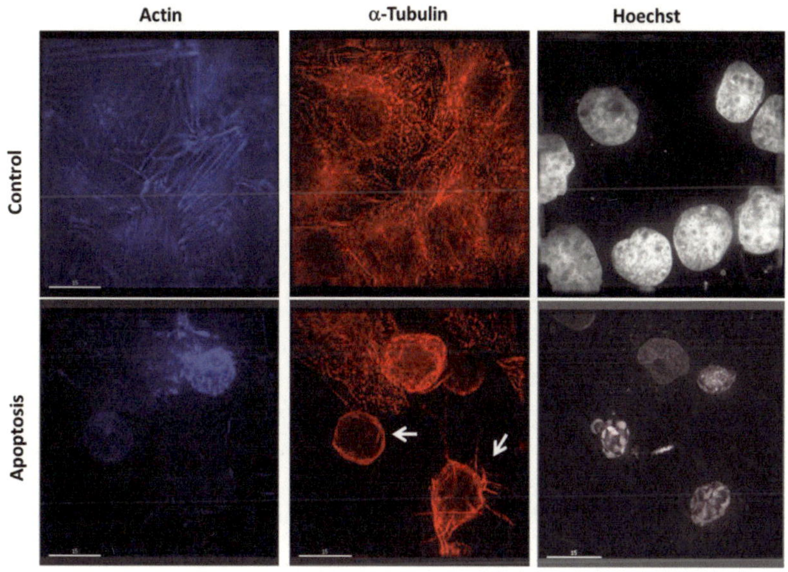

Figure 2. Immunofluorescence microscopy image of control and apoptotic cells in the execution phase. Apoptosis was induced in H460 cells by 10 μM camptothecin treatment for 48 hours. After fixation, control and apoptotic cells were stained with anti α-tubulin (red), and coumarin-phalloidin to visualize actin filaments (blue). Nuclei were revealed by Hoechst staining (blue). Arrows, apoptotic cells with AMN. Bar = 15 μM.

49]. In control interphase cells, microtubules are arranged in long fibres that fill the entire cytosol, growing from a central microtubule organizing centre (MTOC) corresponding to the likely position of the centrosome. In cells undergoing apoptosis, this radial network disappears and is replaced by a cortical arrangement of microtubules corresponding to the AMN observed by immunofluorescence of fixed cells. Initially, interphase microtubules are depolymerized while cells rounded up in the early stages of the execution phase of apoptosis. However, microtubules are soon reorganized beneath plasma membrane with a characteristic cortical localization.

Under physiological conditions, cytoskeleton proteins support plasma membrane integrity. Therefore, changes in the cytoskeletal components beneath plasma membrane can increase membrane permeability. During the execution phase of apoptosis, both the cell cortical actin network and intermediate filaments which support plasma membrane become depolymerized. Therefore, apoptotic microtubules are the only remaining cytoskeletal component supporting plasma membrane and cell shape during apoptosis. The organization of AMN beneath plasma membrane suggests that tubulin repolymerization in the execution phase of apoptosis may have a protective role, helping to maintain plasma membrane integrity and thus, delaying the transition to secondary necrosis. In fact, AMN is present in all genuinely apoptotic cells but is disrupted in cells undergoing secondary necrosis [49]. Furthermore, AMN disorganization by a short treatment with colchicine, an inhibitor of tubulin polymerization, increases cell permeability and the release of cell content into the culturing medium. In addition to a purely supporting role, AMN disorganization by colchicine treatment may also facilitate the access of caspases to essential proteins localized in plasma membrane and cellular cortex such as calcium channels and fodrin (α II-spectrin) whose cleavage could induce ionic imbalance, cellular collapse and eventually secondary necrosis [50].

5. AMN formation

As mentioned above, formation of AMN is a biphasic process: first, during the early phase of apoptosis, interphase microtubules rapidly depolymerized, but soon after actin and intermediate filaments disassemble they are reorganized in extensive bundles of closely packed new tubulin polymers. The initial microtubule depolymerization phase correlate with the loss of centrosomal γ-tubulin, suggesting that the two events may be interconnected [48]. The mechanisms involved in centrosome disorganization remain unknown. One hypothesis is that pericentriolar proteins can be cleaved by active caspases, but to our knowledge, none of these proteins has been identified as caspase targets [51, 52]. Interestingly, it has been demonstrated that dynein, a microtubule motor protein, is essential for the centrosomal localization of pericentrin and γ-tubulin in living cells [53]. Cytoplasmic dynein function is abolished by caspase cleavage during the execution phase [54]. Therefore, an alternative hypothesis is that dynein hydrolysis reduces the content of pericentrin and γ-tubulin at the centrosome, thereby impairing its capacity to nucleate microtubules.

On the other hand, the mechanism of microtubules reassembly in the execution phase remains uncertain. Although the core centrioles remain essentially intact throughout apoptosis, they are unlikely to direct the formation of the novel apoptotic microtubule array, because they are not assembled with a radial pattern, and instead appear randomly throughout the peripheral cytoplasm. Furthermore, apoptotic microtubules assembly takes place in the absence of γ-tubulin ring complex, suggesting that AMN formation is produced by another unknown mechanism [48, 49]. Although they are tightly packed, apoptotic microtubules are dynamic—assessed by tracking the plus-end protein EB1 by time-lapse imaging [48]—indicating that their polymerization is regulated.

It has been postulated that active caspases may cleave the C-terminal regulatory regions of tubulins which increases their ability to polymerize and thus facilitate the formation of apoptotic microtubules [52, 55]. However, AMN reorganization during the execution phase of apoptosis has been also observed in the presence of caspase inhibitors [49].

In another approach, Jon Lane's group has described that active GTP-bound Ran is necessary for apoptotic microtubule polymerization, and that RanGTP release into the apoptotic cytoplasm triggers microtubule nucleation [56]. They showed that RanGTP-activated spindle assembly factor, TPX2 (targeting protein for Xklp2), escapes from the nucleus during the execution phase and associates with apoptotic microtubule bundles [57]. Furthermore, silencing of TPX2 expression by siRNA impairs apoptotic microtubule polymerization. They propose that apoptotic microtubule polymerization shares several common characteristics with mitotic and meiotic spindle assembly, with a particular dependence upon RanGTP and TPX2 [56]. These findings suggest that apoptotic cells utilize the RanGTPase pathway to promote the reorganization of apoptotic microtubules. In another study, the examination of apoptotic microtubules components has showed that in addition to the expected tubulin subunits, they bind other microtubule-associated proteins (MAP) such as MAP-4. These findings may be interesting for elucidating the role of MAPs in AMN nucleation. Given previous evidences associating MAP4 with microtubule nucleation and stabilization [58, 59], this protein may participate in AMN formation and maintenance during apoptosis.

Although AMN lacks the morphological and functional accuracy of the mitotic/meiotic spindle apparatus, it nevertheless represents a model of regulated non-centrosomal microtubule polymerization, and further accentuates how apoptosis should be viewed as regulated process of cellular death.

6. Apoptotic microtubules delimit an active-caspase–free area in the cellular cortex

AMN indeed may work as physical barrier impeding active caspases to access into the cellular cortex where it can cleave critical proteins involved in plasma membrane integrity [60]. AMN disorganization in apoptotic cells by colchicine, a microtubule depolymerizing agent, allowed caspase-mediated cleavage of plasma membrane and cell cortex and proteins such as focal adhesion kinase (FAK), E-cadherin, α-spectrin, paxilin, Na^+/Ca^{2+} exchanger (NCX), plasma membrane Ca^{2+} ATPase-4 (PMCA-4), β4 integrin and Na^+/K^+ pump subunit β. This caspase-mediated proteolysis was associated with increase cell permeability, calcium and sodium overload and bioenergetics failure that eventually led to secondary necrosis [50]. The essential role of caspase-mediated cleavage of plasma membrane and cortical proteins in plasma membrane permeabilization was demonstrated because the concomitant addition of colchicine and Z-VAD, a pan-caspase inhibitor, blocked protein cleavage and significantly reduced plasma membrane permeability and secondary necrosis.

7. Apoptotic cells with AMN enhance phosphatidylserine exposure and interactions with macrophages

Clearance of apoptotic cells by phagocytes (or efferocytosis) can be divided into four distinct processes: aggregation of phagocytes near apoptotic cells, recognition of apoptotic cells by cell surface bridge molecules and receptors, engulfment of apoptotic cells, and degradation of apoptotic cells within phagocytes [61]. The elimination of apoptotic cells by macrophages reduces the probability of inflammation by ensuring that apoptotic cells are eliminated before the release of intracellular contents into de extracellular medium [62, 63]. Apoptotic cells are recognized by phagocytosis through the externalization of phosphatidylserine in the outer leaflet of the plasma membrane [64]. Phosphatidylserine translocation is an early event of apoptosis, occurring while the plasma membrane remains intact and cells exclude membrane-impermeant dyes [65]. Phosphatidylserine exposure has been reported to be a caspase and energy-dependent process [66, 67], but its mechanism is not completely understood. It has been proposed that a combined effect of activation of a lipid scramblase and downregulation of a phospholipid translocase activity may contribute to phosphatidylserine exposure [68].

In agreement with a role of apoptotic microtubules for proper phosphatidylserine translocation, it has been shown that apoptotic cells with AMN show indeed high expression of phosphatidylserine on the cell surface and increased phagocytosis rate. However, both processes were markedly reduced when AMN was depolymerized by colchicine treatment [60]. Interestingly, phosphatidylserine externalization and phagocytosis of apoptotic cells were restored when AMN was depolymerized in the presence of Z-VAD, suggesting that caspase-dependent degradation of plasma membrane and cellular cortex proteins impairs proper phosphatidylserine externalization and apoptotic cell removal by macrophages. These findings

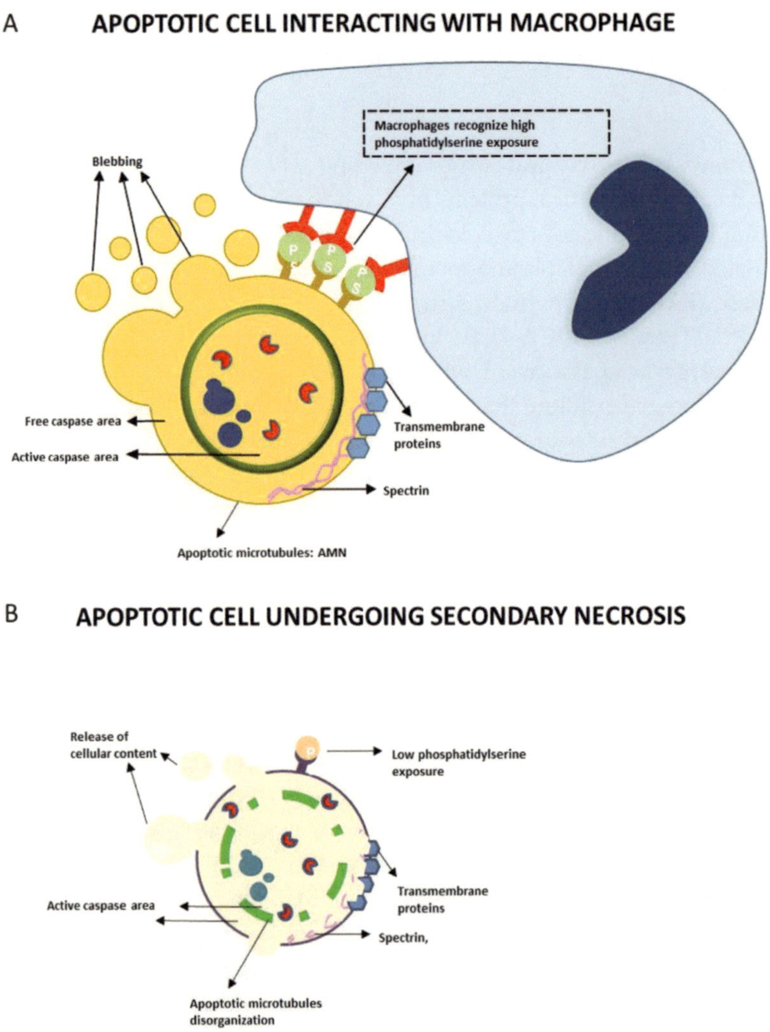

Figure 3. Scheme summarizing the main findings on AMN during the execution phase of apoptosis. (A) Apoptotic cell interacting with macrophage (B) secondary necrotic cell. PS = Phosphatidylserine.

corroborate previous observations showing that after AMN disorganization (by nocodazole treatment) the percentage of macrophages making contacts and engulfing apoptotic cells was significantly reduced compared to apoptotic cells with AMN [48]. The ability of apoptotic cells to stimulate their phagocytosis by macrophages before cell lysis is crucial to prevent the adverse effects (tissue damage and inflammation) associated with secondary necrosis [69] (**Figure 3**).

8. Apoptotic microtubules organization and maintenance depend on high cellular ATP levels and energized mitochondria

Microtubule polymerization is an energy-dependent process because β-tubulin hydrolyzes GTP during polymerization [70]. Therefore, it has been proposed that AMN formation depends on the bioenergetic status of apoptotic cells [71].

ATP levels must be kept high in apoptosis to allow all the energy-dependent processes occurring during the execution phase including AMN formation and maintenance. Thus, in vivo and in vitro experiments have shown that AMN was visualized predominantly in apoptotic cells with polarized/hyperpolarized mitochondria and, on the contrary, was dismantle in apoptotic cells with depolarized mitochondria. These observations suggest that AMN depends on energized mitochondria and high ATP levels [71]. Kinetics examination in pig LLCPK-1α cells expressing GFP-tubulin also showed that AMN was maintained during the execution phase of apoptosis until mitochondria depolarization marked the onset of secondary necrosis. Furthermore, mitochondria depolarization by treatments with uncouplers of mitochondrial oxidative phosphorylation (FCCP) or mitochondrial inhibitors (antimycin, rotenone and oligomycin) induced AMN disassembly associated with enhanced plasma membrane permeability. However, inhibition of glycolysis by 2-deoxyglucose treatment had no effect on mitochondrial polarization or either AMN organization or plasma membrane permeability. In contrast, stabilization of apoptotic microtubules by taxol prevented both mitochondrial depolarization and plasma membrane permeabilization. AMN stabilization also prevented the increased plasma membrane permeability when mitochondria were depolarized by rotenone or FCCP treatment. These results underline the essential role of AMN in plasma membrane integrity during apoptosis.

9. Zombie cells: Stabilization of apoptotic cells

Taken into account that apoptotic cells maintain the integrity of plasma membrane and cellular cortex proteins [60], an innovative method aimed to the temporal stabilization and preservation of apoptotic cells has been developed [72]. This method consists in the treatment of apoptotic cells with a cocktail of taxol, Zn^{2+} and coenzyme Q_{10} (CoQ). This experimental approach has been reported to prevent secondary necrosis for at least 96 h in cell cultures. The rationale for using this stabilizing cocktail is (a) taxol, a microtubule stabilizing agent, prevents AMN depolymerization and the access of active caspases into cellular cortex [73, 74];

(b) Zn^{2+}, a caspase inhibitor, prevents caspase-dependent cleavage of cellular cortex and plasma membrane proteins [75–78]; and (c) CoQ, an antioxidant, that protects against oxidative membrane damage which is increased in apoptotic cells [79].

Stabilized apoptotic cells can be considered as dying cells in which the cellular cortex and plasma membrane are intact or alive. Metaphorically, they can be considered as "living dead" or "zombie cells". Stabilized apoptotic cells retain many of the hallmarks characteristic of apoptotic cells such as cellular cortex and plasma membrane integrity, low intracellular calcium levels, plasma membrane potential, high phosphatidylserine exposure and the ability of being engulfed by phagocytes.

Recently, interest in apoptosis research has increased remarkably for a number of reasons including the technological development of cell cultures and the expansion of new therapeutic strategies. Furthermore, apoptotic cell quantification plays an important role in biomedicine because it is widely used to evaluate the cytotoxic effects of drugs [80]. However, apoptosis determination is often affected by the process of cell manipulation (harvesting, cell centrifugation, cell pipetting…), especially in adherent cell cultures, required for flow cytometry assays. Very often, these manipulations disrupt plasma membrane permeability and leads apoptotic cells to secondary necrosis [81]. As a consequence, reliable apoptosis quantifications are particularly difficult in adherent cell cultures. Stabilization of apoptotic cells before cell harvesting may allow a more accurate and reliable quantification of the actual number of apoptotic cells or the correct determination of biochemical parameters such as mitochondrial membrane potential, intracellular calcium concentration, pH or caspase activity in genuine apoptotic cells.

Currently, apoptotic cells are used for various forms of therapy, especially with the objective of promoting immunological tolerance in recipient individuals [82]. Therefore, stabilization of apoptotic cells before their administration to patients may ensure that apoptotic cells will retain their characteristic features until they are removed by macrophages. The administration of stabilized apoptotic cells can also be of interest for the delivery of proteins (for protein replacement therapy) or drugs to recipient macrophages [83].

There are forms of cell death which by their nature impair the correct formation of AMN (e.g.mitochondrial toxics and cold exposure) [71] and, as a result, apoptotic cells are not able to maintain plasma membrane integrity. Therefore, apoptotic cell stabilization may provide a new approach for preventing the adverse effects of early secondary necrosis.

10. Conclusion

Microtubule cytoskeleton is reformed during the execution phase of apoptosis forming an AMN. AMN is required to maintain plasma membrane integrity and cell morphology during the execution phase of apoptosis. AMN disruption leads cells to secondary necrosis and the release of toxic molecules which can damage neighbour cells. Therefore, AMN formation, preservation or stabilization in apoptosis is essential for tissue homeostasis preventing cell damage and inflammation.

http://dx.doi.org/10.5772/66865

Acknowledgements

This work was supported by FIS PI13/00129 grant, Instituto de Salud Carlos III, Spain and Fondo Europeo de Desarrollo Regional (FEDER-Unión Europea), Proyecto de Investigación de Excelencia de la Junta de Andalucía CTS-5725, and by AEPMI (Asociación de Enfermos de Patología Mitocondrial) and ENACH (Asociación de Enfermedades Neurodegenerativas por Acumulación Cerebral de Hierro).

The authors submitted a patent application describing the stabilization of apoptotic cells.

Author details

Jesús Porcuna Doncel, Patricia de la Cruz Ojeda, Manuel Oropesa-Ávila, Marina Villanueva Paz, Isabel De Lavera, Mario De La Mata, Mónica Álvarez Córdoba, Raquel Luzón Hidalgo, Juan Miguel Suarez Rivero, David Cotán and José Antonio Sánchez-Alcázar*

*Address all correspondence to: jasanalc@upo.es

Andalusian Centre for Developmental Biology, CSIC, University Pablo de Olavide, Seville, Spain

References

[1] Oropesa M, De La Mata M, Maraver JG, Cordero MD, Cotán D, Rodríguez-Hernández A, et al. Apoptotic microtubule network organization and maintenance depend on high cellular ATP levels and energized mitochondria. Apoptosis. 2011;16:404–24.

[2] Kerr JF, Wyllie AH, Currie AR. Apoptosis: a basic biological phenomenon with wide-ranging implications in tissue kinetics. British Journal of Cancer. 1972; 26:239–57.

[3] Fuchs Y, Steller H. Programmed cell death in animal development and disease. Cell. 2014;147:742–58.

[4] Elmore S. Apoptosis: a review of programmed cell death. Toxicologic Pathology. 2007; 35:495–516.

[5] Logue SE, Martin SJ. Caspase activation cascades in apoptosis. Biochemical Society Transactions. 2008;36:1–9.

[6] Krysko DV, D'Herde K, Vandenabeele P. Clearance of apoptotic and necrotic cells and its immunological consequences. Apoptosis. 2006;11:1709–26.

[7] Silva MT. Secondary necrosis: The natural outcome of the complete apoptotic program. FEBS Letters. 2010;584:4491–9.

[8] Poon IKH, Hulett MD, Parish CR. Molecular mechanisms of late apoptotic/necrotic cell clearance. Cell Death and Differentiation. 2010;17:381–97.

[9] Perruche S, Saas P. Immunomodulatory properties of apoptotic cells. Presse Medicale. 2013;42:537–43.

[10] Mills JC, Stone NL, Pittman RN. Extranuclear apoptosis: The role of the cytoplasm in the execution phase. Journal of Cell Biology. 1999;146:703–7.

[11] Ndozangue-Touriguine O, Hamelin J, Brèard J. Cytoskeleton and apoptosis. Biochemical Pharmacology. 2008;76:11–8.

[12] Moss DK, Lane JD. Microtubules: forgotten players in the apoptotic execution phase. Trends in Cell Biology. 2006;16:330–8.

[13] Desouza M, Gunning PW, Stehn JR. The actin cytoskeleton as a sensor and mediator of apoptosis. Bioarchitecture. 2012;2:75–87.

[14] van Engeland M, Kuijpers HJ, Ramaekers FC, Reutelingsperger CP, Schutte B. Plasma membrane alterations and cytoskeletal changes in apoptosis. Experimental Cell Research. 1997;235:421–30.

[15] Sánchez-Alcázar JA, Rodríguez-Hernández n, Cordero MD, Fernández-Ayala DJM, Brea-Calvo G, Garcia K, et al. The apoptotic microtubule network preserves plasma membrane integrity during the execution phase of apoptosis. Apoptosis. 2007;12:1195–208.

[16] Kapur U, Wojcik EM. Follicular neoplasm of the thyroid – vanishing cytologic diagnosis? Diagnostic Cytopathology. 2007;35:525–8.

[17] Fuchs E, Yang Y. Crossroads on cytoskeletal highways. Cell. 1999;98:547–50.

[18] Marceau N, Schutte B, Gilbert S, Loranger A, Henfling MER, Broers JLV, et al. Dual roles of intermediate filaments in apoptosis. Experimental Cell Research. 2007;313:2265–81.

[19] Weerasinghe SVW, Ku N-O, Altshuler PJ, Kwan R, Omary MB. Mutation of caspase-digestion sites in keratin 18 interferes with filament reorganization, and predisposes to hepatocyte necrosis and loss of membrane integrity. Journal of Cell Science. 2014;127:1464–75.

[20] Broers JL, Ramaekers FC. The role of the nuclear lamina in cancer and apoptosis. Advances in Experimental Medicine and Biology. 2014;773:27–48.

[21] Ávila MO, Vega AF, Maraver JG, Paz MV, Lavera ID, Mata MDL, et al. The Apoptotic Microtubule Network During the Execution Phase of Apoptosis. Cytoskeleton (Hoboken). 2015; 72:435–46.

[22] Galjart N, Perez F. A plus-end raft to control microtubule dynamics and function. Current Opinion in Cell Biology. 2002;15:48–53.

[23] Andersen SSL. Spindle assembly and the art of regulating microtubule dynamics by MAPs and Stathmin/Op18. Trends in Cell Biology. 2000;10:261–7.

[24] Carazo-Salas RE, Gruss OJ, Mattaj IW, Karsenti E. Ran-GTP coordinates regulation of microtubule nucleation and dynamics during mitotic-spindle assembly. Nature Cell Biology. 2001;3:228–34.

[25] Lamb NJC, Fernandez A, Watrin A, Labbé JC, Cavadore JC. Microinjection of p34cdc2 kinase induces marked changes in cell shape, cytoskeletal organization, and chromatin structure in mammalian fibroblasts. Cell. 1990;60:151–65.

[26] Ookata K, Hisanaga S, Bulinski JC, Murofushi H, Aizawa H, Itoh TJ, et al. Cyclin B interaction with microtubule-associated protein 4 (MAP4) targets p34 eda kinase to microtubules and is a potential regulator of M-phase microtubule dynamics. Cell. 1995;128:849–62.

[27] Golsteyn RM. Cdk1 and Cdk2 complexes (cyclin dependent kinases) in apoptosis: A role beyond the cell cycle. Cancer Letters. 2005;217:129–38.

[28] Mills JC, Lee VM, Pittman RN. Activation of a PP2A-like phosphatase and dephosphorylation of tau protein characterize onset of the execution phase of apoptosis. Journal of Cell Science. 1998;111(Pt 5):625–36.

[29] Jiang Y. Regulation of the cell cycle by protein phosphatase 2A in Saccharomyces cerevisiae. Microbiology and Molecular Biology Reviews: MMBR. 2006;70:440–9.

[30] Gourlay CW, Ayscough KR. The actin cytoskeleton in ageing and apoptosis. FEMS Yeast Research. 2005;5:1193–8.

[31] Levkau B, Herren B, Koyama H, Ross R, Raines EW. Caspase-mediated cleavage of focal adhesion kinase pp125FAK and disassembly of focal adhesions in human endothelial cell apoptosis. The Journal of Experimental medicine. 1998;187:579–86.

[32] Coleman, Olson. Rho GTPase signalling pathways in the morphological changes associated with apoptosis. Cell Death and Differentiation. 2002;9:493–504.

[33] Julian L, Olson MF. Rho-associated coiled-coil containing kinases (ROCK): structure, regulation, and functions. Small GTPases. 2014;5:e29846.

[34] Sebbagh M, Renvoizé C, Hamelin J, Riché N, Bertoglio J, Bréard J. Caspase-3-mediated cleavage of ROCK I induces MLC phosphorylation and apoptotic membrane blebbing. Nature Cell Biology. 2001;3:346–52.

[35] Maekawa M, Ishizaki T, Boku S, Watanabe N, Fujita a, Iwamatsu a, et al. Signaling from Rho to the actin cytoskeleton through protein kinases ROCK and LIM-kinase. Science (New York, NY). 1999;285:895–8.

[36] Coleman ML, Sahai Ea, Yeo M, Bosch M, Dewar a, Olson MF. Membrane blebbing during apoptosis results from caspase-mediated activation of ROCK I. Nature Cell Biology. 2001;3:339–45.

[37] Dong LQ, Landa LR, Wick MJ, Zhu L, Mukai H, Ono Y, et al. Phosphorylation of protein kinase N by phosphoinositide-dependent protein kinase-1 mediates insulin signals to the

actin cytoskeleton. Proceedings of the National Academy of Sciences of the United States of America. 2000;97:5089–94.

[38] Takahashi M, Mukai H, Toshimori M, Miyamoto M, Ono Y. Proteolytic activation of PKN by caspase-3 or related protease during apoptosis. Proceedings of the National Academy of Sciences. 1998;95:11566–71.

[39] Bokoch GM. Caspase-mediated activation of PAK2 during apoptosis: proteolytic kinase activation as a general mechanism of apoptotic signal transduction? Cell Death and Differentiation. 1998;5:637–45.

[40] Geng YJ, Azuma T, Tang JX, Hartwig JH, Muszynski M, Wu Q, et al. Caspase-3-induced gelsolin fragmentation contributes to actin cytoskeletal collapse, nucleolysis, and apoptosis of vascular smooth muscle cells exposed to proinflammatory cytokines. European Journal of Cell Biology. 1998;77:294–302.

[41] Gu Y, Rosenblatt J. New emerging roles for epithelial cell extrusion. Current Opinion Cell Biology. 2012;24(6):865–70.

[42] Rosenblatt J, Raff MC, Cramer LP. An epithelial cell destined for apoptosis signals its neighbors to extrude it by an actin- and myosin-dependent mechanism. Current Biology. 2001;11(23):1847–57.

[43] Gu Y, Forostyan T, Sabbadini R, Rosenblatt J. Epithelial cell extrusion requires the sphingosine-1-phosphate receptor 2 pathway. Journal of Cell Biology. 2011;193(4):667–76.

[44] Kuipers D, Mehonic A, Kajita M, Peter L, Fujita Y, Duke T, et al. Epithelial repair is a two-stage process driven first by dying cells and then by their neighbours. Journal of Cell Science. 2014;127(Pt 6):1229–41.

[45] Monier B, Gettings M, Gay G, Mangeat T, Schott S, Guarner A, et al. Apico-basal forces exerted by apoptotic cells drive epithelium folding. Nature. 2015;518(7538):245–8.

[46] Teng X, Toyama Y. Apoptotic force: active mechanical function of cell death during morphogenesis. Development, Growth & Differentiation. 2011;53(2):269–76.

[47] Monier B, Suzanne M. The morphogenetic role of apoptosis. Current Topics in Developmental Biology. 2015;114:335–62.

[48] Moss DK, Betin VM, Malesinski SD, Lane JD. A novel role for microtubules in apoptotic chromatin dynamics and cellular fragmentation. Journal of Cell Science. 2006;119(Pt 11):2362–74.

[49] Sanchez-Alcazar JA, Rodriguez-Hernandez A, Cordero MD, Fernandez-Ayala DJ, Brea-Calvo G, Garcia K, et al. The apoptotic microtubule network preserves plasma membrane integrity during the execution phase of apoptosis. Apoptosis. 2007;12(7):1195–208.

[50] Schwab BL, Guerini D, Didszun C, Bano D, Ferrando-May E, Fava E, et al. Cleavage of plasma membrane calcium pumps by caspases: a link between apoptosis and necrosis. Cell Death and Differentiation. 2002;9(8):818–31.

[51] Fischer U, Janicke RU, Schulze-Osthoff K. Many cuts to ruin: A comprehensive update of caspase substrates. Cell Death and Differentiation. 2003;10(1):76–100.

[52] Gerner C, Frohwein U, Gotzmann J, Bayer E, Gelbmann D, Bursch W, et al. The Fas-induced apoptosis analyzed by high throughput proteome analysis. Journal of Biological Chemistry. 2000;275(50):39018–26.

[53] Young A, Dictenberg JB, Purohit A, Tuft R, Doxsey SJ. Cytoplasmic dynein-mediated assembly of pericentrin and gamma tubulin onto centrosomes. Molecular Biology of the Cell. 2000;11(6):2047–56.

[54] Lane JD, Vergnolle MA, Woodman PG, Allan VJ. Apoptotic cleavage of cytoplasmic dynein intermediate chain and p150(Glued) stops dynein-dependent membrane motility. Journal of Cell Biology. 2001;153(7):1415–26.

[55] Adrain C, Duriez PJ, Brumatti G, Delivani P, Martin SJ. The cytotoxic lymphocyte protease, granzyme B, targets the cytoskeleton and perturbs microtubule polymerization dynamics. The Journal of Biological Chemistry. 2006;281(12):8118–25.

[56] Moss DK, Wilde A, Lane JD. Dynamic release of nuclear RanGTP triggers TPX2-dependent microtubule assembly during the apoptotic execution phase. Journal of Cell Science. 2009;122(Pt 5):644–55.

[57] Wittmann T, Wilm M, Karsenti E, Vernos I. TPX2, A novel xenopus MAP involved in spindle pole organization. Journal of Cell Biology. 2000;149(7):1405–18.

[58] Shiina N, Tsukita S. Regulation of microtubule organization during interphase and M phase. Cell Structure Function. 1999;24(5):385–91.

[59] Tokuraku K, Katsuki M, Nakagawa H, Kotani S. A new model for microtubule-associated protein (MAP)-induced microtubule assembly: The Pro-rich region of MAP4 promotes nucleation of microtubule assembly in vitro. European Journal of Biochemistry. 1999;259 (1):158–66.

[60] Oropesa-Avila M, Fernandez-Vega A, de la Mata M, Maraver JG, Cordero MD, Cotan D, et al. Apoptotic microtubules delimit an active caspase free area in the cellular cortex during the execution phase of apoptosis. Cell Death & Disease. 2013;4:e527.

[61] Erwig LP, Henson PM. Clearance of apoptotic cells by phagocytes. Cell Death and Differentiation. 2008;15(2):243–50.

[62] Savill J, Fadok V. Corpse clearance defines the meaning of cell death. Nature. 2000;407 (6805):784–8.

[63] Voll RE, Herrmann M, Roth EA, Stach C, Kalden JR, Girkontaite I. Immunosuppressive effects of apoptotic cells. Nature. 1997;390(6658):350–1.

[64] Fadok VA, Bratton DL, Frasch SC, Warner ML, Henson PM. The role of phosphatidylserine in recognition of apoptotic cells by phagocytes. Cell Death and Differentiation. 1998;5(7):551–62.

[65] Fink SL, Cookson BT. Apoptosis, pyroptosis, and necrosis: mechanistic description of dead and dying eukaryotic cells. Infection and Immunity. 2005;73(4):1907–16.

[66] Castedo M, Hirsch T, Susin SA, Zamzami N, Marchetti P, Macho A, et al. Sequential acquisition of mitochondrial and plasma membrane alterations during early lymphocyte apoptosis. Journal of Immunology. 1996;157(2):512–21.

[67] Nicotera P, Leist M, Ferrando-May E. Intracellular ATP, a switch in the decision between apoptosis and necrosis. Toxicology Letters. 1998;102-103:139–42.

[68] Moreira ME, Barcinski MA. Apoptotic cell and phagocyte interplay: recognition and consequences in different cell systems. Anais da Academia Brasileira de Ciencias. 2004;76(1):93–115.

[69] Wyllie AH. Glucocorticoid-induced thymocyte apoptosis is associated with endogenous endonuclease activation. Nature. 1980;284(5756):555–6.

[70] Duanmu C, Lin CM, Hamel E. Tubulin polymerization with ATP is mediated through the exchangeable GTP site. Biochimica et Biophysica Acta. 1986;881(1):113–23.

[71] Oropesa M, de la Mata M, Maraver JG, Cordero MD, Cotan D, Rodriguez-Hernandez A, et al. Apoptotic microtubule network organization and maintenance depend on high cellular ATP levels and energized mitochondria. Apoptosis. 2011;16(4):404–24.

[72] Oropesa-Avila M, Fernandez-Vega A, de la Mata M, Garrido-Maraver J, Cotan D, Paz MV, et al. Apoptotic cells subjected to cold/warming exposure disorganize apoptotic microtubule network and undergo secondary necrosis. Apoptosis. 2014;19(9):1364–77.

[73] Schiff PB, Fant J, Horwitz SB. Promotion of microtubule assembly in vitro by taxol. Nature. 1979;277(5698):665–7.

[74] Schiff PB, Horwitz SB. Taxol stabilizes microtubules in mouse fibroblast cells. Proceedings of the National Academy Science of the U S A. 1980;77(3):1561–5.

[75] Huber KL, Hardy JA. Mechanism of zinc-mediated inhibition of caspase-9. Protein Science: A Publication of the Protein Society. 2012;21(7):1056–65.

[76] Perry DK, Smyth MJ, Stennicke HR, Salvesen GS, Duriez P, Poirier GG, et al. Zinc is a potent inhibitor of the apoptotic protease, caspase-3. A novel target for zinc in the inhibition of apoptosis. The Journal of Biological Chemistry. 1997;272(30):18530–3.

[77] Smith AF, Longpre J, Loo G. Inhibition by zinc of deoxycholate-induced apoptosis in HCT-116 cells. Journal of Cellular Biochemistry. 2012;113(2):650–7.

[78] Stennicke HR, Salvesen GS. Biochemical characteristics of caspases-3, -6, -7, and -8. The Journal of Biological Chemistry. 1997;272(41):25719–23.

[79] Bentinger M, Brismar K, Dallner G. The antioxidant role of coenzyme Q. Mitochondrion. 2007;7(Suppl):S41–50.

[80] Darzynkiewicz Z, Bruno S, Del Bino G, Gorczyca W, Hotz MA, Lassota P, et al. Features of apoptotic cells measured by flow cytometry. Cytometry. 1992;13(8):795–808.

[81] van Engeland M, Ramaekers FC, Schutte B, Reutelingsperger CP. A novel assay to measure loss of plasma membrane asymmetry during apoptosis of adherent cells in culture. Cytometry. 1996;24(2):131–9.

[82] Saas P, Kaminski S, Perruche S. Prospects of apoptotic cell-based therapies for transplantation and inflammatory diseases. Immunotherapy. 2013;5(10):1055–73.

[83] Perez B, Paquette N, Paidassi H, Zhai B, White K, Skvirsky R, et al. Apoptotic cells can deliver chemotherapeutics to engulfing macrophages and suppress inflammatory cytokine production. The Journal of Biological Chemistry. 2012;287(19):16029–36.

Chapter 9

How are Dynamic Microtubules Stably Tethered to Human Chromosomes?

Duccio Conti, Madeleine Hart, Naoka Tamura,
Roshan Shrestha, Asifa Islam and Viji M. Draviam

Additional information is available at the end of the chapter

http://dx.doi.org/10.5772/intechopen.68321

Abstract

During cell division, microtubules capture and pull chromosomes apart into two equal sets. Without the establishment of proper chromosome-microtubule attachment, microtubules cannot impart the pulling forces needed to separate sister chromatid pairs. How are chromosomes captured along microtubule walls? How is the attachment of chromosomes to dynamic microtubule-ends achieved and monitored? We discuss these key questions by considering the roles of kinetochore-bound microtubule regulating proteins and also the complex regulatory loops of kinases and phosphatases that control chromosome-microtubule attachment and ensure the accurate segregation of chromosomes.

Keywords: chromosome segregation, microtubules, mitosis, kinetochore, mitotic spindle

1. Introduction

When a human cell prepares to divide, its microtubule cytoskeletal network disassembles and reassembles to form a bipolar structure—the mitotic spindle. Microtubules of the mitotic spindle capture chromosomes and segregate the DNA into two equal sets (**Figure 1**). To ensure accurate segregation, the proper attachment between the chromosome and microtubule is important. Chromosome-microtubule attachment is facilitated through a submicron-sized macromolecular structure called the kinetochore. The kinetochore appears as a three-layered structure in electron microscopy: an outer layer that contacts microtubules, a middle layer or interzone and an inner plate assembled on centromeric chromatin. In humans, multiple microtubules engage with the kinetochore, and kinetochore-bound microtubules are bundled into k-fibers. Understanding how microtubules are tethered to kinetochores is not only fascinating but also clinically important. We could exploit some of the redundant regulatory mechanisms in chromosome-microtubule attachment to specifically target properties of cancer cells displaying chromosomal instability (reviewed in Ref. [1]).

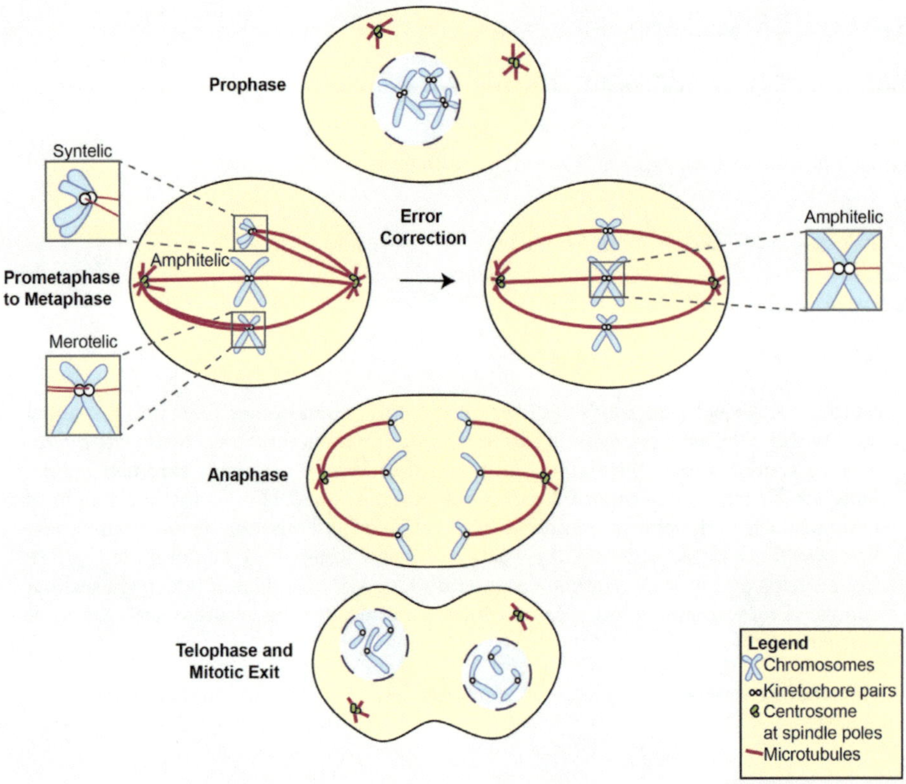

Figure 1. Proper kinetochore-microtubule attachment is needed for the accurate segregation of chromosomes: in prophase, the nuclear envelope breaks down and the kinetochore is poised for microtubule capture. Attachments between microtubules from opposing spindle poles to chromosomes must be bioriented: one kinetochore of a pair is tethered to one spindle pole and the other sister kinetochore of the pair is attached to the opposing pole. Errors in biorientation—syntelic or merotelic attachments—are resolved through the error correction process during prometaphase and metaphase stages. Only after all the kinetochores have achieved biorientation, cells initiate anaphase and pull sister kinetochores apart into two equal sets.

Human kinetochores become available for capture by microtubules soon after nuclear envelope breakdown, when chromosomes are first exposed to the cytoplasm. Kinetochores are initially captured along the lateral walls of microtubules and then tethered to microtubule-ends [2–4], by a multi-step process called end-on conversion. Such a change in the *plane* of kinetochore-microtubule attachment is an important event. Only when kinetochores are tethered to the ends of microtubules, the growth and shrinkage of microtubules can be translated into pushing or pulling forces that move chromosomes. In addition to the proper *plane* of kinetochore-microtubule attachment, proper *orientation* of the attachment is also crucial: one sister kinetochore of a pair must be attached to one spindle pole and the other sister kinetochore of the pair must be attached to the opposing spindle pole. This *orientation* of chromosome-microtubule attachment is called biorientation or an amphitelic attachment (**Figure 1**). When *all* kinetochores are attached in an amphitelic fashion, the cell initiates anaphase, allowing sister chromatids to be synchronously pulled apart into two daughter sets.

To ensure that cells do not prematurely initiate anaphase, the status of kinetochore-microtubule (KT-MT) attachment is continuously monitored by a complex, evolutionarily conserved Spindle Assembly Checkpoint (referred to as SAC) mechanism. Whether the SAC monitors the physical geometry of KT-MT attachment or the outcome of the correct attachment geometry (intra or inter-kinetochore tension) is not fully resolved (reviewed in Refs. [5, 6]).

In this chapter, we focus on events that occur at the kinetochore-microtubule interface. First, we introduce the key molecular components that form the kinetochore-microtubule interface, with a focus on microtubule-associated proteins that bridge the kinetochore and microtubule polymer. Second, we discuss how the plane and orientation of kinetochore-microtubule interaction is correctly established. Finally, we review our current knowledge of phosphorylation-dephosphorylation cycles that control KT-MT attachments and in turn ensure the accurate segregation of chromosomes.

2. The kinetochore-microtubule interface

Over 100 proteins are now known to form the human kinetochore (reviewed in Ref. [7]). The major constituent at the outer kinetochore surface that is responsible for the connection between microtubules and chromosomes is the 10-subunit KMN network, made up of Knl1, Mis12, and Ndc80 complexes. Each of these complexes are composed of proteins, which are evolutionarily conserved from yeasts to humans: the Knl1 complex consists of Zwint1 and KNL1 (hSpc105), the Mis12 complex consists of Mis12, Dsn1, Nnf1 and Nsl1, and the Ndc80 complex consists of Ndc80 (HEC1), Nuf2, Spc24, and Spc25 (reviewed in Ref. [7]). Positioned at the outer face of the KMN complex is the Ndc80-Nuf2 complex, which makes direct contact with microtubules [8, 9].

The KMN network of proteins is recruited to the kinetochore in early mitosis [10–13]. KMN recruitment is mediated by the multi-subunit CCAN—Constitutive Centromere Associated Network—that forms the core structure of centromeres (specialized chromatin regions) [14]. A recent biochemical reconstitution study showed that the KMN network and a seven-subunit CCAN complex, the CHIKMLN complex, along with CENP-A (Histone H3 variant) bound nucleosomes, is sufficient to bridge the DNA and microtubules [15]. Thus, the KMN network bridges the inner centromeric DNA-bound proteins with the microtubule polymer. In addition, the KMN network of proteins acts as a platform for several proteins that monitor and control kinetochore-microtubule attachment status: regulators of the spindle assembly checkpoint (Mad1, Mad2, Bub1, Bub3, BubR1, RZZ complex, Mps1, Aurora-B), microtubule associated nonmotor proteins (for example, CLASP, CLIP170, EB1, APC, ch-TOG, SKA complex, CENP-F and Astrin-SKAP complex), and motor proteins (CENP-E, Dynein/Dynactin complex) are all recruited in a KMN-dependent fashion [12, 13, 16–23]. Therefore, the KMN network is considered as an essential core component of the kinetochore.

3. Mechanisms that ensure proper chromosome-microtubule attachment

3.1. The plane of KT-MT attachment and the end-on conversion process

During early mitosis, the outer kinetochore surface appears expanded [2], and this expansion facilitates the initial interaction between the kinetochore and microtubule wall (lateral attachment). In addition, kinetochore-derived microtubules can interact with spindle-pole-derived microtubules, facilitating kinetochore attachment onto lateral walls of microtubules [24]. The steps involving the conversion of lateral attachments to end-on attachments take place during prometaphase and are collectively termed the end-on conversion process. This multi-step end-on conversion process happens similarly in humans and in yeast, where it was first described (reviewed in Ref. [25]), with a few differences. The process in human cells involves: (a) initial interaction of a kinetochore with the lateral surface of microtubules; (b) transport of the kinetochore toward centrosomes or spindle equator; (c) physical change in the plane of kinetochore interaction, from microtubule-wall to microtubule-end; (d) selective destabilization of laterally engaging kinetochore-microtubules; (e) transient stabilization of end-on attached microtubules; and (f) maintenance of stable end-on attachments after the biorientation of sister kinetochores (**Figure 2**).

The initial attachment between the kinetochore and microtubules, which occurs on the lateral microtubule walls, is an immature one. Laterally attached kinetochores can undergo movement toward and away from the spindle pole; these movements are motor dependent and microtubule-dynamics independent [26–28]. Laterally attached chromosomes should be transported away from the spindle pole and congressed at the spindle equator. This antipoleward transport is, in part, due to an ejection force that pushes chromosomes outward from the spindle pole [29]. Antipoleward forces needed to transport laterally attached kinetochores can also be facilitated by (i) electrostatic repulsion between negatively charged chromosome arms and neighboring negatively charged astral microtubule-ends [30] and (ii) kinetochore or chromosome-associated kinesin motors (reviewed in Ref. [31]).

A well-understood example is the kinetochore-bound kinesin motor protein, CENP-E [32], which facilitates the tethering of kinetochores to the lateral wall of microtubules and slides chromosomes towards the microtubule plus-end [26], providing antipoleward drag force. The CENP-E motor is brought to the kinetochore by a core kinetochore protein, CENP-Q [33] (part of the CENP-O/MCM21R complex [34]). CENP-E facilitates end-on conversion in at least two ways: first, CENP-E prolongs the lifetime of laterally attached kinetochores and thus enables efficient end-on conversion [4]. Second, by transporting kinetochores to the spindle equator, CENP-E ensures that kinetochores are not trapped behind the spindles poles, which would make biorientation geometrically impossible and lead to futile end-on conversion cycles.

http://dx.doi.org/10.5772/intechopen.68321

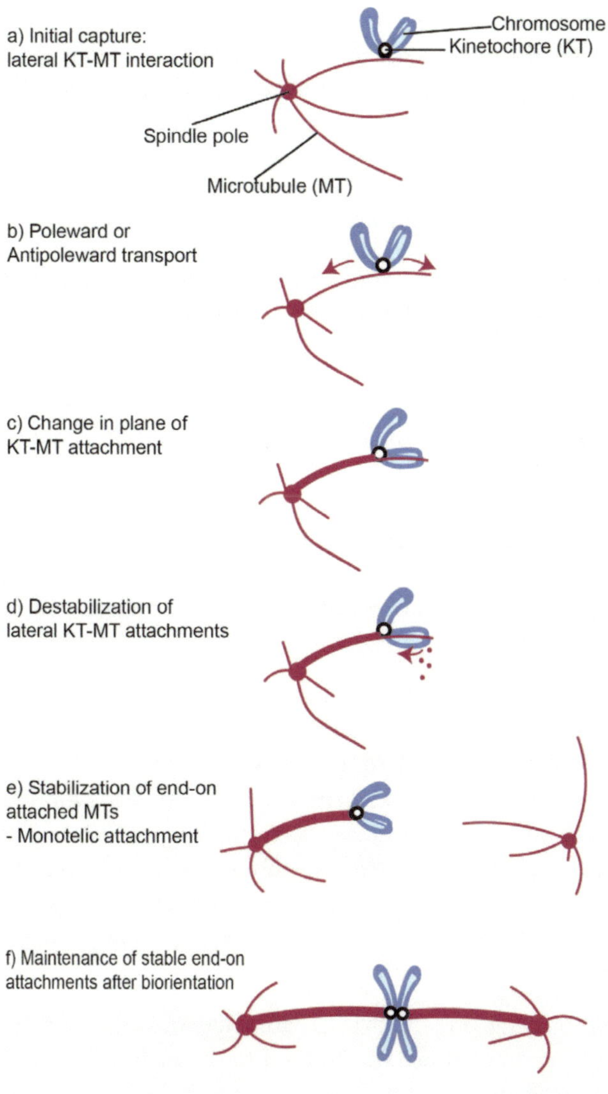

Figure 2. The end-on conversion process: Kinetochores are first captured along lateral walls of microtubules, then are tethered to the ends of microtubules, through multiple steps guided by the kinetochore and microtubule-associated proteins. The end-on conversion process enables the formation of mature end-on tethered kinetochores. In parallel, the error correction process ensures that monotelic attachments are converted into amphitelic attachments (KT-MT, kinetochore-microtubule).

Following end-on conversion, mechanisms that actively remove laterally interacting microtubules are crucial for the maintenance of end-on attachments. The microtubule depolymerizing kinesin MCAK, needed for proper kinetochore-microtubule attachments (reviewed in Ref. [35]), removes the ends of laterally interacting ends of microtubules and aids in the maintenance of a mature end-on attached kinetochore [4].

Whether microtubule wall *versus* end interactions are dominantly controlled by distinct sets of proteins in human cells remains unknown. It is clear that several Microtubule-Associated Proteins (MAPs) are recruited selectively to the immature kinetochore-microtubule interface in prometaphase, but not the mature kinetochore-microtubule interface in metaphase [36, 37]. How MAPs recognize and switch between mature and immature kinetochore states is an exciting unexplored area, which may shed light on the molecular basis underpinning the change in the plane of kinetochore-microtubule interaction.

3.2. Proteins that distinguish ends and walls of microtubules

To develop possible paradigms for how kinetochore proteins may distinguish microtubule walls *versus* microtubule-ends, we will first look at how microtubule associated proteins distinguish microtubule walls from ends.

EB1 differentiates the plus end from the rest of the microtubule wall because of its high affinity for GTP-tubulin, which is enriched at the growing end of microtubules [38–40]. Electrostatic interaction between the C-terminal tail of EB1 and microtubule walls contributes to targeting EB1 to the plus end of microtubules [41]. In addition, EB1 preferentially associates with tubulins along the lateral surface of microtubules, facilitating closure of the open protofilaments and in turn promoting the growth of microtubule ends [42]. Thus, EB1 is believed to directly identify the structural features and chemical signals of the growing end of microtubules.

At the kinetochore-microtubule interface of a mature microtubule-end tethered kinetochore, at least six well-established microtubule-binding proteins are present: Ndc80/HEC1 of the KMN network, SKAP of the Astrin-SKAP complex, the microtubule stabilizer CENP-F and microtubule polymerizers, chTOG, CLASP and the SKA complex. Ndc80 recognizes both the tubulin intradimer and interdimer interface and forms oligomeric arrays along microtubule walls [9, 43]. Similarly, SKAP recognizes intra-tubulin sites [18], enabling a direct link for the Astrin-SKAP complex between the kinetochore and microtubule wall. In addition, SKAP directly interacts with EB1 through its S-X-I-P motif [44] and could potentially link the kinetochore specifically to the growing microtubule-end.

The microtubule-end binding protein chTOG1 interacts directly with Ndc80 [23]. chTOG1 has two TOG domains and could use one to bind to the microtubule end, and the other to present an $\alpha\beta$-tubulin protein for addition to the microtubule-end, thereby promoting microtubule assembly (reviewed in [45]). Similarly, the TOG domains in CLASP1 can interact with tubulin dimers and regulate microtubule rescue (reviewed in Ref. [46]). Although the TOG domains of both chTOG1 and CLASPs bind conserved $\alpha\beta$-tubulin interacting sites, chTOG1 can directly associate with microtubules and CLASPs can be brought to microtubule ends *via* EB1's S-X-I-P motif (reviewed in [16]). In contrast, CENP-F is thought to have a strong binding to depolymerizing microtubules, as one of its two microtubule-binding domains binds strongly to bent microtubules [47–49]. In summary, the kinetochore-microtubule interface recruits a variety of MAPs that can selectively bind to either microtubule walls or ends. Regulated recruitment of kinetochore-bound MAPs can be a fundamentally important way to direct microtubule dynamics and stability.

3.3. Biorientation and the error correction process

The end-on conversion and the error correction process have nonoverlapping tasks. The end-on conversion process facilitates the maturation of kinetochore-microtubule attachment by converting immature lateral attachments into end-on attachments. In contrast, the error correction process acts on kinetochores that are attached to microtubules emanating from incorrect spindle poles. For instance, an erroneous syntelic attachment occurs when both sister chromatids are attached to microtubules from the same spindle pole; an erroneous merotelic attachment occurs when one sister chromatid is attached to microtubules from both spindle poles (**Figure 1**). These erroneous attachments are primarily resolved by an evolutionarily conserved mitotic kinase, Aurora B and to some extent by other kinases; Mps1, Chk1 and Haspin kinases (reviewed in Refs. [50, 51]). When attachments are immature (lateral) or incorrect (nonbiorientated), proteins of the spindle assembly checkpoint are recruited to the kinetochore, generating an inhibitory signal that delays anaphase onset. Our current understanding of how the spindle assembly checkpoint senses attachment status and ensures amphitelic attachments is briefly discussed below.

3.4. The spindle assembly checkpoint (SAC) monitors chromosome-microtubule attachment status

The establishment of amphitelic kinetochore-microtubule attachment (biorientation; see Section 3) is crucial for the cell to accurately segregate its chromosomes in anaphase. For this reason, the status of attachment has to be continuously checked at the kinetochore-microtubule interface by SAC proteins.

The SAC was discovered through genetic screens in yeast, which revealed the majority of the spindle checkpoint proteins: Mad1, Mad2, Bub1, and Bub3 [52, 53]. Other checkpoint proteins Mps1 and BubR1 were later identified [54, 55]. While the above six proteins are generally agreed as evolutionarily conserved SAC proteins from yeasts to humans, several additional proteins have been implicated in monitoring attachment status in human cells (reviewed in Refs. [56, 57]).

In general, two models have been proposed to explain how kinetochores ensure amphitelic attachments (reviewed in Refs. [58, 59]). In the first model—the tension model—only when sister kinetochores are pulled in opposing directions, is there either inter-kinetochore or intra-kinetochore stretching established, which separate the enzymes at the inner or outer kinetochore from their substrates at outer or inner kinetochore, respectively. This tension model is elegant in that it relies on microtubule-pulling induced spatial separation of enzyme-substrate interactions [60]. However, it cannot fully explain how sister kinetochores are allowed to biorientate in the first place, if nonbiorientated attachments are continuously detached.

An alternative model questions the tension model and emphasizes the "stability" of attachment, *per se*, as the sensor for the checkpoint. The attachment stability model relies on evidence for direct competition between microtubule occupancy and checkpoint protein recruitment at the outer kinetochore [61–63].

Because both tensions at kinetochore and microtubule attachment stability depend on each other [64], it is likely the two mechanisms feed into each other to relay signals to the SAC.

The SAC can also monitor the plane of kinetochore-microtubule attachment. The spindle checkpoint proteins Mad2, Mad1, and Mps1 are all present on immature lateral kinetochores [4, 61], but at reduced levels compared to a completely detached kinetochore [61]. The significance of this quantitative difference in checkpoint protein recruitment extent is not fully understood. One possibility, apart from influencing checkpoint signaling strength, could be that different levels of the checkpoint proteins, particularly kinases, facilitate distinct stages of KT-MT attachment.

4. Phosphoregulation of kinetochore-microtubule attachment status

A kinetochore can remain attached, or actively detach, from microtubules tethered to it. The affinity of a kinetochore for microtubules is finely tuned by phosphorylation-dephosphorylation events. Over the years, a number of kinases and phosphatases have been identified as important players in establishing attachment, geometry stabilization and error correction. This section details the role and regulation of principal kinases and phosphatases that monitor and control kinetochore-microtubule attachments through phosphorylation-dephosphorylation cycles.

4.1. Kinases that control kinetochore-microtubule attachments

Several mitotic kinases are recruited to the kinetochore, which, through phosphorylation, dynamically control the localization and function of other kinetochore and microtubule-associated proteins. We have chosen to discuss a few major kinases to illustrate the role of phosphorylation in monitoring and regulating kinetochore-microtubule attachment status (for a detailed review of kinases, we refer to Ref. [36]).

4.1.1. Aurora B kinase

Aurora B is a major regulator of kinetochore-microtubule interaction; it has a dual role in sensing biorientation errors and maintaining the SAC (reviewed in Refs. [50, 65]). Aurora B forms a complex along with INCENP, Survivin and Borealin, termed the Chromosome Passenger Complex or CPC [66–68]. Until anaphase, the CPC is mainly targeted to the inner centromeric region of kinetochores; recently, CPC subpools and active Aurora B has been shown to localise to the outer kinetochore and on microtubules close to the kinetochore-microtubule interface [19, 69–72].

Aurora B forms a gradient of phosphorylation at kinetochores, which suggests its mechanism of action on substrates to be diffusive [60, 73]. To influence kinetochore-microtubule attachment, Aurora-B phosphorylates various members of the KMN network, including Ndc80, that directly interact with the microtubule [74–76]. Recent studies on the establishment of mature end-on attachments in budding yeast indicate a role for Aurora-B in phosphorylating the microtubule associated protein complex, Dam1, and thus promoting lateral kinetochores [77]. Whether Aurora-B similarly controls the plane of kinetochore-microtubule attachment in other models is not known.

4.1.2. Mps1 kinase

Mps1 is an upstream regulator of the SAC. It is recruited to the kinetochore *via* the microtubule-binding domain of the Ndc80 complex, allowing the kinase to "sense" kinetochore-microtubule attachment status [22, 61–63]. Once at kinetochores, Mps1 phosphorylates KNL1 on its repeated MELT motif allowing the recruitment of the Bub1-Bub3-BubR1 module, further strengthening the checkpoint signal (reviewed in Ref. [78]). Dynamic localization of Mps1 kinase is essential for proper kinetochore-microtubule attachment [79]. Thus, although Mps1 is primarily responsible for SAC initiation, it indirectly facilitates the formation of kinetochore-microtubule attachments.

4.1.3. Plk1 and Cdk1

Plk1 and Cdk1 are major mitotic kinases responsible for dramatic cellular reorganization during cell division and are recruited to kinetochores in early mitosis [80–82]. Phosphoproteomic studies revealed Cdk1's ability to phosphorylate a high number of substrates involved in chromosome segregation [83, 84]. In line with its broad range of substrates, Cdk1 controls several crucial events for chromosome segregation, including the building of the spindle, assembly of kinetochore and functioning of the SAC [85–89]. Like Aurora B, Cdk1 is able to modulate microtubule dynamics by acting directly on microtubule-end associated proteins (for example, EB2 [90]). Inactivation of Cdk1 ultimately stabilizes kinetochore-microtubule attachments in anaphase [91] and this is crucial for the proper segregation of chromosomes.

Plk1 activity is essential for prometaphase to metaphase transition in human cells [92, 93]. Phosphoproteomic studies show several substrates of Plk1 among spindle and kinetochore proteins [94–96]. Through phosphorylation, Plk1 directly controls the function of several microtubule-associated proteins: Plk1-mediated phosphorylation promotes the activity of the microtubule depolymerizing kinesin, MCAK [97]. Plk1-mediated phosphorylation can also enhance the association of CLIP-170 with Caesin Kinase 2 and is needed for timely formation of kinetochore-microtubule attachment [98]. To initiate and maintain SAC signal, Plk1 acts on multiple checkpoint components: first, Plk1 acts in synergy with Mps1 kinase by phosphorylating the MELT repeats on KNL1 (reviewed in Ref. [99]). In addition, Plk1 phosphorylates Sds22 to recruit PP1 (Protein Phosphatase 1, introduced in Section 4.2) to the outer kinetochore [100]. Plk1 binds directly to Bub1 kinase to help sustain the SAC signal [101, 102]. In conclusion, Plk1 simultaneously regulates both kinetochore-microtubule interaction and checkpoint signaling.

4.1.4. Bub1 kinase

Bub1, as part of a complex including Bub3 and Mad3/BubR1, is recruited to kinetochores through KNL1 MELT repeats of the KMN network (reviewed in Ref. [99]). In addition to serving as a platform for recruiting Plk1 onto KNL1 [102], Bub1 kinase safeguards sister chromatid cohesion through the Sgo1/PP2A-B56 pathway [103, 104]. Bub1 is also required for the kinetochore recruitment of the RZZ checkpoint protein complex (composed of Rod, Zwilch and Zw10 proteins) [105]. Thus, Bub1 kinase acts as a massive scaffold protein for SAC signaling initiation—a role in addition to its enzymatic activity in sustaining SAC signaling.

4.2. Mitotic phosphatases that influence kinetochore-microtubule attachments

4.2.1. PP1

Protein phosphatase PP1 controls both normal mitotic timing and chromosome segregation accuracy (reviewed in Ref. [106–108]). During mitosis, PP1 localizes specifically at the kinetochore, in addition to spindle microtubules and the cell cortex [109]. PP1 docking proteins share a consensus motif (RVxF motif [110]): KNL1, CENP-E, RepoMan, Sds22 and SKA complex all recruit PP1 to the kinetochore during different stated of kinetochore-microtubule attachment [70, 111–115]. PP1 at the outer-kinetochore is thought to counteract Aurora B for both silencing of SAC and congressing chromosomes (reviewed in [59, 65]). Whether recruiting PP1 to multiple sites of the kinetochore is simply to enable network robustness or to manage a step-wise control of KT-MT attachments is unclear.

4.2.2. PP2A

Protein phosphatase PP2A is targeted to several regions of the kinetochore through its regulatory subunit B56, which is essential for maintaining stable kinetochore-microtubule attachments [116]. At the inner centromere, PP2A is recruited through Sgo1 and Sgo2, where it prevents untimely sister chromatid separation [104, 117, 118]. At the outer kinetochore PP2A directly interacts with BubR1; thus associating with the KMN network, through Knl1 [85, 119–121]. BubR1-bound PP2A-B56 counteracts Aurora B's role in destabilizing kinetochore-microtubule attachments; Sgo1-bound PP2A-B56 is also involved in this role, suggesting a degree of redundancy [85, 116, 119, 120, 122]. PP2A-B56 activity, when bound to Sgo1, is negatively regulated by Bod1 protein and the presence of the latter is important for sister chromatid cohesion [123]. By bringing PP2A to the centromeric region, Sgo1 that is sensitive to intra/inter-kinetochore tension can directly help sense different kinetochore-microtubule attachment states [124]. The various forms of B56 associated with PP2A (positioned at inner centromere and outer kinetochore) are thought to fine-tune the otherwise broad gradient-like action of Aurora-B kinase.

4.2.3. PP4

Protein phosphatase PP4 is a new addition to the list of kinetochore regulating phosphatases important for kinetochore assembly, SAC maintenance, and chromosome congression [125]; its recruitment mode and targets are less clear.

4.3. Higher order regulatory networks controlling KT-MT attachment

4.3.1. Kinetochore-bound kinases form a network to self-regulate localization and activation

Kinetochore-bound kinases are part of a very complex network in which they modulate one another's localization and enzymatic activity [126]. For a detailed overview on the topic, we refer readers to Refs. [59, 65]. Here, we use Aurora-B as an example to illustrate the complex control network that kinetochore-bound kinases are exposed to (**Figure 3**).

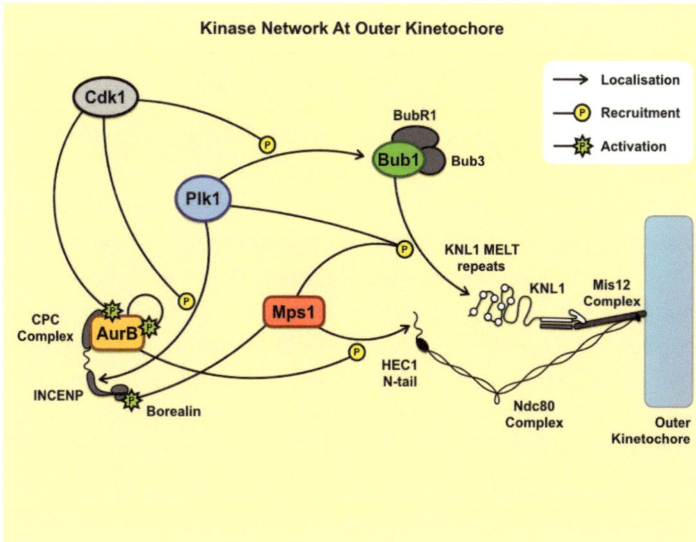

Figure 3. Phospho states are used to monitor and control microtubule attachment status: several kinases act on the KMN network at the outer kinetochore. Recruitment and activation of kinases are finely controlled through a network of phosphorylation-dephosphorylation cycles, allowing rapid monitor and dynamic control of attachment status. Phosphorylation controlled events (recruitment and activation) alone are shown here.

The centromeric enrichment of the CPC complex can be influenced by Cdk1, which phosphorylates CPC components Survivin in fission yeast and Borealin in human cells [127]. Mps1 can also promote rapid accumulation of Aurora B at the centromere [128]. Aurora B-mediated phosphorylation of the HEC1 N-terminus tail allows the recruitment of the checkpoint kinase Mps1 to kinetochores, which, together with Plk1, phosphorylates KNL1 MELT repeats to recruit other checkpoint proteins including Bub1, Bub3, BubR1 and Mad1, Mad2 and Cdc20 (reviewed in Ref. [65]). Moreover, Aurora B modulates its own activity by phosphorylating *in trans* both its kinase domain and INCENP, which results in an enhanced catalytic activity and an increased concentration of the kinase at kinetochores (reviewed in Refs. [50, 65]). In subcellular areas where Aurora-B is highly enriched, with the aid of the positive feedback loop, Aurora-B can thus overcome counteraction by phosphatases and establish areas of high activity [129]. However, when Aurora-B gradient lowers in intensity, many other kinases are able to finely tweak Aurora-B activity locally.

4.3.2. Regulatory loops linking phosphatases as a network

Unlike kinases that usually recognize specific consensus motifs, phosphatases are promiscuous with respect to their substrates. The presence of multiple binding sites for phosphatases at kinetochores could therefore allow precise, spatially restricted counteraction of kinase activity. Moreover, B55 and B56 regulatory subunits bear a highly conserved RVxF motif—a docking domain for PP1—and this puts PP2A and PP1 regulation into a complex loop [130]. In addition, the two phosphatases PP2A and PP1 are part of a KNL1-based feedback loop, which controls both their kinetochore localization and, in turn counteracts Aurora B and Mps1 kinases for silencing the SAC [131, 132]. These complex regulatory loops among phosphatases, together with the loops between kinases, have made it very difficult to dissect upstream and downstream players influencing KT-MT attachment status.

In this final section, we present an overarching model of the key regulatory mechanisms that kinetochores employ to control microtubule interaction, through phosphorylation-dephosphorylation cycles. Kinases and phosphatases are present within the kinetochore scaffold and also the kinetochore-microtubule interface. Kinases regulate one another's activity and localization by building a regulatory network (**Figure 3**), in addition to their direct action on proteins, responsible for kinetochore-microtubule attachment. This regulatory network, coupled to phosphatase-mediated counteraction, is essential to temporally and spatially modulate kinetochore-microtubule attachment. As a general tendency, phosphorylation is associated with lower affinity of the outer kinetochore substrates (e.g., HEC1 and KNL1) to the microtubule, leading to a sustained SAC signal. Nevertheless, kinases and phosphatases can also directly stimulate or silence the checkpoint, respectively, by controlling the dynamic localization of other SAC proteins. Thus, a self-regulating turnover of proteins allows the fine-tuning of the SAC signal in close relationship to the state of kinetochore-microtubule attachments.

5. Concluding remarks and future prospects

Exploring the molecular basis for how cells tether chromosomes to dynamic microtubules has highlighted the importance of phosphorylation-dependent dynamic changes in recruitment and activation of kinetochore and microtubule-associated proteins. In general, these studies provide a conceptual framework for how bulky structures may be precisely moved within the cell. They also shed light on complex regulatory networks that operate at the microtubule end, explaining how cells harness energy stored within the microtubule cytoskeleton.

Our understanding of how microtubules capture and tether onto chromosomes has expanded through the identification of kinetochore proteins which assemble onto the chromatin, the characterization of protein interactions between the outer face of the kinetochore and the microtubule lattice, and last but not the least, the elucidation of how kinetochore-microtubule attachments control the recruitment of checkpoint kinases and phosphatases that finally silence the signal that delays anaphase onset. There is still work needed to understand how kinetochores distinguish between the walls and ends of microtubules; the precise molecular steps that allow separated sister kinetochores to latch onto a disassembling microtubule, without retriggering the checkpoint; whether the feedback loops used by the checkpoint to monitor attachment status can also be used to influence attachment status; and finally, how generally applicable are the mitotic mechanisms to human meiotic systems, where chromosomes remain stably tethered to microtubules for years.

By exploring redundant and nonredundant molecular steps of the highly conserved chromosome segregation process, we progress toward exploiting the mechanism for therapeutic purposes, either to develop biomarkers or drug targets to tackle chromosomal instability in cancers.

Acknowledgements

The authors thank Dr Kozo Tanaka (Tohoku University, Japan) and members of the Draviam lab for helpful feedback and comments. This work was supported by an MRC PhD studentship MR/K50127X/1; (RG70550 PCAG/264 TASK1) to DC and a QMUL PhD studentship to MH.

Author details

Duccio Conti[1,4#], Madeleine Hart[4#], Naoka Tamura[3], Roshan Shrestha[2], Asifa Islam[4] and Viji M. Draviam[4*]

*Address all correspondence to: v.draviam@qmul.ac.uk

1 Department of Genetics, University of Cambridge, Cambridge, UK

2 National Cancer Institute, NIH Bethesda, USA

3 Barts Cancer Institute, Queen Mary University of London, UK

4 School of Biological and Chemical Sciences, Queen Mary University of London, UK

#These authors contributed equally

References

[1] Tanaka K, Hirota T. Chromosomal instability: A common feature and a therapeutic target of cancer. Biochim Biophys Acta - Rev Cancer. 2016;1866(1):64–75.

[2] Magidson V, O'Connell CB, Loncarek J, Paul R, Mogilner A, Khodjakov A. The spatial arrangement of chromosomes during prometaphase facilitates spindle assembly. Cell. 2011;146(4):555–67.

[3] Tanaka K, Mukae N, Dewar H, van Breugel M, James EK, Prescott AR, et al. Molecular mechanisms of kinetochore capture by spindle microtubules. Nature. 2005;434(7036): 987–94.

[4] Shrestha RL, Draviam VM. Lateral to end-on conversion of chromosome-microtubule attachment requires kinesins cenp-e and MCAK. Curr Biol [Internet]. The Authors; 2013;23(16):1514–26. Available from: http://dx.doi.org/10.1016/j.cub.2013.06.040

[5] Nezi L, Musacchio A. Sister chromatid tension and the spindle assembly checkpoint. Curr Opin Cell Biol. 2009;21(6):785–95.

[6] Khodjakov A, Pines J. Centromere tension: a divisive issue. Nat Cell Biol [Internet]. 2010;12(10):919–23. Available from: http://dx.doi.org/10.1038/ncb1010-919

[7] Musacchio A, Desai A. A Molecular View of Kinetochore Assembly and Function. Biology (Basel) [Internet]. 2017;6(1):5. Available from: http://www.mdpi.com/2079-7737/6/1/5

[8] Cheeseman IM, Chappie JS, Wilson-Kubalek EM, Desai A. The Conserved KMN Network Constitutes the Core Microtubule-Binding Site of the Kinetochore. Cell. 2006;127(5):983–97.

[9] Wei RR, Al-Bassam J, Harrison SC. The Ndc80/HEC1 complex is a contact point for kinetochore-microtubule attachment. Nat Struct Mol Biol. 2007;14(1):54–9.

[10] Pagliuca C, Draviam VM, Marco E, Sorger PK, De Wulf P. Roles for the conserved Spc105p/Kre28p complex in kinetochore-microtubule binding and the spindle assembly checkpoint. PLoS One. 2009;4(10):e7640.

[11] Meraldi P, Draviam VM, Sorger PK. Timing and checkpoints in the regulation of mitotic progression. Dev Cell. 2004;7(1):45–60.

[12] Obuse C, Iwasaki O, Kiyomitsu T, Goshima G, Toyoda Y, Yanagida M. A conserved Mis12 centromere complex is linked to heterochromatic HP1 and outer kinetochore protein Zwint-1. Nat Cell Biol. 2004;6(11):1135–41.

[13] Kiyomitsu T, Obuse C, Yanagida M. Human Blinkin/AF15q14 Is Required for Chromosome Alignment and the Mitotic Checkpoint through Direct Interaction with Bub1 and BubR1. Dev Cell. 2007;13(5):663–76.

[14] Hori T, Fukagawa T. Establishment of the vertebrate kinetochores. Chromosom Res. 2012;20(5):547–61.

[15] Weir JR, Faesen AC, Klare K, Petrovic A, Basilico F, Fischböck J, et al. Insights from biochemical reconstitution into the architecture of human kinetochores. Nature [Internet]. Nature Publishing Group; 2016;537(7619):249–53. Available from: http://dx.doi.org/10.1038/nature19333

[16] Tamura N, Draviam VM. Microtubule plus-ends within a mitotic cell are "moving platforms" with anchoring, signalling and force-coupling roles. Open Biol [Internet]. 2012;2(11):120132–120132. Available from: http://www.pubmedcentral.nih.gov/articlerender.fcgi?artid=3513837&tool=pmcentrez&rendertype=abstract%5Cnhttp://rsob.royalsocietypublishing.org/cgi/doi/10.1098/rsob.120132

[17] Nagpal H, Fukagawa T. Kinetochore assembly and function through the cell cycle. Chromosoma. 2016. p. 645–59.

[18] Friese A, Faesen AC, Huis In 't Veld PJ, Fischböck J, Prumbaum D, Petrovic A, et al. Molecular requirements for the inter-subunit interaction and kinetochore recruitment of SKAP and Astrin. Nat Commun [Internet]. 2016;7:11407. Available from: http://www.nature.com.docelec.univ-lyon1.fr/ncomms/2016/160420/ncomms11407/full/ncomms11407.html

[19] Caldas G V, DeLuca KF, DeLuca JG. KNL1 facilitates phosphorylation of outer kinetochore proteins by promoting Aurora B kinase activity. J Cell Biol. 2013;203(6):957–69.

[20] Caldas G V, Lynch TR, Anderson R, Afreen S, Varma D, DeLuca JG. The RZZ complex requires the N-terminus of KNL1 to mediate optimal Mad1 kinetochore localization in human cells. Open Biol [Internet]. 2015;5(11):150160–150160. Available from: http://rsob.royalsocietypublishing.org/content/5/11/150160.abstract

[21] Zhang G, Kelstrup CD, Hu X-W, Kaas Hansen MJ, Singleton MR, Olsen J V., et al. The Ndc80 internal loop is required for recruitment of the Ska complex to establish end-on microtubule attachment to kinetochores. J Cell Sci. 2012;125(13):3243–53.

[22] Nijenhuis W, Von Castelmur E, Littler D, De Marco V, Tromer E, Vleugel M, et al. A TPR domain-containing N-terminal module of MPS1 is required for its kinetochore localization by Aurora B. J Cell Biol. 2013;201(2):217–31.

[23] Miller MP, Asbury CL, Biggins S. A TOG protein confers tension sensitivity to kinetochore-microtubule attachments. Cell [Internet]. Elsevier Inc.; 2016;165(6):1428–39. Available from: http://dx.doi.org/10.1016/j.cell.2016.04.030

[24] Kitamura E, Tanaka K, Komoto S, Kitamura Y, Antony C, Tanaka TU. Kinetochores Generate Microtubules with Distal Plus Ends: Their Roles and Limited Lifetime in Mitosis. Dev Cell [Internet]. Elsevier; 2010;18(2):248–59. Available from: http://dx.doi.org/10.1016/j.devcel.2009.12.018

[25] Tanaka TU. Kinetochore-microtubule interactions: steps towards bi-orientation. EMBO J [Internet]. Nature Publishing Group; 2010;29(24):4070–82. Available from: http://dx.doi.org/10.1038/emboj.2010.294

[26] Kapoor TM, Lampson MA, Hergert P, Cameron L, Cimini D, Salmon E, et al. Chromosomes Can Congress to the Metaphase Plate Before Biorientation. Science (80-). 2006;311(5759):388–91.

[27] Alexander SP, Rieder CL. Chromosome motion during attachment to the vertebrate spindle: Initial saltatory-like behavior of chromosomes and quantitative analysis of force production by nascent kinetochore fibers. J Cell Biol. 1991;113(4):805–15.

[28] Cai S, O'Connell CB, Khodjakov A, Walczak CE. Chromosome congression in the absence of kinetochore fibres. Nat Cell Biol [Internet]. 2009;11(7):832–8. Available from: http://dx.doi.org/10.1038/ncb1890

[29] Rieder CL, Davison EA, Jensen LCW, Cassimeris L, Salmon ED. Oscillatory movements of monooriented chromosomes and their position relative to the spindle pole result from the ejection properties of the aster and half-spindle. J Cell Biol. 1986;103(2):581–91.

[30] Gagliardi LJ, Shain DH. Chromosome congression explained by nanoscale electrostatics. Theor Biol Med Model [Internet]. 2014;11:12. Available from: http://www.pubmedcentral.nih.gov/articlerender.fcgi?artid=3936865&tool=pmcentrez&rendertype=abstract

[31] Cross RA, McAinsh A. Prime movers: the mechanochemistry of mitotic kinesins. Nat Rev Mol Cell Biol [Internet]. Nature Publishing Group; 2014;15(4):257–71. Available from: http://dx.doi.org/10.1038/nrm3768

[32] Yen TJ, Li G, Schaar BT, Illya S, Cleveland D. CENP-E is a putative kinetochore motor that accumulates just before mitosis. Nature. 1992;359(6395):536–9.

[33] Bancroft J, Auckland P, Samora CP, McAinsh AD. Chromosome congression is promoted by CENP-Q- and CENP-E-dependent pathways. J Cell Sci [Internet]. 2015;128(1):171–84. Available from: http://www.ncbi.nlm.nih.gov/pubmed/25395579

[34] Mcainsh AD, Draviam VM, Toso A, Sorger PK. The human kinetochore proteins Nnf1R and Mcm21R are required for accurate chromosome segregation. EMBO J. 2006;25(17):4033–49.

[35] Ems-McClung SC, Walczak CE. Kinesin 13s in Mitosis: Key Players in the Spatial and Temporal Organization of Spindle Microtubules. Semin cell Dev Biol. 2010;21(3):276–82.

[36] Funabiki H, Wynne DJ. Making an effective switch at the kinetochore by phosphorylation and dephosphorylation. Chromosoma. 2014;122(3):135–58.

[37] Manning AL, Samuel F, Maffini S, Correia-melo C, Maiato H, Compton DA. CLASP1, astrin and Kif2b form a molecular switch that regulates kinetochore-microtubule dynamics to promote mitotic progression and fidelity. EMBO J [Internet]. Nature Publishing Group; 2010;29(20):3531–43. Available from: http://dx.doi.org/10.1038/emboj.2010.230

[38] Dimitrov A, Quesnoit M, Moutel S, Cantaloube I, Poüs C, Perez F. Detection of GTP-tubulin conformation in vivo reveals a role for GTP remnants in microtubule rescues. Science. 2008;322(5906):1353–6.

[39] Maurer SP, Bieling P, Cope J, Hoenger A, Surrey T. GTP γ S microtubules mimic the growing microtubule end structure recognized by end-binding proteins (EBs). Proc Natl Acad Sci. 2011;108(10):3988–93.

[40] Maurer SP, Fourniol FJ, Bohner G, Moores CA, Surrey T. EBs recognize a nucleotide-dependent structural cap at growing microtubule ends. Cell. 2012;149(2):371–82.

[41] Buey RM, Mohan R, Leslie K, Walzthoeni T, Missimer JH, Menzel A, et al. Insights into EB1 structure and the role of its C-terminal domain for discriminating microtubule tips from the lattice. Mol Biol Cell [Internet]. 2011;22(16):2912–23. Available from: http://www.ncbi.nlm.nih.gov/pubmed/21737692%5Cnhttp://www.pubmedcentral.nih.gov/articlerender.fcgi?artid=PMC3154886

[42] Vitre B, Coquelle F, Heichette C, Garnier C, Chretien D, Arnal I. EB1 regulates microtubule dynamics and tubulin sheet closure in vitro. Nat Cell Biol. 2008;10(4):415–21.

[43] Alushin GM, Ramey VH, Pasqualato S, Ball DA, Grigorieff N, Musacchio A, et al. The Ndc80 kinetochore complex forms oligomeric arrays along microtubules. Nature [Internet]. Nature Publishing Group; 2010;467(7317):805–10. Available from: http://dx.doi.org/10.1038/nature09423

[44] Tamura N, Simon JE, Nayak A, Shenoy R, Hiroi N, Boilot V. A proteomic study of mitotic phase-specific interactors of EB1 reveals a role for SXIP-mediated protein interactions in anaphase onset. Biol Open. 2015;4:155–69.

[45] Al-bassam J, Chang F. Regulation of Microtubule Dynamics by TOG-domain proteins XMAP215/Dis1 and CLASP. Trends Cell Biol. 2011;21(10):604–14.

[46] Brouhard GJ, Rice LM. The contribution of αβ-tubulin-tubulin curvature to microtubule dynamics. J Cell Biol. 2014;207(3):323–34.

[47] Musinipally V, Howes S, Alushin GM, Nogales E. The Microtubule Binding Properties of CENP-E's C-Terminus and CENP-F. J Mol Biol [Internet]. 2013;425(22):4427–41. Available from: http://dx.doi.org/10.1016/j.jmb.2013.07.027

[48] Mcintosh JR, Toole EO, Zhudenkov K, Morphew M, Schwartz C, Ataullakhanov FI, et al. Conserved and divergent features of kinetochores and spindle microtubule ends from five species. J Cell Biol. 2013;200(4):459–74.

[49] Volkov VA, Grissom PM, Arzhanik VK, Zaytsev A V., Renganathan K, McClure-Begley T, et al. Centromere protein F includes two sites that couple efficiently to depolymerizing microtubules. J Cell Biol. 2015;209(6):813–28.

[50] Carmena M, Wheelock M, Funabiki H, Earnshaw WC. The Chromosomal Passenger Complex (CPC): From Easy Rider to the Godfather of Mitosis. Nat Rev Mol Cell Biol. 2012;13(12):789–803.

[51] Higgins JMG. Haspin: A newly discovered regulator of mitotic chromosome behavior. Chromosoma. 2010;119(2):137–47.

[52] Li R, Murray AW. Feedback Control of Mitosis in Budding Yeast. Cell. 1991;66:519–31.

[53] Hoyt MA, Totis L, Roberts BT. S. cerevisiae genes required for cell cycle arrest in response to loss of microtubule function. Cell. 1991;66:507–17.

[54] Weiss E, Winey M. The Saccharomyces cerevisiae spindle pole body duplication gene MPS1 is part of a mitotic checkpoint. J Cell Biol. 1996;132(1–2):111–23.

[55] Cahill DP, Lengauer C, Yu J, Riggins GJ, Willson JK, Markowitz SD, et al. Mutations of mitotic checkpoint genes in human cancers. Nature [Internet]. 1998;392(6673):300–3. Available from: http://www.ncbi.nlm.nih.gov/pubmed/9521327%5Cnhttp://www.nature.com/nature/journal/v392/n6673/pdf/392300a0.pdf

[56] Murray AW. A brief history of error. Nat Cell Biol [Internet]. Nature Publishing Group; 2011;13(10):1178–82. Available from: http://dx.doi.org/10.1038/ncb2348

[57] Murray AW. Don't Make Me Mad, Bub! Dev Cell. 2012;22(6):1123–5.

[58] Cheerambathur DK, Desai A. Linked In: Formation and Regulation of Microtubule Attachments During Chromosome Segregation. Curr Opin Cell Biol. 2014;26(1):113–22.

[59] Musacchio A. The Molecular Biology of Spindle Assembly Checkpoint Signaling Dynamics. Curr Biol [Internet]. Elsevier; 2015;25(20):R1002–18. Available from: http://dx.doi.org/10.1016/j.cub.2015.08.051

[60] Liu D, Vader G, Vromans MJM, Lampson MA, Lens SMA. Sensing Chromosome Bi-Orientation by Spatial Separation of Aurora B Kinase from Kinetochore Substrates. Science. 2009;323(March):1350–3.

[61] Hiruma Y, Sacristan C, Pachis ST, Adamopoulos A, Kuijt T, Ubbink M, et al. CELL DIVISION CYCLE. Competition between MPS1 and microtubules at kinetochores regulates spindle checkpoint signaling. Science [Internet]. 2015;348(6240):1264–7. Available from: http://www.ncbi.nlm.nih.gov/pubmed/26068855

[62] Ji Z, Gao H, Yu H. Kinetochore attachment sensed by competitive Mps1 and microtubule binding to Ndc80C. Science (80-) [Internet]. 2015;348(6240):1260–4. Available from: http://www.sciencemag.org/cgi/doi/10.1126/science.aaa4029

[63] Zhu T, Dou Z, Qin B, Jin C, Wang X, Xu L, et al. Phosphorylation of microtubule-binding protein hec1 by mitotic kinase aurora b specifies spindle checkpoint kinase mps1 signaling at the kinetochore. J Biol Chem. 2013;288(50):36149–59.

[64] Akiyoshi B, Sarangapani KK, Powers AF, Nelson CR, Reichow SL, Arellano-Santoyo H, et al. Tension directly stabilizes reconstituted kinetochore-microtubule attachments. Nature [Internet]. Nature Publishing Group; 2010;468(7323):576–9. Available from: http://dx.doi.org/10.1038/nature09594

[65] London N, Biggins S. Signalling dynamics in the spindle checkpoint response. Nat Rev Mol Cell Biol. 2014;15(11):736–47.

[66] Gassmann R, Carvalho A, Henzing AJ, Ruchaud S, Hudson DF, Honda R, et al. Borealin: A novel chromosomal passenger required for stability of the bipolar mitotic spindle. J Cell Biol. 2004;166(2):179–91.

[67] Jeyaprakash AA, Klein UR, Lindner D, Ebert J, Nigg EA, Conti E. Structure of a Survivin-Borealin-INCENP Core Complex Reveals How Chromosomal Passengers Travel Together. Cell. 2007;131(2):271–85.

[68] Vanoosthuyse V, Prykhozhij S, Hardwick KG. Shugoshin 2 Regulates Localization of the Chromosomal Passenger Proteins in Fission Yeast Mitosis. Mol Biol Cell. 2007;18(May):1657–69.

[69] Klein UR, Nigg EA, Gruneberg U. Centromere Targeting of the Chromosomal Passenger Complex Requires a Ternary Subcomplex of Borealin, Survivin, and the N-Terminal Domain of INCENP. Mol Biol Cell. 2006;17(June):2547–58.

[70] Posch M, Khoudoli GA, Swift S, King EM, Deluca JG, Swedlow JR. Sds22 regulates aurora B activity and microtubule–kinetochore interactions at mitosis. J Cell Biol. 2010;191(1):61–74.

[71] Campbell CS, Desai A. Tension Sensing by Aurora B Kinase is Independent of Survivin-Based Centromere Localization. Nature. 2013;497(7447):118–21.

[72] Banerjee B, Kestner CA, Stukenberg PT. EB1 enables spindle microtubules to regulate centromeric recruitment of Aurora B. J Cell Biol. 2014;204(6):947–63.

[73] Wang E, Ballister ER, Lampson MA. Aurora B dynamics at centromeres create a diffusion-based phosphorylation gradient. J Cell Biol. 2011;194(4):539–49.

[74] Welburn JPI, Vleugel M, Liu D, Iii JRY, Lampson MA, Fukagawa T, et al. Aurora B phosphorylates spatially distinct targets to differentially regulate the kinetochore-microtubule interface. Mol Cell. 2011;38(3):383–92.

[75] DeLuca JG, Gall WE, Ciferri C, Cimini D, Musacchio A, Salmon ED. Kinetochore Microtubule Dynamics and Attachment Stability Are Regulated by Hec1. Cell. 2006;127(5):969–82.

[76] Ciferri C, Pasqualato S, Screpanti E, Varetti G, Santaguida S, Dos Reis G, et al. Implications for Kinetochore-Microtubule Attachment from the Structure of an Engineered Ndc80 Complex. Cell. 2008;133(3):427–39.

[77] Kalantzaki M, Kitamura E, Zhang T, Mino A, Novák B. Kinetochore – microtubule error correction is driven by differentially regulated interaction modes. Nat cell. 2015;17(4):421–33.

[78] Hervas-Aguilar A, Millar JBA. Mph1/MPS1 checks in at the kinetochore. Cell Cycle [Internet]. Taylor & Francis; 2016;1–2. Available from: http://dx.doi.org/10.1080/15384101.2016.1159888

[79] Dou Z, Liu X, Wang W, Zhu T, Wang X, Xu L, et al. Dynamic localization of Mps1 kinase to kinetochores is essential for accurate spindle microtubule attachment. Proc Natl Acad Sci [Internet]. 2015;112(33):E4546–55. Available from: http://www.pnas.org/lookup/doi/10.1073/pnas.1508791112

[80] Arnaud L, Pines J, Nigg EA. GFP tagging reveals human Polo-like kinase 1 at the kinetochore/centromere region of mitotic chromosomes. Chromosoma. 1998;107(6–7):424–9.

[81] Chen Q, Zhang X, Jiang Q, Clarke PR, Zhang C. Cyclin B1 is localized to unattached kinetochores and contributes to efficient microtubule attachment and proper chromosome alignment during mitosis. Cell Res. 2008;18(2):268–80.

[82] Sunkel CE, Glover DM. polo, a mitotic mutant of Drosophila displaying abnormal spindle poles. J Cell Sci. 1988;89(Part 1):25–38.

[83] Pagliuca FW, Collins MO, Lichawska A, Zegerman P. Quantitative Proteomics Reveals the Basis for the Biochemical Specificity of the Cell Cycle Machinery. Mol Cell. 2011;43(3):406–17.

[84] Jiménez JL, Hegemann B, Hutchins JRA, Peters J-M, Durbin R. A systematic comparative and structural analysis of protein phosphorylation sites based on the mtcPTM database. Genome Biol. 2007;8(5):R90.

[85] Kruse T, Zhang G, Sofie M, Larsen Y, Lischetti T, Streicher W, et al. Direct binding between BubR1 and B56 – PP2A phosphatase complexes regulate mitotic progression. J Cell Sci. 2013;126:1086–92.

[86] Draviam VM, Orrechia S, Lowe M, Pardi R, Pines J. The localization of human cyclins B1 and B2 determines CDK1 substrate specificity and neither enzyme requires MEK to disassemble the Golgi apparatus. J Cell Biol. 2001;152(5):945–58.

[87] Gascoigne KE, Cheeseman IM. CDK-dependent phosphorylation and nuclear exclusion coordinately control kinetochore assembly state. J Cell Biol. 2013;201(1):23–32.

[88] Robellet X, Thattikota Y, Wang F, Wee TL, Pascariu M, Shankar S, et al. A high-sensitivity phospho-switch triggered by Cdk1 governs chromosome morphogenesis during cell division. Genes Dev. 2015;29(4):426–39.

[89] Maia ARR, Garcia Z, Kabeche L, Barisic M, Maffini S, Macedo-Ribeiro S, et al. Cdk1 and Plk1 mediate a CLASP2 phospho-switch that stabilizes kinetochore-microtubule attachments. J Cell Biol. 2012;199(2):285–301.

[90] Iimori M, Watanabe S, Kiyonari S, Matsuoka K, Sakasai R, Saeki H, et al. Phosphorylation of EB2 by Aurora B and CDK1 ensures mitotic progression and genome stability. Nat Commun [Internet]. Nature Publishing Group; 2016;7:11117. Available from: http://dx.doi.org/10.1038/ncomms11117

[91] Vázquez-Novelle MD, Sansregret L, Dick AE, Smith CA, Mcainsh AD, Gerlich DW, et al. Cdk1 inactivation terminates mitotic checkpoint surveillance and stabilizes kinetochore attachments in anaphase. Curr Biol. 2014;24(6):638–45.

[92] Ahonen LJ, Kallio MJ, Daum JR, Bolton M, Manke IA, Yaffe MB, et al. Polo-like kinase 1 creates the tension-sensing 3F3/2 phosphoepitope and modulates the association of spindle-checkpoint proteins at kinetochores. Curr Biol. 2005;15(12):1078–89.

[93] Lenart P, Petronczki M, Steegmaier M, Fiore B Di, Lipp JJ, Hoffmann M, et al. The Small-Molecule Inhibitor BI 2536 Reveals Novel Insights into Mitotic Roles of Polo-like Kinase 1. Curr Biol. 2007;17:304–15.

[94] Dou Z, von Schubert C, Körner R, Santamaria A, Elowe S, Nigg EA. Quantitative mass spectrometry analysis reveals similar substrate consensus motif for human Mps1 kinase and Plk1. PLoS One. 2011;6(4):e18793.

[95] Hegemann B, Hutchins JRA, Hudecz O, Novatchkova M, Rameseder J, Sykora MM, et al. Systematic phosphorylation analysis of human mitotic protein complexes. Sci Signal [Internet]. 2011;4(198):rs12. Available from: http://www.pubmedcentral.nih.gov/articlerender.fcgi?artid=4206221&tool=pmcentrez&rendertype=abstract

[96] Santamaria A, Wang B, Elowe S, Malik R, Zhang F, Bauer M, et al. The Plk1-dependent phosphoproteome of the early mitotic spindle. Mol Cell Proteomics. 2011;10(1):M110.004457.

[97] Shao H, Huang Y, Zhang L, Yuan K, Chu Y, Dou Z, et al. Spatiotemporal dynamics of Aurora B-PLK1-MCAK signaling axis orchestrates kinetochore bi-orientation and faithful chromosome segregation. Sci Rep [Internet]. Nature Publishing Group; 2015;5(October 2014):12204. Available from: http://dx.doi.org/10.1038/srep12204

[98] Li H, Liu XS, Yang X, Wang Y, Wang Y, Turner JR, et al. Phosphorylation of CLIP-170 by Plk1 and CK2 promotes timely formation of kinetochore-microtubule attachments. Embo j [Internet]. 2010;29(17):2953–65. Available from: http://www.ncbi.nlm.nih.gov/pubmed/20664522

[99] Faesen AC, Musacchio A. The (Phospho) Needle in the (MELT) Haystack. Mol Cell. 2015;57(5):765–6.

[100] Duan H, Wang C, Wang M, Gao X, Yan M, Akram S, et al. Phosphorylation of PP1 regulator Sds22 by PLK1 ensures accurate chromosome segregation. J Biol Chem. 2016;291(40):21123–36.

[101] Jia L, Li B, Yu H. The Bub1-Plk1 kinase complex promotes spindle checkpoint signalling through Cdc20 phosphorylation. Nat Commun [Internet]. 2016;7:10818. Available from: http://www.pubmedcentral.nih.gov/articlerender.fcgi?artid=4773433&tool=pmcentrez&rendertype=abstract

[102] Qi W, Tang Z, Yu H. Phosphorylation- and polo-box-dependent binding of Plk1 to Bub1 is required for the kinetochore localization of Plk1. Mol Biol Cell [Internet]. 2006;17(8):3705–16. Available from: http://www.pubmedcentral.nih.gov/articlerender.fcgi?artid=1525235&tool=pmcentrez&rendertype=abstract

[103] Tang Z, Sun Y, Harley SE, Zou H, Yu H. Human Bub1 protects centromeric sister-chromatid cohesion through Shugoshin during mitosis. Proc Natl Acad Sci U S A. 2004;101(52):18012–7.

[104] Tang Z, Shu H, Qi W, Mahmood NA, Mumby MC, Yu H. PP2A Is Required for Centromeric Localization of Sgo1 and Proper Chromosome Segregation. Dev Cell. 2006;10(5):575–85.

[105] Zhang G, Lischetti T, Hayward DG, Nilsson J. Distinct domains in Bub1 localize RZZ and BubR1 to kinetochores to regulate the checkpoint. Nat Commun [Internet]. 2015;6:7162. Available from: http://www.nature.com.proxy.library.uu.nl/ncomms/2015/150602/ncomms8162/full/ncomms8162.html#affil-auth

[106] Wurzenberger C, Gerlich DW. Phosphatases: providing safe passage through mitotic exit. Nat Rev Mol Cell Biol [Internet]. 2011;12(8):469–82. Available from: http://www.ncbi.nlm.nih.gov/pubmed/21750572

[107] Trinkle-Mulcahy L, Lamond AI. Mitotic phosphatases: no longer silent partners. Curr Opin Cell Biol. 2006;18(6):623–31.

[108] Maldonado M, Kapoor TM. Moving right along: How PP1 helps clear the checkpoint. Dev Cell. 2011;20(6):733–4.

[109] Trinkle-Mulcahy L, Andrews PD, Wickramasinghe S, Sleeman J, Prescott A, Lam YW, et al. Time-lapse imaging reveals dynamic relocalization of PP1gamma throughout the mammalian cell cycle. Mol Biol Cell. 2003;14(1):107–17.

[110] Heroes E, Lesage B, Görnemann J, Beullens M, Van Meervelt L, Bollen M. The PP1 binding code: A molecular-lego strategy that governs specificity. FEBS Journal. 2013. p. 584–95.

[111] Kim Y, Holland AJ, Lan W, Cleveland DW. Aurora kinases and protein phosphatase 1 mediate chromosome congression through regulation of CENP-E. Cell. 2010;142(3):444–55.

[112] Rosenberg JS, Cross FR, Funabiki H. KNL1/Spc105 recruits PP1 to silence the spindle assembly checkpoint. Curr Biol. 2011;21(11):942–7.

[113] Meadows JC, Shepperd LA, Vanoosthuyse V, Lancaster TC, Sochaj AM, Buttrick GJ, et al. Spindle checkpoint silencing requires association of PP1 to both Spc7 and kinesin-8 motors. Dev Cell. 2011;20(6):739–50.

[114] Liu D, Vleugel M, Backer CB, Hori T, Fukagawa T, Cheeseman IM, et al. Regulated targeting of protein phosphatase 1 to the outer kinetochore by KNL1 opposes Aurora B kinase. J Cell Biol. 2010;188(6):809–20.

[115] Sivakumar S, Janczyk P, Qu Q, Brautigam CA, Stukenberg PT, Yu H, et al. The human SKA complex drives the metaphase-anaphase cell cycle transition by recruiting protein phosphatase 1 to kinetochores. Elife. 2016;5:e12902.

[116] Foley E a., Maldonado M, Kapoor TM. Formation of stable attachments between kinetochores and microtubules depends on the B56-PP2A phosphatase. Nat Cell Biol [Internet]. 2011;13(10):1265–71. Available from: http://dx.doi.org/10.1038/ncb2327

[117] Kitajima TS, Ohsugi M, Ellenberg J. Complete kinetochore tracking reveals error-prone homologous chromosome biorientation in mammalian oocytes. Cell. 2011;146(4):568–81.

[118] Liu H, Rankin S, Yu H. Phosphorylation-enabled binding of SGO1-PP2A to cohesin protects sororin and centromeric cohesion during mitosis. Nat Cell Biol [Internet]. 2013;15(1):40–9. Available from: http://www.pubmedcentral.nih.gov/articlerender.fcgi?artid=3531828&tool=pmcentrez&rendertype=abstract

[119] Suijkerbuijk SJE, Vleugel M, Teixeira A, Kops GJPL. Integration of Kinase and Phosphatase Activities by BUBR1 Ensures Formation of Stable Kinetochore-Microtubule Attachments. Dev Cell. 2012;23(4):745–55.

[120] Xu P, Raetz E a, Kitagawa M, Virshup DM, Lee SH. BUBR1 recruits PP2A via the B56 family of targeting subunits to promote chromosome congression. Biol Open [Internet]. 2013;2(5):479–86. Available from: http://www.pubmedcentral.nih.gov/articlerender.fcgi?artid=3654266&tool=pmcentrez&rendertype=abstract

[121] Wang J, Wang Z, Yu T, Yang H, Virshup DM, Kops GJPL, et al. Crystal structure of a PP2A B56-BubR1 complex and its implications for PP2A substrate recruitment and localization. Protein Cell. 2016;7(7):516–26.

[122] Meppelink A, Kabeche L, Vromans MJM, Compton DA, Lens SMA. Shugoshin-1 Balances Aurora B Kinase Activity via PP2A to Promote Chromosome Bi-orientation. Cell Rep. 2015;11(4):508–15.

[123] Porter IM, Schleicher K, Porter M, Swedlow JR. Bod1 regulates protein phosphatase 2A at mitotic kinetochores. Nat Commun [Internet]. 2013;4:2677. Available from: http://www.ncbi.nlm.nih.gov/pmc/articles/PMC3826647/pdf/ncomms3677.pdf

[124] Marston AL. Shugoshins: tension-sensitive pericentromeric adaptors safeguarding chromosome segregation. Mol Cell Biol [Internet]. 2015;35(4):634–48. Available from: /pmc/articles/PMC4301718/?report=abstract

[125] Lipinszki Z, Lefevre S, Savoian MS, Singleton MR, Glover DM, Przewloka MR. Centromeric binding and activity of Protein Phosphatase 4. Nat Commun [Internet]. 2015;6:5894. Available from: http://www.nature.com/ncomms/2015/150106/ncomms6894/full/ncomms6894.html

[126] Vigneron S, Prieto S, Bernis C, Labbe J-C, Castro A, Lorca T. Kinetochore localization of spindle checkpoint proteins: who controls whom? Mol Biol Cell [Internet]. 2004;15(10):4584–96. Available from: http://www.hubmed.org/display.cgi?uids=15269280

[127] Tsukahara T, Tanno Y, Watanabe Y. Phosphorylation of the CPC by Cdk1 promotes chromosome bi-orientation. Nature [Internet]. 2010;467(7316):719–23. Available from: http://dx.doi.org/10.1038/nature09390

[128] van der Waal MS, Saurin AT, Vromans MJ, Vleugel M, Wurzenberger C, Gerlich DW, et al. Mps1 promotes rapid centromere accumulation of Aurora B. EMBO Rep [Internet]. 2012;13(9):847–54. Available from: http://www.ncbi.nlm.nih.gov/pmc/articles/PMC3432816/pdf/embor201293a.pdf

[129] Zaytsev A V., Segura-Pena D, Godzi M, Calderon A, Ballister ER, Stamatov R, et al. Bistability of a coupled aurora B kinase-phosphatase system in cell division. Elife. 2016;5(January):e10644.

[130] Grallert A, Boke E, Hagting A, Hodgson B, Connolly Y, Griffiths JR, et al. A PP1-PP2A phosphatase relay controls mitotic progression. Nature [Internet]. 2015;517(7532):94–8. Available from: http://www.pubmedcentral.nih.gov/articlerender.fcgi?artid=4338534&tool=pmcentrez&rendertype=abstract

[131] Espert A, Uluocak P, Bastos RN, Mangat D, Graab P, Gruneberg U. PP2A-B56 opposes Mps1 phosphorylation of Knl1 and thereby promotes spindle assembly checkpoint silencing. J Cell Biol. 2014;206(7):833–42.

[132] Nijenhuis W, Vallardi G, Teixeira A, Kops GJPL, Saurin AT. Negative feedback at kinetochores underlies a responsive spindle checkpoint signal [Internet]. Nature. 2014. p. 1257–64. Available from: http://www.nature.com/ncb/journal/vaop/ncurrent/full/ncb3065.html

Section 2

Cytoskeleton: Function and Disease

Chapter 10

Muscle Fibers Lacking Desmin in the Extraocular Muscles: A Paradigm Shift

Fatima Pedrosa Domellöf

Additional information is available at the end of the chapter

http://dx.doi.org/10.5772/66928

Abstract

The extraocular muscles are highly specialized muscles responsible for the complex movements of the eyeball. They differ from other skeletal muscles in many respects, including fundamental components of the contractile apparatus and the extracellular matrix. Using immunohistochemistry and a battery of well-characterized antibodies, we have investigated the composition of the cytoskeleton of their myofibers with respect to desmin, vimentin, and nestin. In the adult and fetal human extraocular muscles, a subgroup of the slow tonic muscle fibers is lacking desmin. These fibers, which are multiply innervated, show a normal myofibrillar arrangement, maintained mitochondrial distribution, and sarcolemma integrity. Desmin, the most abundant intermediate filament protein in muscle, has been considered a ubiquitous protein in skeletal muscle fibers where it links adjacent myofibrils and the myofibrillar network to the sarcolemma, the mitochondria and the membrane of the nuclei. The functional implications of the lack of desmin remain to be determined, but these findings represent a paradigm shift, as desmin has been regarded a ubiquitous protein of the cytoskeleton of muscle fibers.

Keywords: desmin, human, extraocular muscle, myosin heavy chain slow tonic, multiple innervation

1. Introduction

The extraocular muscles differ significantly from other muscles in the body, including their structural composition, physiological properties, and response to disease [1–3]. Because the extraocular muscles are more resistant to a number of diseases than the other muscles [3], it has been hypothesized that a better understanding of their composition may provide useful clues to develop strategies to face the challenge posed by muscle dystrophies and other neuromuscular diseases. We have studied the composition of the human extraocular muscles with respect to the myosin heavy chains (MyCHs) [4], the major determinants of contraction velocity and force; SERCA-1 and SERCA-2 (sarco/endoplasmic endoreticulum calcium ATPase) [5], two important calcium transportation proteins; laminins [6], major components of the basement membrane and a number of proteins relevant for the neuromuscular junctions, e.g., gangliosides [7]. Because data on the composition of the cytoskeleton in the human extraocular muscles were lacking and because defects in the desmin gene cause myopathy and cardiomyopathy, we recently also investigated the distribution of the major intermediate filament proteins, desmin, vimentin, and nestin in the human adult extraocular muscles, using immunohistochemistry in serial sections and in 1-μm thick sections [8].

Desmin has been regarded as a ubiquitous protein in skeletal muscle fibers, being the first muscle-specific protein detected during development [9]. Desmin is the most abundant intermediate filament protein in muscle fibers, forming a three-dimensional scaffold along the entire muscle fiber. Desmin links adjacent myofibrils at the Z-discs, and it links the most peripheral myofibrils to the sarcolemma and to the membrane of the nuclei. It also links the myofibrils to other organelles, such as the mitochondria [10–12].

In spite of being the first muscle-specific protein expressed early during muscle development and being so abundant in mature muscle fibers, desmin is not strictly necessary for myogenesis or muscle fiber maturation (reviewed in Refs. [13, 14]). Desmin knockout (KO) mice develop rather normally, but they develop a cardiomyopathy and/or a skeletal myopathy that predominantly affects highly used muscles, such as the diaphragm. Early muscle differentiation appears normal but with time signs of pathology develop. The affected myofibers show misaligned myofibrils with the loss of anchorage to the sarcolemma and mitochondrial abnormalities. Data from the desmin KO animals indicate that desmin is essential for the integrity of highly used skeletal muscle and for proper mitochondrial morphology and positioning but that it is not essential for muscle development. In contrast, these desmin KO animals show affected regeneration capacity, with adipocyte accumulation and abnormal neuromuscular junction morphology. Furthermore, in this animal model, desmin seems to be necessary for proper excitation-contraction coupling and respiratory function [13, 14].

Patients with defects in the desmin gene develop progressive skeletal myopathy and cardiomyopathy, although the age of onset of the disease may be rather high. Desmin-related myopathies, which may be familial or sporadic, are generally characterized by the formation of abnormal desmin aggregates and have heterogeneous clinical hallmarks, which may include neuropathy and smooth muscle involvement, in addition to skeletal muscle and heart symptoms [13].

The complexity of actions performed by the extraocular muscles is reflected in their distinct fiber type composition and structural organization that differ substantially from ordinary skeletal muscles [1, 2]. The extraocular muscles are among the fastest muscles in the human body, yet they are extremely fatigue-resistant. Their muscle fibers are organized into two layers: the orbital layer, closest to the orbital wall, and the global layer, facing the bulb. The muscle fibers in the orbital layer have a smaller diameter and do not extend the full length of the muscle. The muscle fibers in the extraocular muscles have small diameters, are loosely arranged in a bed of connective tissue, are richly supplied by capillaries, and belong to very small muscle units.

The fiber type composition of the human extraocular muscles is very complex [4, 15]. Typically, the muscle fibers in the human extraocular muscles contain several myosin heavy chain (MyHC) isoforms, including isoforms not typically found in mature skeletal muscle such as MyHC embryonic and fetal, MyHC alpha cardiac, MyHC extraocular, and MyHC slow tonic and the composition of the muscle fibers varies along their length [4, 15]. A particular fiber type contains MyHC slow tonic, and it is multiply innervated. In the orbital layer, these fibers also have a twitch motor endplate in the middle portion, whereas those in the global layer seem to be tonic along their entire length (reviewed in Refs. [1, 2]).

The gene expression profile of the extraocular muscles differs fundamentally from that of the limb muscles with respect to metabolic pathways, structural components, developmental and regeneration markers, by over 300 genes [16]. Indeed, the extraocular muscles are classified as a separate muscle allotype (a class of muscles), in contrast to the other allotypes: (i) the limb and trunk muscles and (ii) the masticatory muscles [17].

The extraocular muscles also differ from the other muscles in the human body by their particular response to disease [3]. The extraocular muscles are selectively affected in autoimmune disorders such as myasthenia gravis, Miller-Fisher syndrome, and Grave's ophthalmopathy. In contrast, they are strikingly spared in various muscular dystrophies such as Duchene and Becker muscular dystrophy, limb-girdle, and congenital muscular dystrophies that are devastating for the other muscles in the body, by causing severe weakness, loss of ambulation, and early death [18].

Here, the present state of knowledge regarding the composition of the cytoskeleton of human extraocular muscles with respect to desmin, vimentin, and nestin is reviewed [8].

2. Lack of desmin in fibers of the human extraocular muscles

The vast majority of muscle fibers in the extraocular muscles contain desmin (**Figure 1**), as expected, but the muscle fibers containing MyHC slow tonic show different staining patterns concerning desmin, irrespective of whether they are present in the orbital or global layers [8]. Some of these muscle fibers apparently lack desmin completely (**Figure 1**), both at the subsarcolemmal level and in between myofibrils, as they are completely unstained by a battery of mono and polyclonal antibodies against desmin.

In thin sections (1 μm, **Figure 2**), a normal myofibrillar organization of these muscle fibers is apparent and antibodies against MyHC slow tonic and/or MyHC slow label them, but they are completely unlabeled by different antibodies against desmin. Three additional staining patterns are present among the muscle fibers containing MyHC slow tonic in the human adult extraocular muscles: (i) muscle fibers that show desmin only subsarcolemmally; (ii) muscle fibers with desmin staining present both between myofibrils and below the sarcolemma, but with extremely weak staining intensity and (iii) muscle fibers showing a typical desmin cytoskeleton pattern, as shown in the other muscle fibers (**Figure 2**). Moreover, the lack of desmin in the myofibers containing MyHC slow tonic is already present in fetal human extraocular muscles at 16 and 18 weeks of gestation. The pattern of distribution of mitochondria is maintained in all adult muscle fibers [8].

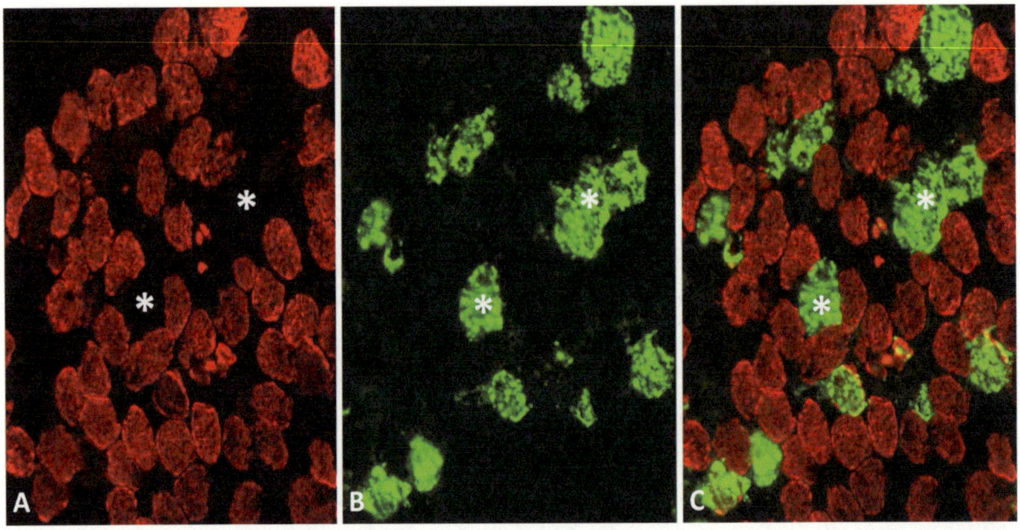

Figure 1. Cross-section (5 μm) of adult human extraocular muscle double-stained for desmin (A, red) and MyHC slow (B, green). The merged picture of A and B is shown in C. Examples of muscle fibers lacking desmin are marked with an asterisk.

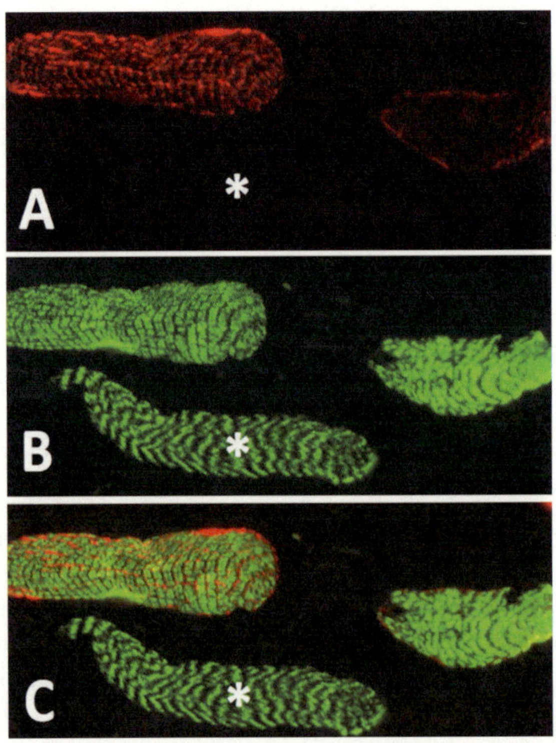

Figure 2. Thin section (1 μm) of adult human extraocular muscle double-stained for desmin (A, red) and MyHC slow tonic (B, green), showing a muscle fiber lacking desmin (asterisk) both in between myofibrils and under the sarcolemma. A merged picture with both stainings is shown in C. The typical distribution of desmin is seen in the other two muscle fibers shown. Notice that all three muscle fibers contain MyHC slow tonic.

Nestin does not appear to substitute for desmin in the desmin-negative muscle fibers, although nestin is present in a very large number of normal adult muscle fibers in the human extraocular muscles. Vimentin is not present in any muscle fibers in the adult human extraocular muscles [8].

More recently, the absence of desmin has also been reported in slow muscle fibers of the uvula and palatopharyngeal muscles [19]. These fibers contain MyHC slow but lack MyHC slow tonic and they are also unstained with antibodies against the C-terminal of dystrophin. Similarly to the extraocular muscles, nestin and vimentin do not compensate for the lack of desmin in these muscle fibers [19]. The uvula and palatopharyngeal muscles are highly specialized and unusual in that they lack a firm bony attachment at least at one end and the authors suggest that the lack of desmin in a subgroup of the fibers in these two muscles likely reflects specialization to meet the complex requirements of oropharyngeal functions [19].

The presence of normal adult muscle fibers lacking desmin is a paradigm shift as this cytoskeletal protein has been regarded as ubiquitous in normal skeletal muscle fibers. Because the absence or presence of only trace amounts of desmin in muscle fibers of the human extraocular muscles is confirmed with a polyclonal antibody and two different monoclonal antibodies, in both 5 and 1 μm sections, it is not likely that the findings are due to differences in epitope availability.

The absence of desmin in KO animals primarily leads to pathological changes in highly used muscles [13, 14, 20]. The extraocular muscles are also highly used muscles, but it may be so that the muscle fibers containing MyHC slow tonic and lacking desmin are used in a different pattern. These muscle fibers are typically multiply innervated, have a sustained mode of contraction, and they have been suggested to play a role in proprioception [9]. The fact that (i) no abnormalities in the pattern of mitochondria distribution are present in the fibers lacking desmin in the human extraocular muscles, along with (ii) a maintained myofibrillar organization, (iii) maintained fiber integrity, and (iv) the early establishment of this pattern during development [8] strongly suggest that the lack of desmin is indeed a special feature of these muscle fibers. These findings really represent a paradigm shift, as desmin can no longer be considered ubiquitous in muscle fibers and its ascribed fundamental role anchoring the mitochondria apparently does not apply to all muscle fibers.

Desmin is a crucial cytoskeletal protein in the vast majority of muscle fibers, and it is proposed to be fundamental for maintenance of fiber integrity, force transduction, and mechanochemical signaling as it links the contractile elements both to the sarcolemma and to the nuclear membrane [14, 21]. However, it is difficult to speculate on possible physiological consequences of the absence of desmin on these slow tonic muscle fibers of the human extraocular muscles, and further studies correlating structural properties and physiological behavior are required.

3. Conclusions

Against all previous knowledge, desmin is lacking in a subgroup of normal myofibers in the human eye muscles [8]. Furthermore, nestin does not compensate for the lack of desmin, but an important number of muscle fibers normally contains nestin in these adult muscles.

The finding of normal muscle fibers lacking desmin in adult and fetal human extraocular muscle represents a paradigm shift as desmin has been regarded as ubiquitous in muscle fibers. The presence of nestin in normal mature muscle fibers further emphasizes that the extraocular muscles differ fundamentally from the other muscles in the body.

Acknowledgements

Supported by grants from the Swedish Research Council, the County Council of Västerbotten (Cutting Edge Medical Research and Central ALF) and Stiftelsen Kronprinsessan Margaretas Arbetsnämnd för Synskadade.

Author details

Fatima Pedrosa Domellöf

Address all correspondence to: fatima.pedrosa-domellof@umu.se

Department of Clinical Sciences, Ophthalmology, Umeå University, Umeå, Sweden

References

[1] Sadeh M. Extraocular muscles. In: Engel AG, Franzini-Armstrong C, editors. Myology. New York: McGraw-Hill; 2004. pp. 119–127. DOI: 10.1002/ana.410360536

[2] McLoon LK, Christiansen SP. Extraocular muscles: Extraocular muscle anatomy. In: Dartt DA, Besharse JC, Dana R, editors. Encyclopedia of the Eye, Vol. 2. Oxford: Academic Press; 2010. pp. 89–98. DOI: 10.1016/b978-0-12-374203-2.00274-8

[3] Pedrosa Domellöf F. Extraocular muscles: Extraocular muscle involvement in disease. In: Dartt DA, Besharse JC, Dana R, editors. Encyclopedia of the Eye, Vol. 2. Oxford: Academic Press; 2010. pp. 99–104. DOI: 10.1016/B978-0-12-374203-2.00279-7

[4] Kjellgren D, Thornell LE, Pedrosa-Domellöf F. Myosin heavy chain isoforms in human extraocular muscles. Invest Ophthalmol Vis Sci. 2003;**44**:1419–1425. DOI: 10.1167/iovs.02-0638

[5] Kjellgren D, Stal P, Larsson L, Furst D, Pedrosa-Domellof F. Uncoordinated expression of myosin heavy chains and myosin-binding protein C isoforms in human extraocular muscles. Invest Ophthalmol Vis Sci. 2006;**47**:4188–4193. DOI:10.1167/iovs.05-1496

[6] Kjellgren D, Thornell LE, Virtanen I, Pedrosa-Domellof F. Laminin isoforms in human extraocular muscles. Invest Ophthalmol Vis Sci. 2004;**45**:4233–4239. DOI:10.1167/iovs.04-0456

[7] Liu JX, Willison HJ, Pedrosa-Domellof F. Immunolocalization of GQ1b and related gangliosides in human extraocular neuromuscular junctions and muscle spindles. Invest Ophthalmol Vis Sci. 2009;**50**:3226–3232. DOI: 10.1167/iovs.08-3333

[8] Janbaz AH, Lindström M, Liu JX, Pedrosa Domellöf F. Intermediate filaments in the human extraocular muscles. Invest Ophthalmol Vis Sci. 2014;**55**:5151–5159. DOI: 10.1167/iovs.14-14316

[9] Furst DO, Osborn M, Weber K. Myogenesis in the mouse embryo: differential onset of expression of myogenic proteins and the involvement of titin in myofibril assembly. J Cell Biol. 1989;**109**:517–527. DOI: 10.1083/jcb.109.2.517

[10] Small JV, Furst DO, Thornell LE. The cytoskeletal lattice of muscle cells. Eur J Biochem. 1992;**208**:559–572. DOI: 10.1111/j.1432-1033.1992.tb17220.x

[11] Tokuyasu KT, Dutton AH, Singer SJ. Immunoelectron microscopic studies of desmin (skeletin) localization and intermediate filament organization in chicken skeletal muscle. J Cell Biol. 1983;**96**:1727–1735. DOI: 10.1083/jcb.96.6.1727

[12] Fuchs E, Weber K. Intermediate filaments: structure, dynamics, function, and disease. Ann Rev Biochem 1994;**63**:345–382. DOI: 10.1146/annurev.bi.63.070194.002021

[13] Paulin D, Li Z. Desmin: a major intermediate filament protein essential for the structural integrity and function of muscle. Exp Cell Res. 2004;**301**:1–7. DOI: 10.1016/j.yexcr.2004.08.004

[14] Capetanaki Y, Bloch RJ, Kouloumenta A, Mavroidis M. Psarras S. Muscle intermediate filaments and their links to membranes and membranous organelles. Exp Cell Res. 2007;**313**:2063–2076. DOI: 10.1016/j.yexcr.2007.03.033

[15] Wasicky R, Ziya-Ghazvini F, Blumer R, Lukas JR, Mayr R. Muscle fiber types of human extraocular muscles: a histochemical and immunohistochemical study. Invest Ophthalmol Vis Sci. 2000;**41**:980–990. Apparently there is no DOI for this article. This is the link to the article: http://iovs.arvojournals.org/article.aspx?articleid=2123010

[16] Fischer D, Bakay M, Gorospe JR, Kjellgren D, Pedrosa-Domellöf F, Hoffman E, Khurana TS. Definition of the unique human extraocular muscle allotype by expression profiling. Physiol Genomics. 2005;**22**:283–291. DOI: 10.1152/physiolgenomics.00158.2004

[17] Hoh JF, Hughes S, Hoy JF. Myogenic and neurogenic regulation of myosin gene expression in cat jaw-closing muscles regenerating in fast and slow limb muscle beds. J Mus Res Cell Motil. 1988;**9**:59–72. DOI: 10.1007/bf01682148

[18] Emery AEH. The Muscular Dystrophies. Oxford: Oxford University Press; 2001. 330 p. DOI: 10.1016/S0960-8966(03)00006-3 DOI:10.1016/S0960-8966%2803%2900006-3

[19] Shah F, Berggren D, Holmlund T, Levring Jäghagen E, Stål P. Unique expression of cytoskeletal proteins in human soft palate muscles. J Anat. 2016;**228**:487–494. DOI: 10.1111/joa.12417

[20] Carlsson L, Thornell LE. Desmin-related myopathies in mice and man. Acta Physiol Scand. 2001;**171**:341–348. DOI: 10.1046/j.1365-201x.2001.00837.x

[21] Goldfarb LG, Olive M, Vicart P, Goebel HH. Intermediate filament diseases: desminopathy. Adv Exp Med Biol. 2008;**642**:131–164. DOI: 10.1007/978-0-387-84847-1_11

Chapter 11

The Actomyosin Network and Cellular Motility: A S100A4 Regulatory View into the Process

Stephane R. Gross

Additional information is available at the end of the chapter

http://dx.doi.org/10.5772/66940

Abstract

Cell migration is a fundamental process responsible for numerous physiological and physiopathological conditions such as inflammation, embryogenesis and cancer. This central aspect of cell biology has seen quantum leaps in our understanding of the coordinated regulations, both spatial and temporal of numerous cytoskeletal proteins and their orchestrations. At the molecular level, this dynamic cellular process can be naively summarised as an engineered cycle composed of three distinct phases of (1) formation of cellular protrusion to initiate contact followed by (2) the adhesion with the external environment/cell-extracellular established connection and (3) the actomyosin force generation to consequently remodel the cytoskeleton. A prominent factor that regulates cellular motility is S100A4, a protein that has received constant attention for its significant role in cellular migration. Consequently, and in order to focus further the impact of this work, the present chapter aims to review some of the actomyosin proteins/complexes that have been demonstrated to be crucial players of the cyclic migration process but are also S100A4 interactors. In doing so, this chapter aims to capture a picture of how expression of this small, calcium-binding protein may, in essence, remodel at different levels the actin organisation and fulfil the motility engineered cycle of protrusion, attachments and contractions.

Keywords: S100A4, Actin, Arp2/3, formin, tropomyosin, myosin, Rho-GTPAses, Rhotekin

1. Introduction

Cellular motility has been an essential cellular phenomenon throughout phylogeny that has allowed organisms to survive, adapt and prosper in different environments. It is engrained in the chemoattraction and nutrient-seeking mechanisms in protozoa such as *Dictyostelium discoideum* [1]; whilst in metazoan, it is found to be a key concept for physiological regulations during all aspects of life. For instance, cellular migration in the early stages of gastrulation allows the coordinated movements of progenitor cells for the subsequent development of the different layers of precursor tissues and organs [2, 3]. Equally important is the profound effects cellular migration occupies in the process of healing during wound closure and/or tissue regeneration undertaken by tissues of the mesenchyme or epithelium [4]. Cell motility also plays essential functions during all stages of the immune response, from the development of mature effector cells, to endothelium trans-crossing and phagocytosis [5–7]. Given the indispensable roles of cellular migration in these events, and others, it is therefore not surprising to learn that loss of functions of many actin-regulating genes result in embryonic lethality or severe immunodeficiency syndromes [8].

Other than these physiological conditions, cellular motility is essential in regulating some of the physiopathological steps seen in disease. As example, it is well-documented that cellular migration is one of the prominent factors involved in the later stages of carcinogenesis and the subsequent phases of metastasis [9–11]. Cancer cell dissemination is clearly dependent upon the ability of migratory tumour cells to evade away from their initial niche, leading to the colonisation and formation of distant secondary lesions in the body [12].

At the molecular level, cell migration requires the coordinated regulations, both spatial and temporal of numerous cytoskeletal proteins, to orchestrate the dynamic cellular processes needed for cells to acquire movement. In this context, the actin cytoskeleton and the closely linked myosin network play essential functions [13, 14]. The process of cellular motility can be summarised as an engineered cycle composed of three distinct phases which are, (1) formation of cellular protrusion in the forms of lammelipodia and fillopodia to initiate contact and adhesion with the external environment, (2) regulation of cell-extracellular matrix established connections, usually integrin-dependent, and (3) force generation by the actomyosin network which will control both the structure and organisation of the motile architecture [15]. I provide here a brief overview of some of the different elements and protein complexes that are regulated during this migratory cycle, focusing primarily on specific components of the actomyosin complexes.

A group of low-molecular weight polypeptides that has been demonstrated to have key functions in remodelling the overall actin cytoskeletal network is the S100 protein family [16]. Composed of approximately 25 members, the presence of the majority of these in different cellular systems, both in vivo and in vitro has been associated with significant changes in cellular migration. One of the most prominent members of this family to have been linked to regulate cellular motility is S100A4, a protein that has received constant attention for its significant role in promoting cancer metastasis [16–18]. Consequently, and in order to emphasis the impact of this work and strengthened its delivery, I have concentrated our attentions on actomyosin proteins/complexes that have both been demonstrated to be crucial players of the migration process but also S100A4 interactors. In doing so, this chapter aims to capture a picture of how expression of this small, calcium-binding protein may in essence remodel at different levels the actin organisation and fulfil the motility engineered cycle of protrusion, attachments and contractions.

2. The actomyosin machinery in cellular migration

Motility can be seen as a lone activity where a single cell may migrate (also known as amoeboid or mesenchymal migration [19, 20]) or is referred to as collective, if this effort is the result of concerted effort undertaken by numerous cells, either in sheet or clusters [9]. Equally important is the cell physiognomy that will be regulated in the process. Mesenchymal motility as seen during fibroblast migration leads to cellular characteristic of a more elongated spindle-like shape. In this type of migration, an actin-rich leading edge can be observed, where extension of the front leading edge is driven by actin polymerisation [21]. In amoeboid migration, cells adopt a more rounded morphology and rely on the contraction-based membrane blebbing and enriched levels of myosin II at the cell rear [22]. Both of these migratory processes have been shown to play important roles in both physiological and pathological events.

The complexity of the different types of cell migration that can be used is mirrored by the number of different molecular pathways that are available to orchestrate these processes. Among them, however, the remodelling of the actin cytoskeleton and its organisation stands as an irreplaceable circuitry which is undisputably common to all. At the molecular level, this network is considered to provide the platform where physical forces will be exerted. Pushing forces generated by the assembly of filamentous actin (F-actin) will encourage the formation of cellular protrusions, such as filopodia, lamellipodia, blebbing and the most recently characterised invadopodia [23–25]. These changes in actin polymerisation and their dynamics will be directly responsible for reshaping and remodelling the underling plasma membranes.

2.1. Cellular protrusions and regulators

Actin polymerisation. The actin filaments are considered to be the backbone of cellular protrusion, providing the physically necessary special platform that will provide sufficient force to deform the plasma membrane. Their overall organisation relies primarily on their polymerisation from monomeric globular actin (G-actin) into long arrays. This process is regulated by numerous partners but the core regulator lies in ATP hydrolysis to promote actin molecule recognition and bonding between two monomers. When ATP bound G-actin is hydrolysed, the newly created ADP+Pi G-actin structure can form stable filaments. Binding of the nucleotide takes place in the high-affinity binding site located in the deep upper inter-domain cleft of actin (**Figure 1**). The presence of a cleft around exposed subdomains II and IV results in the polarisation of the monomeric structure and is referred to as the pointed end (**Figure 1**). The other exposed side, composed of subdomains I and III is known as the barbed end [26] and constitute the major binding site for most actin binding proteins ([27], **Figure 1**). This is a very important distinction which will result in sticking difference in behavioural characteristics in both G-actin and F-actin, of which polarised polymerisation is only one aspect.

In the early stages of assembly, also known as nucleation, actin protomers aggregate in an energetically unfavourable process to form a dimer that is more likely to dissociate. Addition of another subunit stabilises the complex and represents the nucleus, a state where actin polymerisation is now more favourable than dissociation (**Figure 1**). The association of monomers

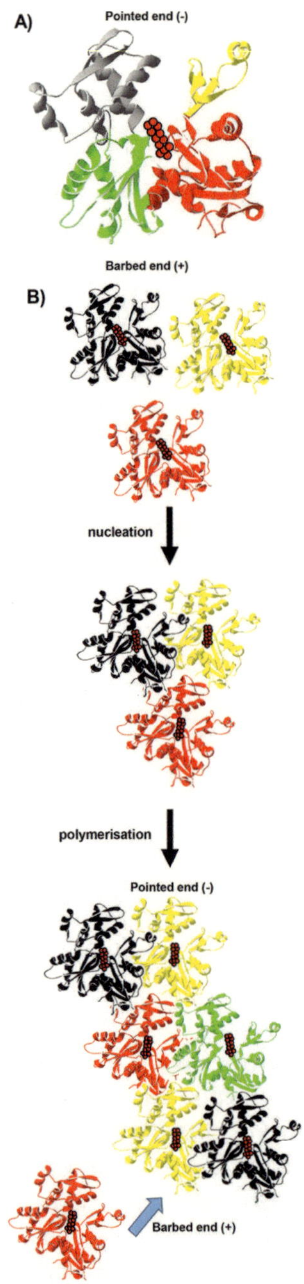

Figure 1. Actin structure and cartoon of F-actin polymerisation. (A) G-actin monomer at 1.54-Å resolution bound to ADP (PDB code 1J6Z) by Otterbein et al. [165] obtained from striated rabbit muscle tissue. Subdomain I (red, residues 1–32, 70–144 and 338–374), subdomain II (yellow, residues 3369), subdomain III (green, residues 145–180 and 270–337) and subdomain IV (grey, residues 181–269) are highlighted, resulting in the orientation of the actin molecule with the pointed end (– end) and the nucleotide cleft in the upper part, and the barbed end (+ end) in the lower part. (B) Process of actin polymerisation highlighting the steps of nucleus formation and filament formation. Please note this is a schematic representation which does not illustrate the current model of actin polymerisation initially proposed by Holmes et al. [166] suggesting that actin filaments are structured as a two right handed long pitch helices of head to tail bound actin subunits or a single left handed short pitch helix with consecutive lateral subunits staggered with respect to another by half a monomer length.

into a trimeric structure is seen as the rate limiting step of the whole polymerisation process as it is reversible where monomers can easily dissociate [28–30]. It is during the stage of nucleation that addition of further actin subunits is supported at both ends. Once the nucleus and newly added monomers have been locked into position by conformational changes, the process of elongation begins and the addition of actin molecules at the barbed end of the filament can be seen, resulting in the formation of structural polarised complexes (**Figure 1**). Whilst G-actin subunits can self-assemble, this process only occurs if the concentration of actin exceeds a critical concentration.

Within cells, a growing number of binding partners, or actin-binding proteins, will act both antagonistically and agonistically to regulate the polymerisation process. Some factors will act as nucleators, such as formins and Arp2/3, facilitating the process through providing a scaffold structure which encourages de novo assembly. Others will regulate the overall structure of filaments through their remodelling in larger structures. Examples provided here will control the cross-linked state of actin filaments through the involvement of bundling regulators such as the tropomyosins and to an extent myosins. Involvements of all these factors, as well as many others that are too numerous to be listed, here, will be responsible for the remodelling of the actin cytoskeleton into different substructures seen during cell migration (**Figure 2**).

When grown in a 2D environment, cells will encourage the formation of differential planar filamentous actin, in the form of filipodia/microvilli or sheet-like structures referred to as lamellipodia (**Figure 2**, [31, 32]). Whilst the former act as sensory organelles that enable cells to probe their local environment, through the formation of thin extensions that are mainly made of long, unbranched bundles, the latter is viewed as the main driving force for locomotion, through the organisation of short branched actin networks (**Figure 2**). In both instances, however, regulation of F-actin polymerisation, especially at their barbed end is essential, in order to control their elongation in the direction of the plasma membrane and is thought to require nucleation-promoting factors where both formins and Arp2/3 have been shown to play key functions (**Figure 2**, [31, 33]).

Formins. The family of formins, encoded by 15 different genes in mammals represent a cluster of large multi-domain proteins, grouped in eight different subfamilies, that regulate actin nucleation and polymerisation, primarily at the barbed end [34, 35]. Their nucleation abilities are regulated by signature regions of the proteins, the formin homology domains 1 and 2 (FH1/FH2), located at the C-terminus (**Figure 3**). Although a clear picture as to how formins nucleate the assembly of actin filament is still under investigation, the C-terminal region has been demonstrated to be a key regulator as it recruits actin monomers in the presence of profilin. The FH2 domain also plays key function during the polymerisation of F-actin as it allows addition of large amounts of actin subunits at the barbed end [36]. This continuous tracking results from alternate contact of the two halves of the FH2 domain with the two most terminal actin subunits in the filament, allowing the sliding of the whole formin molecule through an open/closed conformation as the subunits, remaining bound as subunits are added [37–39]. For some formins, activation is also controlled through the release of the head to tail auto-inhibition as well as through the movement of proteins away from the leading edge [40]. For such formins, classified as diaphanous-related formins (DRF) [41], comprised

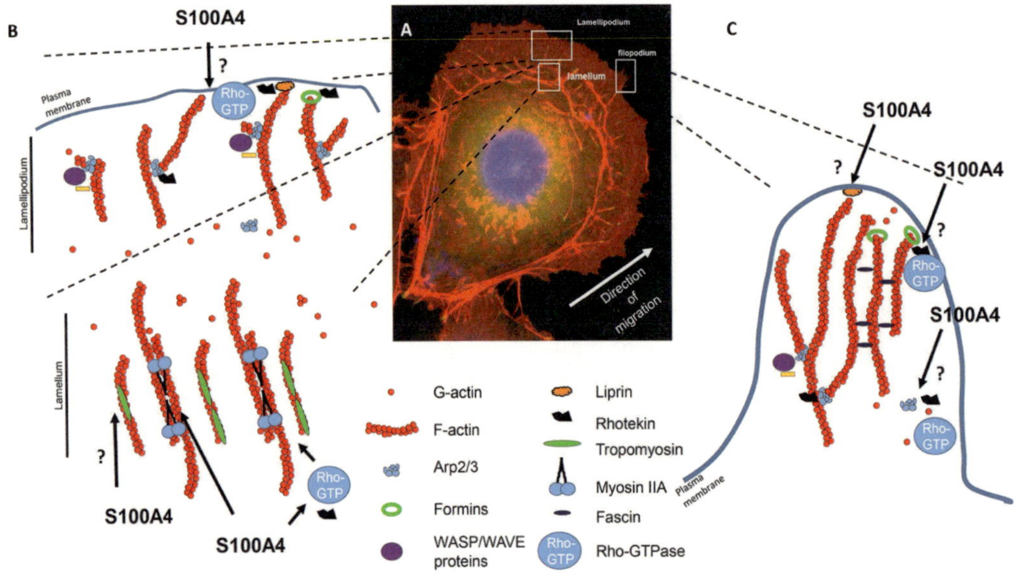

Figure 2. S100A4 interacting actomyosin complexes and their simplified organisation in the different protrusions of a migrating cell. (A) Actin and focal adhesion organization in a HeLa migrating cell. Staining for F-actin using Phalloidin-rhodamin (red) and paxillin with antibodies coupled to FITC (green) in a migrating HeLa cell. In this image, the actin mediated structures of the filopodium and lamellipodium/lamellum are distinctly visible at the leading edge of the cell. B and C present models for the lamellipodium/lamellum and filopodium and the respective molecular organisation within, focusing on the proteins presented in this chapter. (B) A simplified model for lamellipodium/lamellum formation. In the lamellipodium, the Arp2/3 complex via activation by WASP/WAVE complex interacts with actin filaments resulting in the nucleation of new actin filaments from the side of existing filaments. Formin proteins are also found at the barbed end of filaments. Limprin and the Rho-GTPase-Rhotekin complexes could get regulated by S100A4 to promote lamellipodia protrusions. In the lamellum, tropomyosin wrapping around the actin filaments prevents interactions with other actin binding proteins. NMMIIA regulates retrograde flow in the lamellum. At the interface of the lamellipodium–lamellum, actin is depolymerised. Interactions of S100A4 with tropomyosins and the NMMII complexes have been reported and could result in significant changes in their overall organisation. (C) A simplified model for filopodia formation. In this diagram, actin polymerisation promoted by the Arp2/3 complex leads to the branching and extension of nascent individual actin filaments in the filopodium. Recruitment of the formins to this location promotes the elongation of the filaments through the addition of actin monomers at the barbed end. Other actin bundling proteins such as fascin regulates filopodia stability through the clustering of actin filaments. Both the Limprin and the Rho-GTPase-Rhotekin complexes could be regulated by S100A4 to control filopodial protrusions.

of 4 families; diaphanous (Dia including mDia), Dishevelled associated activators of morphogenesis (Daam), formin-like proteins (FMNL) and FH1/FH2 domain-containing proteins (FHOD) in mammals, the auto-inhibitory mechanism relies on the folding of the N-terminal portion, containing domains FH3, which physically obstructs the diaphanous autoregulatory domain (DAD) at the C-terminus and prevents it to interact with actin molecules (**Figure 3**, [42]). Binding of the Rho-GTP to the formin polypeptide in the GBD (GTPase binding domain) region is thought to result, at least in part, to the displacement of the masking DAD region away from the FH3 domain [43]. Molecular mechanisms to explain this process are currently being investigated. The relocalisation of formin to the leading edge is also a key concept to control their activities. Membrane relocalisation has been reported to be performed primarily by Rho-GTPases through their binding to the GBD [44]. Other studies have also revealed that

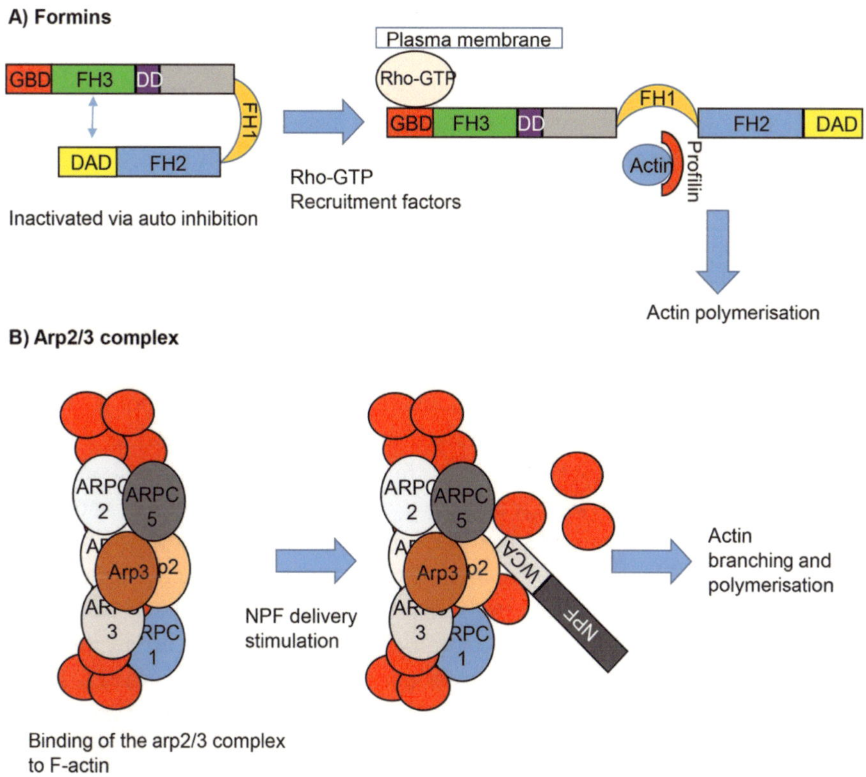

Figure 3. Cartoon showing some of the regulation steps for different actin nucleating proteins. (A) Activation of the Diaphanous-related formin. Autoinhibition of the actin nucleating ability is due to the interaction of the C-terminal Diaphanous auto-regulation domain (DAD) with the N-terminal FH3 (Formin homology) domain. Rho-GTP Binding of the GTP bound Rho to the GTPase binding domain (GBD) region is thought to lead to a partial displacement of the DAD as well as relocalisation of the complex, resulting in the unfolding of the protein and the relieve of the autoinhibition (DD dimerization domain). (B) Cartoon representation of the Arp2/3 complex and nucleation of a branched filament. The Arp2/3 complex initially binds to the pointed end of the mother F-actin. Binding of the WCA domain of a nucleation promoting factor (NPF) to exposed regions of Arp2 and Arp3 allows the delivery of actin monomer and initiate the polymerisation of a nascent branched filament as elegantly demonstrated [52].

the FH3 and DD (dimerization domain) regions on mDia also mediate its membrane localisation [45, 46], indicating that other proteins capable of interaction with such domains could be efficient regulators. The liprin family have been suggested to possess such properties and have been put forward as another series of proteins which may affect formin cellular functions [47].

Arp2/3 complex. Another regulator of actin nucleation and polymerisation that plays a critical role in the process of formation of lamellipodia and filopodia structures is the 220kDa Arp2/3 factor [48]. Composed of seven different subunits (ARPC1-5, Arp2 and Arp3), this complex promotes the formation of newly formed actin filament from the sides of existing filaments, forming a 70° side-branched network from pre-existing filaments [49, 50] (**Figure 3**). This property is predominantly the result of a striking similarity between the Arp2 and Arp3 proteins and that of monomeric actin molecules [51], providing a mimicking dimer that serves as a cooperative docking for other actin subunits and in doing so, accelerates the nucleation process and thereby reduces the rate of the limiting step at this stage [52]. Whilst all components of this hetero-heptamer are critical for the generation of newly formed actin arrays from the pointed end, albeit with distinct functions, the Arp2 and Arp3 proteins are seen as the principal components responsible for establishing the initial base of the newly assembled filament [52]. The other components, especially ARPC1, are mainly involved in the binding to the mother filament [53, 54] (**Figure 3**). Interestingly, weak basal activity of the purified Arp2/3 complex in promoting actin nucleation and branch formation [55, 56] highlights its intrinsic association with other regulators [57]. Activation of the Arp2/3 complex is regulated by different complexes at distinct cellular locations. Whilst Arp2/3 is controlled by the WAVE regulatory complex in a Rac-GTPase pathway in lamellipodia, the Wiskott-Aldrich syndrome protein family (WASP), downstream of Cdc42, is predominantly implicated with the regulation of Arp2/3 in filopodia [58]. By all accounts, these nucleation-promoting factors (NPF) stimulate Arp2/3 mediated-nucleation through a WCA domain found at the C-terminus (**Figure 3**). It is thought that the WH2 region within the WCA domain is responsible for binding and therefore delivering the actin monomer, whilst the CA sequence promotes binding to the exposed regions of both Arp2 and Arp3 [59]. It is the clustering of the different subunits, along with the newly added actin molecule that encourages formation of a new nucleus and further actin polymerisation, resulting in the elongation of 70° side-branched network. Since the NPF family has been continuously expanding, it is now subcategorised into five groups including WASp and neural Wiskott-Aldrich syndrome protein (N-WASP), three SCAR/WAVE proteins and the recently identified factors WASH, WHAMM and JMY [60].

Taken all together, actin polymerization at the leading edge is a vital process for cellular migration, through the orchestrations of events that will ultimately lead to different cellular protrusion events. In this section, different actin polymerization factors and their functions (Arp2/3 and Formin) were briefly explored. One should remember that this is only a preferential view in regards to their potential involvements through a S100A4-dependent process and that numerous other regulators not mentioned here play equally vital roles in the process of actin remodelling and cellular migration.

Away from the leading edge and the protrusions of the lamellipodia and filopodia, the array of filamentous actin is seen to exist as more bundling rather than the branched sheets reported previously, mainly due to the interaction of different actin-binding proteins. This contractile network is seen as a unique structural complex, spatially posterior to the lamellipodium, and is referred to as the lamellum [61].

2.2. Lamellum and cellular contractions

In the spatial arrangement of the lamellum, filaments are organised in different structures, known as stress, dorsal and ventral fibres; they are the result of interaction of the actin filaments with different partners (**Figure 2**). It is in this context and primarily through the control of the tropomyosin and myosins that contractile forces are exerted to manipulate their overall organisation. Generation of such tensile forces is provided by the myosin network, mainly non-muscle myosin II (NMMII) which is responsible for the majority of the morphological and architectural reorganisations that promote cell movement.

Tropomyosin. The tropomyosin (tpm) family is composed of four separate genes, TPM1-4 which can be further subdivided, due to different alternative splicing and post-translational modifications, resulting in the presence of more than 40 tpm products [62, 63]. Interestingly, these isoforms have been shown to interact differentially with actin filaments, ensuing biophysical and dynamic property changes, as well as different subpopulations occurring in different locations and in abundance [64]. It is unclear today, how these association-promoting mechanisms are regulated over time and space, to result in such highly selective and discriminatory organisation [65, 66], but all interactions necessitate dimerization as well as head to tail contact between individual complexes to form continuous actin/tropomyosin filaments [67]. The formation of these highly selective complexes is thought to seclude, or at least regulate the interactions of other actin-binding proteins with these actin filaments, therefore playing a major role in determining the functions of different filaments [68, 69]. For instance, the absence of tropomyosin in the leading edge is thought to be a predominant factor that allows specific branching of the actin network, since different isoforms have been shown to compete and inhibit the actin polymerisation of Arp2/3, at least in vitro [52, 70]. Equally important is the fact that tropomyosin has been implicated in the regulation and recruitments of NMMII in stress fibre formation [68], regulating both elasticity and stiffness [71]

The overall organisation of the actin cytoskeleton can also be dictated by actin bundling and contractile motor proteins. Binding of individual filaments, actin cross-linking and motor proteins allow the formation of thicker, linear and either paralleled or antiparallel filamentous F-actin networks that can be found in all subcellular localisations. In the lamellum, the class II non-muscle myosin family has been shown to be a key regulator, participating in the bundling of actin filaments and generating mechanical forces, which result in filaments sliding and/or contractions [72, 73].

Non-muscle Myosin II family. The myosin II family, which encompasses a group of 34 different isoforms, are expressed in all eukaryotes, except plants with 15 genes corresponding to the myosin II cluster. These myosin II motor proteins are exclusively expressed in non-muscles cells and can therefore be referred to as non-muscle myosin II (NMMII). In its fully formed state, the NMMII complex corresponds to a 525 kDa structure composed of six non-covalently associated polypeptides. The backbone of this is a homodimeric myosin heavy chain containing a head domain and a long coiled-coil rod domain, separated by a neck area. Two essential light chains and two regulatory light chains bind to this backbone [74]. The N-terminal head portion of the heavy chain is globular in structure and possesses the actin-binding domains as well as the ATPase activity which is required for movement towards the plus end of the actin filament, thereby inducing sliding between filaments and force generation. In contrast, the long coiled-coil C-terminal part of this protein is essential for dimerization and further assembly of one hexamer to another thereby forming a multimeric network of bipolar NMMII with motor domains positioned at both ends of the filaments. Bipolar filaments of NMMII formation are the result of electrostatic interactions between these C-terminal helical tails [75] and are essential for its cellular functions. Stability of these NMMII filaments is controlled by phosphorylation of the myosin heavy chain [76, 77] or by interaction with proteins that recognize the C-terminal helical tail region.

NMMII isoforms in the form of NMMIIA, NNMIIB and NMMIIC are found in most, but not all human cells and most mammals. NMMIIA and NMMIIB are expressed at similar levels in endothelial and epithelial cells. There is, however, little or no NMMIIC present in these cell subsets although it is known to be much more prominent in cells of lung and nervous tissues. Their location appears to be cell-specific and will be regulated differentially depending on the type of cell migration. Perhaps their expression patterns further reflects their function, since NMMIIA and NMMIIB play fundamental roles in mediating cell shape and matrix interactions during migration, whilst a clear role of NMMIIC in this process is still missing. As a consequence, during mesenchymal motility, NMMIIA is localised throughout the cells, in cellular protrusions and in both the lamellipodia and lamella of migrating cells [78] where, with Arp2/3-dependent actin polymerisation in the former, it results in actomyosin contraction and the retrograde flow of F-actin towards the cell body [61]. NMMIIB is predominantly found in the central and rear regions of the cell but not at the cell front [79].

Whilst formation of protrusions is a key element of cellular motility, it is equally critical that these newly created extensions also adhere and attach to the substratum, as their inability to do so result in their rearward movements in waves due to cellular tension [80]. Nascent adhesions are the first observable adhesive structures established in the lamellipodium [72, 81]. Their transient nature will force them to either disassemble quickly or mature into larger complexes known as focal adhesions which reside at the boundary of the lamellipodium and lamellum [82]. Actomyosin contraction is the main regulator that controls nascent adhesion enlargement [72]. This initial interaction with the extracellular environment will in turn induces activation of downstream effectors transmitted to the plasma membrane to encourage further integrin adhesion and clustering through their activation as well as changes in confirmation of extracellular matrix proteins [83]. Following on from this integrin activation, other downstream cellular pathways are also instigated; these lead to the recruitment, in a force dependent manner, of numerous framework and adaptor proteins which associate to form an adhesome [84]. Indeed mechanical stretch generated during this process induces changes in structural protein configurations, thereby encouraging binding of certain factors to other partners, as is the case for paxillin, vinculin, talin and actin, three essential component of the adhesome/adhesion complex [85, 86] (**Figure 4**).

NMMII cross-linking and contractile functions to remodel the actin cytoskeleton are regulated by phosphorylation of (1) regulatory light chains (RLC), (2) heavy chains (HC) and (3) myosin's ability to assemble into filaments. The RLC polypeptide can be specifically phosphorylated at different regulatory sites. The most prominent site is Ser19, leading, when phosphorylated, to a significant increase in ATPase activity of the head domain in the presence of actin [87]. Phosphorylation sites in the C-terminus of the myosin heavy chain have also been identified, although the true implications of such changes in ATPase activity remain unclear. Regulation of the reversible phosphorylation of specific residues is obtained through the activation of different kinases and phosphatases, depending on the residues considered. Whilst phosphorylation at residue Ser19 will be regulated by MLCK (myosin light chain kinase) and MYPT1 (myosin phosphatase target subunit 1), the most prominent examples, respectively [88, 89], the C-terminal region of the myosin heavy chain is directly targeted by kinases such as TRPM7 (transient receptor potential cation channel subfamily M member 7) and PKC (protein kinase C) [77, 90].

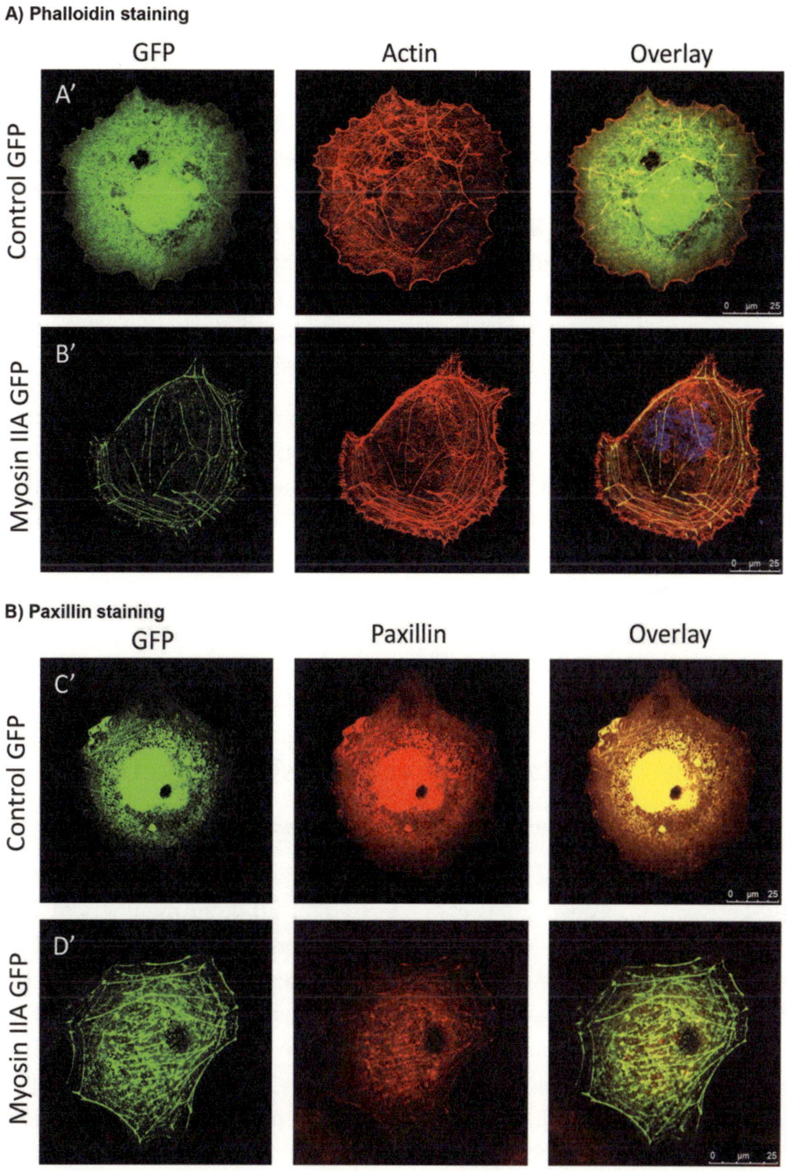

Figure 4. Myosin IIA expression in Cos7 cells regulates actin fiber formation and focal adhesion maturation. (A) Regulation of the actin organisation in Cos7 cells in the presence of NMMIIA. Cos7 cells were seeded into 24 well plates and left to incubate for 24 h. Cells were washed and fresh medium without antibiotics were added into each well 4 h prior to adding transfection mixture containing either the empty PeGFP-C3 plasmid (green cells labelled as A') or the PeGFP plasmid expressing NMMIIA (green cells labelled B'). Following a 48-h incubation, cells were fixed, permeabilised and stained either for actin (Phalloidin-rhodamine (red)) prior to mounting and viewing by epifluorescence microscopy. Note that the expression of NMMIIA leads to significant increases in stress fibres formation (B' where both NMMIIA and actin colocalise). (B) Focal adhesion organisation in Cos7 cells in the presence of NMMIIA. Cells grown as above were transfected with an empty PeGFP-C3 plasmid (green cells labelled as C') or the PeGFP plasmid expressing NMMIIA (green cells labelled D') prior to fixing, permeabilisation and immunostaining for paxillin using a mouse anti-paxillin primary antibody and a secondary anti-mouse rabbit Alexa-568 secondary antibody (red), prior to mounting and viewing by epifluorescence microscopy. Note that the expression of NMMIIA leads to significant increases in formation of paxillin cluster at the end of the myosin fibres (D' where paxillin foci are seen).

Compelling evidence also demonstrates that phosphorylation and interactions with the C-terminal region of NMMII regulate assembly of the complex into filaments. Such post-translational modifications also result in a change in overall structure. In its fully dephosphorylated form, the compact NMMII complex adopts an asymmetric state, unable to polymerise which is relaxed into an extended conformation following phosphorylation of RLC [91]. Phosphorylation of the C-terminal region of NMMII by kinases such as TRPM7 and PKC has also been shown to interfere with its abilities to form filaments [92]. Finally, it is predominantly through changes at the C-terminus, via the coiled-coil domain that NMMII assembly into filaments is controlled. Truncation experiments have demonstrated that the C-terminal region containing an ACD (assembly competence domain), as well as the C-terminal tail piece, are both important to promote correct assembly of NMMII into parallel and anti-parallel filaments [93, 94]

Another key regulator of motility is the cell's ability to orchestrate the different contraction and tension forces that are necessary to promote movement. In this section, lamellum components such as those of the NMMII and tropomyosin networks have been briefly reviewed. The overarching purpose here being once again to focalise the reader's mind on certain complexes, which are known interactors for the S100A4 proteins and could be key players in the process of motility observed when this protein is aberrantly expressed. Other pivotal factors that regulate motility via contractions have not been mentioned here but have been addressed in numerous other reviews and chapters [23, 95–101].

3. S100A4 protein regulation of the actomyosin cytoskeleton

Initially, referred to as mts1, FSP1, metastatin, p9Ka, PEL98, calvasculin, 42A and placental calcium-binding protein, S100A4 is a low-molecular weight acidic protein belonging to the S100 family. This small polypeptide is characterised, as all other members of this family, by a pair of calcium-binding helix-loop-helix regions referred to as EF-hand calcium-binding domains located at either side of the protein and separated by a hinge region [16, 102]. Over the years, S100A4 has received a large degree of interest in the field of cancer cell biology, since its expression is linked to increased motility and invasion directly promoting metastasis in animals, and it is now considered a potent marker for human metastasis and predictor for poor patient outcomes [17, 103]. Its expression is also observed in non-physiopathological states in different motile cell types in vivo, including those of the immune system (lymphocytes, macrophages and neutrophils) as well as mesenchymal fibroblastic cells. The biological implications of S100A4 expression on cellular migration are well-known [16], but the mechanisms required to attain such phenotypic changes are not fully characterised. In this part of the chapter, I will review its interactions with the different actomyosin components that have been discussed, whether actin nucleating activators, actin binding proteins or myosin regulators, highlighting how their functions are being affected in the presence of the S100A4 protein.

3.1. S100A4 regulates cellular protrusion

Expression of S100A4 has been correlated to significant changes in overall actin organisation and cellular extensions, with extensive increases in lamellipodia and forward protrusions

[104, 105] which, in turn, are thought to result in greatly enhanced cellular motility. Similarly, S100A4 has been found to be enriched at both the leading edge and in pseudopodia of migrating cells [104, 106]. A clear molecular explanation as to how S100A4 can regulate such processes is currently lacking, but different hypotheses have been formulated in view of what is known about its different binding partners and regulatory functions.

Actin interaction. Direct interaction and binding of S100A4 with actin has been demonstrated. This process is obtained, in the presence of calcium ions, at a ratio of 3 to 5:1. This high ratio of S100A4 to actin is intriguing, especially in view of that of other actin-binding proteins, suggesting possible oligomerisation of S100A4 with itself as well as its ability to cross-link actin filaments [107, 108]. Further association of S100A4 with actin stress fibres have been reported both in vitro and in vivo [108, 109], although it is important to highlight that it is unsure whether this interaction is the result of direct protein interactions or is the result of an association of S100A4 with other actin-associated factors. To this date, the biological consequences of such a partnership have not been deciphered and the true role of their interaction remains to be elucidated.

Liprin interaction. S100A4 has been shown to interact with the liprin family of proteins and both can be seen to colocalise at the plasma membrane [110]. Interaction of these two proteins leads to reduction in phosphorylation of the liprin $\beta 1$ molecule. Current thinking indicates that the liprin family of proteins are key promoters of cellular migration [47, 111]. Among possible functions to regulate cytoskeletal remodelling at the leading edge, liprins have been shown to interact with formin proteins and affect the appropriate localisation of the formin mDia to the plasma membrane. Their interaction is modulated by liprin binding to the FH3-DD domains on mDia, preferably when in the open configuration after Rho-GTPase activation (**Figure 3**, [47]). In cells, Liprin $\alpha 1$ has been shown to form a pentameric complex with another isoform, liprin $-\beta 1$ as well as LL5 and ERC1/ELKS proteins. This complex is located primarily at the leading edge of migrating cells and stimulates lamellipodia formation [112] as well as integrin-mediated focal adhesion stability [113]. Other reports further demonstrate the essential functions of liprin on stabilising lamellipodia and invadopodia [114]. The true involvement of S100A4 in the regulation of liprin and/or formin activities is currently missing, but one could hypothesise that the three protein functions may be intertwined. Given that S100A4 expression has been linked to filopodia instability [105], and the critical known role of formins in the process [58], one could for instance speculate and put forward a new formin-liprin-dependent function for this small calcium-binding protein.

Rhotekin interaction. S100A4 has recently been linked to into the Rho-GTPase pathway through its association with the scaffold protein Rhotekin. The Rho-GTPase family has been demonstrated to be a key regulator of the actin cytoskeleton structure and organisation. Initially, their roles appeared to remodel the formation of stress fibres, myosin activities and focal adhesion formation through modulations of its two main effectors ROCK and mDia [115, 116]. Activation of ROCK, a Rho-associated coiled- coil serine/threonine kinase, would result in phosphorylation of myosin phosphatases, which, in turn, would lead to increase phosphorylation of RLC and subsequent contractibility of NMMIIA (as discussed above). Actin polymerisation through the mDia axis would also be stimulated by ROCK and Rho-GTPase [116, 117], resulting in actin filament and ultimately fibre formation. More recent analysis clearly demonstrates that Rho activity is a key in the process of restructuring the leading edge in migrating cells, particularly RhoA [118–121]. Activation of RhoA has further been demonstrated to take place within a short distance of the leading edge, regulating cellular protrusion in the process, preceding the activation of other GTPases such as Rac1 and Cdc42 [120]. A ternary complex consisting of RhoA, Rhotekin and S100A4 has been shown to play a key role in hijacking this complex towards the leading edge to stimulate membrane ruffling *in lieu* of stress fibres [122]. S100A4 binding specifically to the Rho-binding domain of Rhotekin does not interfere with the association of the scaffolding protein with RhoA, and instead, leads to further activation of the complex, although the direct effectors of this process to fully explain how this enhances lamellae formation remain unknown [122].

NMMIIA interactions. The best studied S100A4 interacting partner is undoubtedly the NMMIIA heavy chain [92, 123–126]. Structural models have revealed that the S100A4 dimer binds both the C-terminal region of the coiled-coil ACD1 domain as well as the N-terminal part of the disordered tailpiece [127, 128]. Such interaction is thought to induce unwinding and destabilisation of the coiled-coil region, potentially disrupting intermolecular interaction between myosin molecules and subsequent disassembly and losses in contractibility [123, 128, 129].

Recent evidence indicates that the retrograde flow exerted by NMMII impedes extension of the leading edge and subsequent rates of protrusions because of reduced actin polymerisation [130, 131]. Therefore, a reduction in contractile forces, as observed when NMMII activity in cells is inhibited by blebbsistatin, leads to significant decreases in actin retrograde flow in the lamellum. This reduced flow leads, in turn, to subsequent increases in actin clustering into bundles at the lamellipodium–lamellum interface, and increased leading edge extension [132, 133]. The opposite experiment where activation of the NMMII complex is achieved through a Eph receptor/RhoA/ROCK signalling pathway was found to inhibit lamellipodial extension [134].

It is therefore rational to suggest that one of the protrusion-promoting abilities of S100A4 may be directly linked to its ability to interfere with formation of myosin fibres and their role in protrusion formation. Experimental support for a S100A4-NMMIIA role in cellular protrusion was obtained when S100A4 overexpression in tumour cells led to significant increase in leading edge protrusive activity [104]. This change in phenotype was shown indirectly to be NMMIIA dependent since the exogenous addition of antibody targeted towards the NMMIIA-binding site mimicked the cellular behaviour [104]. Interestingly, a reverse experiment also confirmed a clear role for S100A4 in this process, since S100A4 (−/−) macrophages demonstrated large amounts of over assembled NMMIIA filaments, leading to significant lamellipodia instability and reduced persistence [135]. Because NMMIIA has been clearly demonstrated to be the preferential cytoskeletal target for S100A4 to date, it is expected that their partnership may also influence the overall organisation of the actomyosin network in the lamellipodium and lamellum where the majority of the NMMIIA pools are located.

Whilst S100A4 expression in cells has undoubtedly been linked to regulation of cellular protrusions, mainly increases in lamellipodia and forward protrusions, and arguably being responsible at least in part for some of the enhanced cellular motility observed, the true molecular processes responsible are yet to be fully characterized. In this section, S100A4 partners such as Rhotekin, actin or liprin possibly found at the leading edge are presented in order to provide possible explanations as to how such biological properties may be regulated and it is expected that future researches will shed new lights into the true mechanisms that are involved in this process.

3.2. S100A4 functions in the lamellum and cellular contractions

Behind the cellular protrusion of the highly dynamic lamellipodium, the actomyosin network contributes to cell migration through contractibility and substrate adhesion [61, 72, 81]. Although direct evidence of S100A4 being localised in the lamella of migrating cells is still lacking, numerous reports have highlighted potential regulatory functions of the S100A4 within this subcellular fraction since its expression has been shown to lead to dramatic changes in numbers and organisation of focal adhesions and actin stress fibres [104, 105, 109, 122, 128, 136]. I will briefly discuss here the different properties that S100A4 encompasses towards a remodelling of this cellular architecture.

Tropomyosin interactions. The ability of S100A4 to bind to tropomyosin has been put forward both in vitro and in vivo [137], but with relatively low affinity if at all [138]. The true consequences of this interaction remain to be fully elucidated in regards to biological cellular consequences.

NMMIIA interactions. As discussed previously, binding of S100A4 leads to significant disassembly of NMMIIA filaments. Mechanical forces exerted by the myosin network have been shown to be key regulators for the growth and maturation of focal adhesions since altering contractility using either inhibitors [139] or knockout studies [79] results in impeded maturation and stabilisation. Although myosin is not physically present in the adhesion, it influences the process through its attachment to actin bundles with which adhesion is associated [79]. Whilst no direct evidence has been presented to demonstrate that S100A4 interacting with the myosin network results in the loss of adhesion of cells with their substratum, there is a compelling number of reports which have linked S10A4 expression to either stress fibre losses or reduction of focal adhesion stability or maturation [104, 105, 109, 122, 128, 136]. Similarly, regulating the assembly status of myosin via phosphorylation of sites in the C-terminal coiled coil and tailpiece regions have also been associated with reorganisation of the cytoskeleton and focal adhesions. Phosphorylation of serine residue S1943 results in disassembly of NMMIIA filaments [76] and leads in vivo to cytoskeletal reorganisation, whilst conversely inhibiting phosphorylation of the C-terminal tail pieces was shown to induce stabilisation of focal adhesion [140]. Other myosin-disassembling factors which bind to the C-terminal coiled coil and tailpieces regions such as S100P or lethal giant larvae have also been shown to regulate focal adhesions [141–143].

This final section briefly summarizes the best characterized and known S100A4 interactor, namely NMMII, in view of their high binding affinity, as well as by the number of reports highlighting their association. Yet again, whilst we are gathering further understanding related to their association and the different regions of the proteins that are responsible, a true model as to how their interactions regulate cellular motility remains elusive both theoretically and more importantly experimentally.

4. Conclusion

Controlling the actin cytoskeleton and the motility process is a key function, that when going awry, leads to significant pathological conditions. Not surprisingly, mutations or aberrant expression of all actin interacting proteins listed in this section have been liked to diseases, thought to be the result of significantly reduced cellular motility. Mutations in cytoplasmic actin have been related to autosomal dominant hearing loss [144]. Mutation or molecular mechanisms that result in changes of activity of actin-binding proteins, such as the nucleating facilitating protein complex Arp2/3 have been linked, albeit indirectly, to bacillary dysentery, as their functions are high jacked by the *Shigella* bacterial strains to disseminate in the colonic epithelium [145]. Involvement of the Arp2/3 complex and their regulator WASp have also been shown to lead to immunodeficiency and reduced platelet numbers, as the loss of WASp expression leads to the Wiskott-Aldrich syndrome [146]. Formin has also been linked to pathological conditions such as deafness [147] and immunity deficiencies, at different levels, presumably because of the differential expression of formins isoforms in different cell systems [148]. Mutation in NMMII has also been associated with a multitude of defects [89], with for instance, loss in NMMIIA affecting platelet and leukocyte dysfunction, renal diseases, loss to cataracts formation and neuronal disorders [14]. Equally important is the fact that differential expression of these factors and/or their activations are also key regulator of changes in cell division and migration/invasion, playing a predominant role in tumorigenesis and metastasis [14, 15, 149]. In this context, the high levels of S100A4 expression have also been shown to be significant determinant allowing cellular spreading and distant tumour formation. Beside its role in cellular motility through its interactions with actomyosin components which have been discussed throughout, S100A4 has also been demonstrated to play key roles in both cellular motility through the interactions with other partners. Both the Wnt/· catenin [150] and the AKT/slug [151] pathways have been shown to be capable of regulation by S100A4, leading to changes in cytoskeletal architecture and overall cellular migration. Other signalling pathways such as the PI3K/AKT/mTOR [152] or the NF-kB [153] route have also been shown to be capable of activation by S100A4 leading to significant changes in migration but without directly linking this process to cytoskeletal reorganisation. Another important aspect of S100A4 ability to encourage cellular migration and potential chemotaxis relates to its presence in the extracellular matrix. In the context of this chapter, I have concentrated our efforts to consider cytoplasmic S100A4, where it is found at concentration as high as 10 μM [154]. Traces of the protein have, however, also been reported both in medium of cultured cells [155, 156] and in biological fluids [157, 158], whether physiological or pathological, where it is thought to regulate the activities of metalloproteinases [159] or cellular receptors such as annexin2/plasmin and RAGE [156, 160]. Perhaps not surprisingly, different approaches have been utilised to combat the S100A4 motility and invasion inducing capability through the isolation of inhibitors. The result of this early work has demonstrated that some of these inhibitors regulate the interactions of S100A4 with the actomyosin network. For instance, trifluoperazine, a phenothiazine-based compound, has been reported to block S100A4 ability to depolymerise NMMIIA filaments [161, 162]. The true potential of this compound in cells and possibly in vivo, along with deciphering the mode of actions of other molecules that have been isolated as specific inhibitors of the S100A4 associated cell motility/

http://dx.doi.org/10.5772/66940

invasion/metastasis [163, 164] will pave the way for the development of further drugs that can regulate S100A4 interaction with the actomyosin architecture.

Acknowledgements

I would like to apologise for the numerous studies, which have significantly improved our understanding of actin dependent processes regulating cellular motility and invasion, but could not be included in this work owing to limits on the number of references. Special thanks are given to Prof. Philip S. Rudland (the University of Liverpool) for his critical comments on the chapter and for Prof. David Poyner (Aston University) for his invaluable help with the preparation of the different actin structures used.

Author details

Stephane R. Gross

Address all correspondence to: S.R.Gross@aston.ac.uk

School of Life and Health Sciences, Aston University, Aston Triangle, Birmingham, B4 7ET, UK

References

[1] Nichols, J.M., D. Veltman, and R.R. Kay, *Chemotaxis of a model organism: progress with Dictyostelium*. Curr Opin Cell Biol, 2015. **36**: p. 7–12.

[2] Ridley, A.J., et al., *Cell migration: integrating signals from front to back*. Science, 2003. **302**(5651): p. 1704–9.

[3] Solnica-Krezel, L. and D.S. Sepich, *Gastrulation: making and shaping germ layers*. Annu Rev Cell Dev Biol, 2012. **28**: p. 687–717.

[4] Lambrechts, A., M. Van Troys, and C. Ampe, *The actin cytoskeleton in normal and pathological cell motility*. Int J Biochem Cell Biol, 2004. **36**(10): p. 1890–909.

[5] Ridley, A.J., *Rho GTPase signalling in cell migration*. Curr Opin Cell Biol, 2015. **36**: p. 103–12.

[6] Freeman, S.A. and S. Grinstein, *Phagocytosis: receptors, signal integration, and the cytoskeleton*. Immunol Rev, 2014. **262**(1): p. 193–215.

[7] Yang, H., et al., *Changes of cytoskeleton affect T cell biological behaviors*. Front Biosci (Landmark Ed), 2015. **20**: p. 829–37.

[8] Moulding, D.A., et al., *Actin cytoskeletal defects in immunodeficiency*. Immunol Rev, 2013. **256**(1): p. 282–99.

[9] Mayor, R. and S. Etienne-Manneville, *The front and rear of collective cell migration.* Nat Rev Mol Cell Biol, 2016. **17**(2): p. 97–109.

[10] Frugtniet, B., W.G. Jiang, and T.A. Martin, *Role of the WASP and WAVE family proteins in breast cancer invasion and metastasis.* Breast Cancer (Dove Med Press), 2015. **7**: p. 99–109.

[11] Fife, C.M., J.A. McCarroll, and M. Kavallaris, *Movers and shakers: cell cytoskeleton in cancer metastasis.* Br J Pharmacol, 2014. **171**(24): p. 5507–23.

[12] Friedl, P. and S. Alexander, *Cancer invasion and the microenvironment: plasticity and reciprocity.* Cell, 2011. **147**(5): p. 992–1009.

[13] Vicente-Manzanares, M. and A.R. Horwitz, *Cell migration: an overview.* Methods Mol Biol, 2011. **769**: p. 1–24.

[14] Newell-Litwa, K.A., R. Horwitz, and M.L. Lamers, *Non-muscle myosin II in disease: mechanisms and therapeutic opportunities.* Dis Model Mech, 2015. **8**(12): p. 1495–515.

[15] Gross, S.R., *Actin binding proteins: their ups and downs in metastatic life.* Cell Adh Migr, 2013. **7**(2): p. 199–213.

[16] Gross, S.R., et al., *Joining S100 proteins and migration: for better or for worse, in sickness and in health.* Cell Mol Life Sci, 2014. **71**(9): p. 1551–79.

[17] Davies, B.R., et al., *Induction of the metastatic phenotype by transfection of a benign rat mammary epithelial cell line with the gene for p9Ka, a rat calcium-binding protein, but not with the oncogene EJ-ras-1.* Oncogene, 1993. **8**(4): p. 999–1008.

[18] Mishra, S.K., H.R. Siddique, and M. Saleem, *S100A4 calcium-binding protein is key player in tumor progression and metastasis: preclinical and clinical evidence.* Cancer Metastasis Rev, 2012. **31**(1–2): p. 163–72.

[19] Yilmaz, M. and G. Christofori, *Mechanisms of motility in metastasizing cells.* Mol Cancer Res, 2010. **8**(5): p. 629–42.

[20] Bear, J.E. and J.M. Haugh, *Directed migration of mesenchymal cells: where signaling and the cytoskeleton meet.* Curr Opin Cell Biol, 2014. **30**: p. 74–82.

[21] Pollard, T.D. and G.G. Borisy, *Cellular motility driven by assembly and disassembly of actin filaments.* Cell, 2003. **112**(4): p. 453–65.

[22] Poincloux, R., et al., *Contractility of the cell rear drives invasion of breast tumor cells in 3D Matrigel.* Proc Natl Acad Sci USA, 2011. **108**(5): p. 1943–8.

[23] Ridley, A.J., *Life at the leading edge.* Cell, 2011. **145**(7): p. 1012–22.

[24] Le Clainche, C. and M.F. Carlier, *Regulation of actin assembly associated with protrusion and adhesion in cell migration.* Physiol Rev, 2008. **88**(2): p. 489–513.

[25] Gimona, M., et al., *Assembly and biological role of podosomes and invadopodia.* Curr Opin Cell Biol, 2008. **20**(2): p. 235–41.

[26] Kabsch, W., et al., *Atomic structure of the actin:DNase I complex*. Nature, 1990. **347**(6288): p. 37–44.

[27] Dominguez, R., *Actin-binding proteins—a unifying hypothesis*. Trends Biochem Sci, 2004. **29**(11): p. 572–8.

[28] dos Remedios, C.G., et al., *Actin binding proteins: regulation of cytoskeletal microfilaments*. Physiol Rev, 2003. **83**(2): p. 433–73.

[29] Kasai, M., S. Asakura, and F. Oosawa, *The cooperative nature of G-F transformation of actin*. Biochim Biophys Acta, 1962. **57**: p. 22–31.

[30] Kasai, M., S. Asakura, and F. Oosawa, *The G-F equilibrium in actin solutions under various conditions*. Biochim Biophys Acta, 1962. **57**: p. 13–21.

[31] Skau, C.T. and C.M. Waterman, Specification of architecture and function of actin structures by actin nucleation factors. Annu Rev Biophys, 2015. **44**: p. 285–310.

[32] Blanchoin, L., et al., *Actin dynamics, architecture, and mechanics in cell motility*. Physiol Rev, 2014. **94**(1): p. 235–63.

[33] Bugyi, B. and M.F. Carlier, *Control of actin filament treadmilling in cell motility*. Annu Rev Biophys, 2010. **39**: p. 449–70.

[34] Breitsprecher, D. and B.L. Goode, *Formins at a glance*. J Cell Sci, 2013. **126**(Pt 1): p. 1–7.

[35] Schonichen, A. and M. Geyer, *Fifteen formins for an actin filament: a molecular view on the regulation of human formins*. Biochim Biophys Acta, 2010. **1803**(2): p. 152–63.

[36] Thompson, M.E., et al., *FMNL3 FH2-actin structure gives insight into formin-mediated actin nucleation and elongation*. Nat Struct Mol Biol, 2013. **20**(1): p. 111–8.

[37] Baker, J.L., et al., *Electrostatic interactions between the Bni1p Formin FH2 domain and actin influence actin filament nucleation*. Structure, 2015. **23**(1): p. 68–79.

[38] Paul, A.S. and T.D. Pollard, *Review of the mechanism of processive actin filament elongation by formins*. Cell Motil Cytoskeleton, 2009. **66**(8): p. 606–17.

[39] Goode, B.L. and M.J. Eck, *Mechanism and function of formins in the control of actin assembly*. Annu Rev Biochem, 2007. **76**: p. 593–627.

[40] Kuhn, S. and M. Geyer, *Formins as effector proteins of Rho GTPases*. Small GTPases, 2014. **5**: p. e29513.

[41] Alberts, A.S., *Identification of a carboxyl-terminal diaphanous-related formin homology protein autoregulatory domain*. J Biol Chem, 2001. **276**(4): p. 2824–30.

[42] Bechtold, M., J. Schultz, and S. Bogdan, *FHOD proteins in actin dynamics—a formin' class of its own*. Small GTPases, 2014. **5**(2): p. 11.

[43] Rose, R., et al., *Structural and mechanistic insights into the interaction between Rho and mammalian Dia*. Nature, 2005. **435**(7041): p. 513–8.

[44] Lammers, M., et al., *Specificity of interactions between mDia isoforms and Rho proteins*. J Biol Chem, 2008. **283**(50): p. 35236–46.

[45] Kato, T., et al., *Localization of a mammalian homolog of diaphanous, mDia1, to the mitotic spindle in HeLa cells*. J Cell Sci, 2001. **114**(Pt 4): p. 775–84.

[46] Gorelik, R., et al., *Mechanisms of plasma membrane targeting of formin mDia2 through its amino terminal domains*. Mol Biol Cell, 2011. **22**(2): p. 189–201.

[47] Sakamoto, S., et al., *Liprin-alpha controls stress fiber formation by binding to mDia and regulating its membrane localization*. J Cell Sci, 2012. **125**(Pt 1): p. 108–20.

[48] Swaney, K.F. and R. Li, *Function and regulation of the Arp2/3 complex during cell migration in diverse environments*. Curr Opin Cell Biol, 2016. **42**: p. 63–72.

[49] Goley, E.D. and M.D. Welch, *The ARP2/3 complex: an actin nucleator comes of age*. Nat Rev Mol Cell Biol, 2006. **7**(10): p. 713–26.

[50] Volkmann, N., et al., *Structure of Arp2/3 complex in its activated state and in actin filament branch junctions*. Science, 2001. **293**(5539): p. 2456–9.

[51] Robinson, R.C., et al., *Crystal structure of Arp2/3 complex*. Science, 2001. **294**(5547): p. 1679–84.

[52] Rouiller, I., et al., *The structural basis of actin filament branching by the Arp2/3 complex*. J Cell Biol, 2008. **180**(5): p. 887–95.

[53] Insall, R.H. and L.M. Machesky, *Actin dynamics at the leading edge: from simple machinery to complex networks*. Dev Cell, 2009. **17**(3): p. 310–22.

[54] Beltzner, C.C. and T.D. Pollard, *Identification of functionally important residues of Arp2/3 complex by analysis of homology models from diverse species*. J Mol Biol, 2004. **336**(2): p. 551–65.

[55] Mullins, R.D., J.A. Heuser, and T.D. Pollard, *The interaction of Arp2/3 complex with actin: nucleation, high affinity pointed end capping, and formation of branching networks of filaments*. Proc Natl Acad Sci USA, 1998. **95**(11): p. 6181–6.

[56] Welch, M.D., et al., *Interaction of human Arp2/3 complex and the Listeria monocytogenes ActA protein in actin filament nucleation*. Science, 1998. **281**(5373): p. 105–8.

[57] Rodnick-Smith, M., et al., *Role and structural mechanism of WASP-triggered conformational changes in branched actin filament nucleation by Arp2/3 complex*. Proc Natl Acad Sci USA, 2016. **113**(27): p. E3834–43.

[58] Campellone, K.G. and M.D. Welch, *A nucleator arms race: cellular control of actin assembly*. Nat Rev Mol Cell Biol, 2010. **11**(4): p. 237–51.

[59] Firat-Karalar, E.N. and M.D. Welch, *New mechanisms and functions of actin nucleation*. Curr Opin Cell Biol, 2011. **23**(1): p. 4–13.

[60] Rottner, K., J. Hanisch, and K.G. Campellone, *WASH, WHAMM and JMY: regulation of Arp2/3 complex and beyond*. Trends Cell Biol, 2010. **20**(11): p. 650–61.

[61] Ponti, A., et al., *Two distinct actin networks drive the protrusion of migrating cells*. Science, 2004. **305**(5691): p. 1782–6.

[62] Colote, S., et al., *Evolution of tropomyosin functional domains: differential splicing and genomic constraints*. J Mol Evol, 1988. **27**(3): p. 228–35.

[63] Geeves, M.A., S.E. Hitchcock-DeGregori, and P.W. Gunning, *A systematic nomenclature for mammalian tropomyosin isoforms*. J Muscle Res Cell Motil, 2015. **36**(2): p. 147–53.

[64] Schevzov, G., et al., *Tropomyosin isoforms and reagents*. Bioarchitecture, 2011. **1**(4): p. 135–164.

[65] Gunning, P.W., et al., *Tropomyosin—master regulator of actin filament function in the cytoskeleton*. J Cell Sci, 2015. **128**(16): p. 2965–74.

[66] Johnson, M., D.A. East, and D.P. Mulvihill, *Formins determine the functional properties of actin filaments in yeast*. Curr Biol, 2014. **24**(13): p. 1525–30.

[67] Tobacman, L.S., *Cooperative binding of tropomyosin to actin*. Adv Exp Med Biol, 2008. **644**: p. 85–94.

[68] Bryce, N.S., et al., *Specification of actin filament function and molecular composition by tropomyosin isoforms*. Mol Biol Cell, 2003. **14**(3): p. 1002–16.

[69] Skau, C.T. and D.R. Kovar, *Fimbrin and tropomyosin competition regulates endocytosis and cytokinesis kinetics in fission yeast*. Curr Biol, 2010. **20**(16): p. 1415–22.

[70] Bugyi, B., D. Didry, and M.F. Carlier, *How tropomyosin regulates lamellipodial actin-based motility: a combined biochemical and reconstituted motility approach*. EMBO J, 2010. **29**(1): p. 14–26.

[71] Jalilian, I., et al., *Cell elasticity is regulated by the tropomyosin isoform composition of the actin cytoskeleton*. PLoS One, 2015. **10**(5): p. e0126214.

[72] Choi, C.K., et al., *Actin and alpha-actinin orchestrate the assembly and maturation of nascent adhesions in a myosin II motor-independent manner*. Nat Cell Biol, 2008. **10**(9): p. 1039–50.

[73] Xu, X.S., et al., *During multicellular migration, myosin ii serves a structural role independent of its motor function*. Dev Biol, 2001. **232**(1): p. 255–64.

[74] Mooseker MS, F.B., The structural and functional diversity of the myosin family of actin-based molecular motors. In: Coluccio LM, editor (Netherlands) Springer, 2008: p. 1–34.

[75] Hostetter, D., et al., *Dictyostelium myosin bipolar thick filament formation: importance of charge and specific domains of the myosin rod*. PLoS Biol, 2004. **2**(11): p. e356.

[76] Dulyaninova, N.G., et al., *Regulation of myosin-IIA assembly and Mts1 binding by heavy chain phosphorylation*. Biochemistry, 2005. **44**(18): p. 6867–76.

[77] Clark, K., et al., *TRPM7 regulates myosin IIA filament stability and protein localization by heavy chain phosphorylation*. J Mol Biol, 2008. **378**(4): p. 790–803.

[78] Beach, J.R., et al., *Nonmuscle myosin II isoforms coassemble in living cells*. Curr Biol, 2014. **24**(10): p. 1160–6.

[79] Vicente-Manzanares, M., et al., *Regulation of protrusion, adhesion dynamics, and polarity by myosins IIA and IIB in migrating cells.* J Cell Biol, 2007. **176**(5): p. 573–80.

[80] Burnette, D.T., et al., *A role for actin arcs in the leading-edge advance of migrating cells.* Nat Cell Biol, 2011. **13**(4): p. 371–81.

[81] Alexandrova, A.Y., et al., *Comparative dynamics of retrograde actin flow and focal adhesions: formation of nascent adhesions triggers transition from fast to slow flow.* PLoS One, 2008. **3**(9): p. e3234.

[82] Giannone, G., et al., *Lamellipodial actin mechanically links myosin activity with adhesion-site formation.* Cell, 2007. **128**(3): p. 561–75.

[83] Friedland, J.C., M.H. Lee, and D. Boettiger, *Mechanically activated integrin switch controls alpha5beta1 function.* Science, 2009. **323**(5914): p. 642–4.

[84] Zaidel-Bar, R., et al., *Functional atlas of the integrin adhesome.* Nat Cell Biol, 2007. **9**(8): p. 858–67.

[85] Jiang, G., et al., *Two-piconewton slip bond between fibronectin and the cytoskeleton depends on talin.* Nature, 2003. **424**(6946): p. 334–7.

[86] Aguilar-Cuenca, R., A. Juanes-Garcia, and M. Vicente-Manzanares, *Myosin II in mechanotransduction: master and commander of cell migration, morphogenesis, and cancer.* Cell Mol Life Sci, 2014. **71**(3): p. 479–92.

[87] Somlyo, A.P. and A.V. Somlyo, *Ca^{2+} sensitivity of smooth muscle and nonmuscle myosin II: modulated by G proteins, kinases, and myosin phosphatase.* Physiol Rev, 2003. **83**(4): p. 1325–58.

[88] Adelstein, R.S. and M.A. Conti, *Phosphorylation of platelet myosin increases actin-activated myosin ATPase activity.* Nature, 1975. **256**(5518): p. 597–8.

[89] Betapudi, V., *Life without double-headed non-muscle myosin II motor proteins.* Front Chem, 2014. **2**: p. 45.

[90] Murakami, N., V.P. Chauhan, and M. Elzinga, *Two nonmuscle myosin II heavy chain isoforms expressed in rabbit brains: filament forming properties, the effects of phosphorylation by protein kinase C and casein kinase II, and location of the phosphorylation sites.* Biochemistry, 1998. **37**(7): p. 1989–2003.

[91] Liu, J., et al., *Refined model of the 10S conformation of smooth muscle myosin by cryo-electron microscopy 3D image reconstruction.* J Mol Biol, 2003. **329**(5): p. 963–72.

[92] Dulyaninova, N.G. and A.R. Bresnick, *The heavy chain has its day: regulation of myosin-II assembly.* Bioarchitecture, 2013. **3**(4): p. 77–85.

[93] Ronen, D. and S. Ravid, *Myosin II tailpiece determines its paracrystal structure, filament assembly properties, and cellular localization.* J Biol Chem, 2009. **284**(37): p. 24948–57.

[94] Nakasawa, T., et al., *Critical regions for assembly of vertebrate nonmuscle myosin II.* Biochemistry, 2005. **44**(1): p. 174–83.

[95] Akhshi, T.K., D. Wernike, and A. Piekny, *Microtubules and actin crosstalk in cell migration and division*. Cytoskeleton (Hoboken), 2014. **71**(1): p. 1–23.

[96] Colpan, M., N.A. Moroz, and A.S. Kostyukova, *Tropomodulins and tropomyosins: working as a team*. J Muscle Res Cell Motil, 2013. **34**(3–4): p. 247–60.

[97] Unbekandt, M. and M.F. Olson, *The actin-myosin regulatory MRCK kinases: regulation, biological functions and associations with human cancer*. J Mol Med (Berl), 2014. **92**(3): p. 217–25.

[98] Sackmann, E., *How actin/myosin crosstalks guide the adhesion, locomotion and polarization of cells*. Biochim Biophys Acta, 2015. **1853**(11 Pt B): p. 3132–42.

[99] Goicoechea, S.M., S. Awadia, and R. Garcia-Mata, *I'm coming to GEF you: regulation of RhoGEFs during cell migration*. Cell Adh Migr, 2014. **8**(6): p. 535–49.

[100] Wehrle-Haller, B., *Assembly and disassembly of cell matrix adhesions*. Curr Opin Cell Biol, 2012. **24**(5): p. 569–81.

[101] Olson, M.F. and E. Sahai, *The actin cytoskeleton in cancer cell motility*. Clin Exp Metastasis, 2009. **26**(4): p. 273–87.

[102] Barraclough, R., et al., *Molecular cloning and sequence of the gene for p9Ka. A cultured myo-epithelial cell protein with strong homology to S-100, a calcium-binding protein*. J Mol Biol, 1987. **198**(1): p. 13–20.

[103] Boye, K. and G.M. Maelandsmo, *S100A4 and metastasis: a small actor playing many roles*. Am J Pathol, 2010. **176**(2): p. 528–35.

[104] Li, Z.H. and A.R. Bresnick, *The S100A4 metastasis factor regulates cellular motility via a direct interaction with myosin-IIA*. Cancer Res, 2006. **66**(10): p. 5173–80.

[105] Goh Then Sin, C., et al., *S100A4 downregulates filopodia formation through increased dynamic instability*. Cell Adh Migr, 2011. **5**(5): p. 439–47.

[106] Kim, E.J. and D.M. Helfman, *Characterization of the metastasis-associated protein, S100A4. Roles of calcium binding and dimerization in cellular localization and interaction with myosin*. J Biol Chem, 2003. **278**(32): p. 30063–73.

[107] Flynn, A.M., P.S. Rudland, and R. Barraclough, *Protein interactions between S100A4 (p9Ka) and other cellular proteins identified using in vitro methods*. Biochem Soc Trans, 1996. **24**(3): p. 341S.

[108] Watanabe, Y., et al., *Calvasculin, as a factor affecting the microfilament assemblies in rat fibroblasts transfected by src gene*. FEBS Lett, 1993. **324**(1): p. 51–5.

[109] Mandinova, A., et al., *Distinct subcellular localization of calcium binding S100 proteins in human smooth muscle cells and their relocation in response to rises in intracellular calcium*. J Cell Sci, 1998. **111 (Pt 14)**: p. 2043–54.

[110] Kriajevska, M., et al., *Liprin beta 1, a member of the family of LAR transmembrane tyrosine phosphatase-interacting proteins, is a new target for the metastasis-associated protein S100A4 (Mts1)*. J Biol Chem, 2002. **277**(7): p. 5229–35.

[111] Sakamoto, S., S. Narumiya, and T. Ishizaki, *A new role of multi scaffold protein Liprin-alpha: liprin-alpha suppresses Rho-mDia mediated stress fiber formation*. Bioarchitecture, 2012. **2**(2): p. 43–49.

[112] Shen, J.C., et al., *Inhibitor of growth 4 suppresses cell spreading and cell migration by interacting with a novel binding partner, liprin alpha1*. Cancer Res, 2007. **67**(6): p. 2552–8.

[113] Asperti, C., et al., *Liprin-alpha1 promotes cell spreading on the extracellular matrix by affecting the distribution of activated integrins*. J Cell Sci, 2009. **122**(Pt 18): p. 3225–32.

[114] Astro, V., et al., *Liprin-alpha1, ERC1 and LL5 define polarized and dynamic structures that are implicated in cell migration*. J Cell Sci, 2014. **127**(Pt 17): p. 3862–76.

[115] Burridge, K. and K. Wennerberg, *Rho and Rac take center stage*. Cell, 2004. **116**(2): p. 167–79.

[116] Narumiya, S., M. Tanji, and T. Ishizaki, *Rho signaling, ROCK and mDia1, in transformation, metastasis and invasion*. Cancer Metastasis Rev, 2009. **28**(1–2): p. 65–76.

[117] Tsuji, T., et al., *ROCK and mDia1 antagonize in Rho-dependent Rac activation in Swiss 3T3 fibroblasts*. J Cell Biol, 2002. **157**(5): p. 819–30.

[118] Pertz, O., et al., *Spatiotemporal dynamics of RhoA activity in migrating cells*. Nature, 2006. **440**(7087): p. 1069–72.

[119] O'Connor, K.L., B.K. Nguyen, and A.M. Mercurio, *RhoA function in lamellae formation and migration is regulated by the alpha6beta4 integrin and cAMP metabolism*. J Cell Biol, 2000. **148**(2): p. 253–8.

[120] Machacek, M., et al., *Coordination of Rho GTPase activities during cell protrusion*. Nature, 2009. **461**(7260): p. 99–103.

[121] Kurokawa, K. and M. Matsuda, *Localized RhoA activation as a requirement for the induction of membrane ruffling*. Mol Biol Cell, 2005. **16**(9): p. 4294–303.

[122] Chen, M., A.R. Bresnick, and K.L. O'Connor, *Coupling S100A4 to Rhotekin alters Rho signaling output in breast cancer cells*. Oncogene, 2013. **32**(32): p. 3754–64.

[123] Li, Z.H., et al., *Mts1 regulates the assembly of nonmuscle myosin-IIA*. Biochemistry, 2003. **42**(48): p. 14258–66.

[124] Ford, H.L., et al., *Effect of Mts1 on the structure and activity of nonmuscle myosin II*. Biochemistry, 1997. **36**(51): p. 16321–7.

[125] Ramagopal, U.A., et al., *Structure of the S100A4/myosin-IIA complex*. BMC Struct Biol, 2013. **13**: p. 31.

[126] Kiss, B., et al., *Structural determinants governing S100A4-induced isoform-selective disassembly of nonmuscle myosin II filaments*. FEBS J, 2016. **283**(11): p. 2164–80.

[127] Kiss, B., et al., *Crystal structure of the S100A4-nonmuscle myosin IIA tail fragment complex reveals an asymmetric target binding mechanism*. Proc Natl Acad Sci USA, 2012. **109**(16): p. 6048–53.

[128] Elliott, P.R., et al., *Asymmetric mode of Ca(2)(+)-S100A4 interaction with nonmuscle myosin IIA generates nanomolar affinity required for filament remodeling*. Structure, 2012. **20**(4): p. 654–66.

[129] Badyal, S.K., et al., *Mechanism of the Ca(2)+-dependent interaction between S100A4 and tail fragments of nonmuscle myosin heavy chain IIA*. J Mol Biol, 2011. **405**(4): p. 1004–26.

[130] Cai, Y., et al., *Nonmuscle myosin IIA-dependent force inhibits cell spreading and drives F-actin flow*. Biophys J, 2006. **91**(10): p. 3907–20.

[131] Even-Ram, S., et al., *Myosin IIA regulates cell motility and actomyosin-microtubule crosstalk*. Nat Cell Biol, 2007. **9**(3): p. 299–309.

[132] Lim, J.I., et al., *Protrusion and actin assembly are coupled to the organization of lamellar contractile structures*. Exp Cell Res, 2010. **316**(13): p. 2027–41.

[133] Shih, W. and S. Yamada, *Myosin IIA dependent retrograde flow drives 3D cell migration*. Biophys J, 2010. **98**(8): p. L29–31.

[134] Astin, J.W., et al., *Competition amongst Eph receptors regulates contact inhibition of locomotion and invasiveness in prostate cancer cells*. Nat Cell Biol, 2010. **12**(12): p. 1194–204.

[135] Li, Z.H., et al., *S100A4 regulates macrophage chemotaxis*. Mol Biol Cell, 2010. **21**(15): p. 2598–610.

[136] Takenaga, K., Y. Nakamura, and S. Sakiyama, *Cellular localization of pEL98 protein, an S100-related calcium binding protein, in fibroblasts and its tissue distribution analyzed by monoclonal antibodies*. Cell Struct Funct, 1994. **19**(3): p. 133–41.

[137] Takenaga, K., et al., *Binding of pEL98 protein, an S100-related calcium-binding protein, to nonmuscle tropomyosin*. J Cell Biol, 1994. **124**(5): p. 757–68.

[138] Chen, H., et al., *Binding to intracellular targets of the metastasis-inducing protein, S100A4 (p9Ka)*. Biochem Biophys Res Commun, 2001. **286**(5): p. 1212–7.

[139] Pasapera, A.M., et al., *Myosin II activity regulates vinculin recruitment to focal adhesions through FAK-mediated paxillin phosphorylation*. J Cell Biol, 2010. **188**(6): p. 877–90.

[140] Dulyaninova, N.G., et al., *Myosin-IIA heavy-chain phosphorylation regulates the motility of MDA-MB-231 carcinoma cells*. Mol Biol Cell, 2007. **18**(8): p. 3144–55.

[141] Du, M., et al., *S100P dissociates myosin IIA filaments and focal adhesion sites to reduce cell adhesion and enhance cell migration*. J Biol Chem, 2012. **287**(19): p. 15330–44.

[142] Dahan, I., et al., *The tumor suppressor Lgl1 regulates NMII-A cellular distribution and focal adhesion morphology to optimize cell migration*. Mol Biol Cell, 2012. **23**(4): p. 591–601.

[143] Dahan, I., et al., *The tumor suppressor Lgl1 forms discrete complexes with NMII-A and Par6alpha-aPKCzeta that are affected by Lgl1 phosphorylation*. J Cell Sci, 2014. **127**(Pt 2): p. 295–304.

[144] van Wijk, E., et al., *A mutation in the gamma actin 1 (ACTG1) gene causes autosomal dominant hearing loss (DFNA20/26)*. J Med Genet, 2003. **40**(12): p. 879–84.

[145] Agaisse, H., *Molecular and cellular mechanisms of Shigella flexneri dissemination*. Front Cell Infect Microbiol, 2016. **6**: p. 29.

[146] Imai, K., S. Nonoyama, and H.D. Ochs, *WASP (Wiskott-Aldrich syndrome protein) gene mutations and phenotype*. Curr Opin Allergy Clin Immunol, 2003. **3**(6): p. 427–36.

[147] Lynch, E.D., et al., *Nonsyndromic deafness DFNA1 associated with mutation of a human homolog of the Drosophila gene diaphanous*. Science, 1997. **278**(5341): p. 1315–8.

[148] DeWard, A.D., et al., *The role of formins in human disease*. Biochim Biophys Acta, 2010. **1803**(2): p. 226–33.

[149] Randall, T.S. and E. Ehler, *A formin-g role during development and disease*. Eur J Cell Biol, 2014. **93**(5–6): p. 205–11.

[150] Zhang, K., et al., *S100A4 regulates motility and invasiveness of human esophageal squamous cell carcinoma through modulating the AKT/Slug signal pathway*. Dis Esophagus, 2012. **25**(8): p. 731–9.

[151] Sack, U., et al., *S100A4-induced cell motility and metastasis is restricted by the Wnt/beta-catenin pathway inhibitor calcimycin in colon cancer cells*. Mol Biol Cell, 2011. **22**(18): p. 3344–54.

[152] Wang, H., et al., *Activation of the PI3K/Akt/mTOR/p70S6K pathway is involved in S100A4-induced viability and migration in colorectal cancer cells*. Int J Med Sci, 2014. **11**(8): p. 841–9.

[153] Zhang, J., et al., *S100A4 regulates migration and invasion in hepatocellular carcinoma HepG2 cells via NF-kappaB-dependent MMP-9 signal*. Eur Rev Med Pharmacol Sci, 2013. **17**(17): p. 2372–82.

[154] Malashkevich, V.N., et al., *Structure of Ca^{2+}-bound S100A4 and its interaction with peptides derived from nonmuscle myosin-IIA*. Biochemistry, 2008. **47**(18): p. 5111–26.

[155] Forst, B., et al., *Metastasis-inducing S100A4 and RANTES cooperate in promoting tumor progression in mice*. PLoS One, 2010. **5**(4): p. e10374.

[156] Semov, A., et al., *Metastasis-associated protein S100A4 induces angiogenesis through interaction with Annexin II and accelerated plasmin formation*. J Biol Chem, 2005. **280**(21): p. 20833–41.

[157] Cabezon, T., et al., *Expression of S100A4 by a variety of cell types present in the tumor microenvironment of human breast cancer*. Int J Cancer, 2007. **121**(7): p. 1433–44.

[158] Ambartsumian, N., et al., *The metastasis-associated Mts1(S100A4) protein could act as an angiogenic factor*. Oncogene, 2001. **20**(34): p. 4685–95.

[159] Schmidt-Hansen, B., et al., *Extracellular S100A4(mts1) stimulates invasive growth of mouse endothelial cells and modulates MMP-13 matrix metalloproteinase activity*. Oncogene, 2004. **23**(32): p. 5487–95.

[160] Spiekerkoetter, E., et al., *S100A4 and bone morphogenetic protein-2 codependently induce vascular smooth muscle cell migration via phospho-extracellular signal-regulated kinase and chloride intracellular channel 4*. Circ Res, 2009. **105**(7): p. 639–47, 13 p following 647.

[161] Garrett, S.C., et al., *A biosensor of S100A4 metastasis factor activation: inhibitor screening and cellular activation dynamics*. Biochemistry, 2008. **47**(3): p. 986–96.

[162] Malashkevich, V.N., et al., *Phenothiazines inhibit S100A4 function by inducing protein oligomerization*. Proc Natl Acad Sci USA, 2010. **107**(19): p. 8605–10.

[163] Klingelhofer, J., et al., *Anti-S100A4 antibody suppresses metastasis formation by blocking stroma cell invasion*. Neoplasia, 2012. **14**(12): p. 1260–8.

[164] Sack, U., et al., *Novel effect of antihelminthic Niclosamide on S100A4-mediated metastatic progression in colon cancer*. J Natl Cancer Inst, 2011. **103**(13): p. 1018–36.

[165] Otterbein, L.R., P. Graceffa, and R. Dominguez, *The crystal structure of uncomplexed actin in the ADP state*. Science, 2001. **293**(5530): p. 708–11.

[166] Holmes, K.C., et al., *Atomic model of the actin filament*. Nature, 1990. **347**(6288): p. 44–9.

Chapter 12

Intermediate Filaments as a Target of Signaling Mechanisms in Neurotoxicity

Ariane Zamoner and Regina Pessoa-Pureur

Additional information is available at the end of the chapter

http://dx.doi.org/10.5772/66926

Abstract

In this chapter, we deal with the current knowledge and important results on the cytoskeletal proteins and their differential regulation by kinases/phosphatases and Ca^{2+}-mediated mechanisms in developmental rat brain. We focus on the misregulation of the phosphorylating system associated with intermediate filament proteins of neural cells and its relevance to cell and tissue dysfunction. Taking into account our findings, we propose that intermediate-filament proteins are dynamic structures whose regulation is crucial for proper neural cell function. Given their relevance, they must be regulated in response to extracellular and intracellular signals. The complexity and connection between signaling pathways regulating intermediate-filament dynamics remain obscure. In this chapter, we get light into some kinase/phosphatase cascades downstream of membrane receptors disrupting the dynamics of intermediate filaments and its association with neural dysfunction. However, intermediate filaments do not act individually into the neural cells. Our results evidence the importance of misregulated cytoskeletal crosstalk in disrupting cytoskeletal dynamics and cell morphology underlying neural dysfunction in experimental conditions mimicking metabolic diseases and nongenomic actions of thyroid hormones and as an end point in the neurotoxicity of organic tellurium.

Keywords: intermediate filament, cytoskeleton, cell signaling, calcium, neurotoxicity

1. Introduction

All the cell functions accomplished by the living cell are dependent on a sophisticated network of protein filaments with different compositions, distributions and roles into the cell, forming an integrated meshwork known as the cytoskeleton. However, the most striking feature of the cytoskeleton concerns its ability to respond to signals and conditions to which cells are submitted, taking part of adaptive cell response to different stimuli. The cytoskeleton is an end point of signaling pathways adapting cells to immediate or long-lasting behaviors in healthy and sick organisms.

Cytoskeleton of most animal cells is constituted by three interconnected filament subsystems: microfilaments (MFs), microtubules (MTs) and intermediate filaments (IFs). Compelling evidence from the last decades has brought convincing understanding for the highly regulated and interconnected interactions between the cytoskeletal elements giving support to sculpting and maintaining cell shape and sustaining all kinds of morphological alterations or internal organization, as well as their implications for the behavior of animal cells. **Figure 1** demonstrates the organization of the cytoskeleton in neurons.

A cohort of accessory proteins and signaling machinery regulates the dynamic turnover of the cytoskeleton. Although each type of filament has specific cell distribution, molecular constituents and equilibrium, the coordinated intertwining among the different networks provides the force for a number of coherent processes in response to all kinds of intra- and extracellular stimuli leading responses so decisive as cell survival or death [1].

This chapter initiates with a brief introduction about the structure and function of IFs, emphasizing those from neural cells. However, the main purposes of the chapter are the experimental evidence of our laboratory that the roles of IFs are beyond protection from mechanical and nonmechanical stress. They might be the end point of misregulated-signaling mechanisms in neurotoxic conditions adapting their dynamics, in concert with the other cytoskeletal fibers, to cell survival or death.

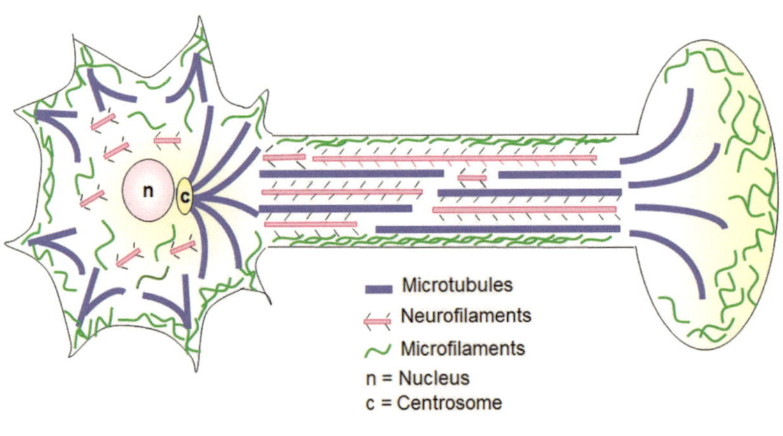

Figure 1. Distribution of cytoskeletal constituents into neurons. Neuronal cytoskeleton is composed by microfilaments, microtubules, and intermediate filaments. The microtubules are nucleated at the centrosome, then released and delivered to either the dendrites or the axon. Neurofilaments are abundant in axons and the spacing of neurofilaments is sensitive to the level of phosphorylation. The microfilaments are dispersed within the cells and they are most abundant near the plasma membrane.

http://dx.doi.org/10.5772/66926

2. Intermediate filaments

2.1. Molecular architecture of intermediate filaments

IFs are flexible, rod-shaped fibers averaging 10 nm in diameter, a size that is intermediate between MFs and MTs. They are ubiquitous constituents of the structural scaffold of the eukaryotic cells and considered mechanical integrators of cytomatrix [2]. These cytoskeletal filaments are widespread expressed in practically all animal cell types and are the most diverse cytoskeletal protein family, encoded by an estimated 70 IF genes in the humans. IFs have been grouped into six sequence homology classes (SHC) according to the degree of sequence identity: acidic keratins (SHC group I); basic keratins (SHC group II); desmin, vimentin and other mesenchymal IF proteins, such as glial fibrillary acidic protein (GFAP) (SHC group III); neurofilament proteins (SHC group IV); and lamins (SHC group V).

IF building blocks are fibrous proteins stabilized by multistranded left-handed coiled coils giving rise to a rope-like structure. Their structures are constituted by a long central α-helical region, also designed rod domain, with a distinct number of equally sized coiled coils forming segments flanked by non-α-helical N-terminal (the head domain) and C-terminal domains (the tail domain). Both head and tail domains are highly varying in size and sequence, thus, the functional and molecular heterogeneity of IF proteins are a consequence of the highly variable non-α-helical end domains of subunits.

The central rod domain of IF subunits is α-helical rod highly charged, with a role in the first phase of IF assembly. By contrast, the head domain enriched in basic amino acids is essential for the formation of tetramers (the polymerization units) and complete IF assembly.

The non-α-helical tail domain can vary drastically between different IF proteins. This domain is not essential for the assembly of cytoplasmic IFs but plays a significant role in filament width control. The functional role of the tail domain is particularly important in the neurofilaments, the neuronal-specific IFs, as discussed below.

Overall, the assembly of subunits giving rise to functional IFs is a complex and multistep process with individual specificities among the different representatives of this molecularly heterogeneous family. Taking into account the *in vitro* polymerization of vimentin, filament assembly starts with the formation of parallel, in-register dimers. These dimers spontaneously associate laterally into antiparallel, half-staggered tetramers. Tetramers aggregate into higher-order oligomers to form unit length filaments (ULFs) that undergo reorganization and elongation by longitudinal annealing to form immature IFs. The final step is radial compaction of the filaments from approximately16 nm to a diameter of 10–12 nm [3].

Different from the other IFs, NFs comprise three subunits with different molecular masses and distributions into the filament. They are formed by light, medium and heavy molecular mass NF triplet proteins (NF-L, NF-M and NF-H), respectively. NF-L can self-assemble forming the core of the filament. NF-M and NF-H are peripherally disposed on the filament, with their long and flexible tails rich in highly charged domains and multiple phosphorylation sites, radially projecting out from the filament backbone when NF-M and NF-H co-assemble with the short-tail NF protein NF-L. Interestingly, NF-H and NF-M by their own are not able to assemble into filaments, but by contrast, self-assembled NF-L yields normal looking 10-nm filaments. These side arms of NF-M and NF-H contain multiple phosphorylation sites regulating the interactions of NFs with each other and with other cytoskeletal structures [4].

2.2. Roles of intermediate filaments in neural cell function

Neurons are highly specialized in the transmission and processing of electrical and chemical signals. A functional nervous system is dependent of a proper axonal array, which in turn is critically dependent upon the organization of the axonal cytoskeleton. Five main subunit proteins form the neuronal specific NFs: the group IV NF-L, NF-M and NF-H triplet proteins, α-internexin and the group III peripherin. Mature filaments are composed of several combinations of these five subunits. In most differentiated neurons, α-internexin expression precedes that of the NF triplet and declines somewhat postnatally, while the expression of the NF triplet sharply rises. Neurofilaments found in perikarya, dendrites and axons differ considerably in their organization and function. Perikarial NFs form a meshwork around the nucleus. In the axons of mature neurons, a large number of longitudinally oriented and phosphorylated NFs play a fundamental role increasing the diameter of myelinated axons and consequently nerve conductivity. Neurofilaments present in dendrites are less abundant and less phosphorylated than those of axons.

Neurofilaments are transported from the cell body, where they are synthesized, to be delivered along the axon by a mechanism called axonal transport. The motors implicated in the anterograde transport are kinesins, while the retrograde transport is mediated in association with dynein, the same motor proteins involved in the fast axonal transport along MTs [4].

The multiple roles of cytoskeletal proteins in the neural cells imply that there is an underlying cytoskeletal pathology associated with several neurodegenerative processes. The major neurodegenerative diseases are characterized by the presence of inclusion bodies in implicated neurons. These inclusion bodies all contain elements of the cytoskeleton. In addition, mutations and/or accumulations of NFs are frequently observed in several human neurodegenerative disorders including amyotrophic lateral sclerosis, Alzheimer disease, Parkinson disease, Charcot-Marie-Tooth, giant axonal neuropathy, neuronal intermediate-filament inclusion disease and diabetic neuropathy [5]. Multiple factors can potentially induce the accumulation of NF, including deregulation of NF gene expression, NF mutations, defective axonal transport, abnormal posttranslational modifications and proteolysis [4]. Beyond their association with neural damage in inherited or age-dependent neurodegenerative diseases, studies from our laboratory indicated that the disruption of NF homeostasis is a response to toxic agents and abnormally accumulated metabolites in rat brain.

Astrocytes are important cytoarchitectural elements of the CNS; however, during the past few years, molecular and functional characterization of astroglial cells indicates that they have a much broader function than only support the neurons in the brain. Compelling evidence supports that astrocytes have specialized functions in inducing and regulating the blood-brain barrier (BBB), glutamate uptake, synaptic transmission, plasticity and metabolic homeostasis of the brain [6]. Astrocytes express 10 different isoforms of glial fibrillary acidic protein

(GFAP), the specific astrocytic IF, together with vimentin, nestin and synemin. However, GFAP is the main IF protein expressed in mature astrocytes, where it helps maintaining mechanical strength, as well as cell shape. However, recent evidence has shown that GFAP plays a role in a variety of additional astrocyte functions, such as cell motility/migration, cell proliferation, glutamate homeostasis, neurite outgrowth and injury/protection [7].

Astrocytes are also involved in a wide range of CNS pathologies, including trauma, ischemia and neurodegeneration. In such situations, the cells change both their morphology and their expression of many genes leading to activation of astroglia, or astrogliosis. It is accepted that the increase of IFs with accompanying cellular hypertrophy and an abnormal apparent increase in the number of astrocytes characterize astrogliosis. However, upregulation of IF proteins, in particular GFAP, but also vimentin and nestin, two IF proteins abundantly expressed in immature astrocytes, is regarded as the hallmark of astrogliosis [7]. In this regard, the most remarkable evidence of the relevance of GFAP in the physiological roles of astrocytes in maintaining normal brain function is Alexander disease, a fatal disorder in which GFAP mutations might compromise the astrocyte stress response [8].

3. Protein phosphorylation in signaling transduction

Phosphorylation is the most widespread type of posttranslational modification of the intracellular signaling proteins. Phosphorylation of proteins occurs within seconds or minutes of a regulatory signal, typically an extracellular signal.

Phosphorylation is an enzymatic process in which the introduction of a phosphoryl group to specific amino acid residues of a protein is catalyzed by protein kinases and the removal of phosphoryl groups is catalyzed by protein phosphatases. For phosphorylation to be useful in the regulation of a protein activity, it is important to be a reversible process, in which the phosphorylated form of the protein could restore its original dephosphorylated form when signal ends, functioning therefore as a molecular switch. The addition of a phosphoryl group to the side chain of a Ser, Thr, or Tyr residue introduces a bulky, charged group into a polar region. The oxygen atoms of a phosphoryl group can hydrogen bond with one or several groups in a protein, commonly the amide groups of the peptide backbone at the α-helix start or the charged guanidinium group of an Arg residue influencing the functionality of the protein [9].

3.1. Phosphorylation of intermediate-filament proteins

Phosphorylation, glycosylation and transglutamination take part in the multiple mechanisms of IF regulation. However, phosphorylation/dephosphorylation is a major regulatory mechanism orchestrating IF dynamics. Phosphorylation sites of IF subunits are located on their head and tail domains and phosphorylation plays a major role in regulating the structural organization and function of these cytoskeletal proteins in a cell- and tissue-specific manner [10].

Amino-terminal phosphorylation regulates the assembly/disassembly equilibrium of type III and IV IFs. Second messenger-dependent protein kinases add phosphate groups on the amino-terminal head domain on GFAP, vimentin and NF-L. Specific phosphorylating sites for cAMP-dependent protein kinase (PKA), Ca^{2+}/calmodulin-dependent protein kinase II (PKCaMII) and protein kinase C (PKC) are associated with IF disassembly; however, the action of the protein phosphatases 1, 2A and 2B (PP1, PP2A and PP2B), respectively, removes phosphate and restores the IF ability to polymerize [11].

Otherwise, the main phosphorylation sites on NF-M and NF-H are located in Lys-Ser-Pro (KSP) repeat regions of the tail domain of these subunits. The KSP repeats are phosphorylated by proline-directed kinases such as Cdk5, the mitogen-activated protein kinases (MAPK) such as Erk1/2, JNK, p38MAPK as well as glycogen synthase kinase 3 (GSK3). Phosphorylation of these KSP sites regulates the interactions of NFs with each other and with other cytoskeletal structures, since the tail domain of NF-M and NF-H protrudes laterally from the filament backbone to form "side-arms" when phosphorylated. These lateral interactions are central in the formation of a cytoskeletal lattice that supports the mature axon. Moreover, carboxyl-terminal phosphorylation of NF-M and NF-H subunits has long been considered to regulate their axonal transport rate and in doing so to provide stability to mature axons [12]. The axonal transport of NFs results from binding to the fast motor proteins kinesin and dynein intermitted with prolonged pauses. It is known that carboxyl-terminal phosphorylation of NF-H progressively restricts the association of NFs with kinesin and stimulates its interaction with dynein. This event could represent one of the mechanisms by which aberrant carboxyl-terminal phosphorylation would slow NF axonal transport. Both the maintenance of axonal caliber and axonal transport are dependent on the adequately phosphorylated NF subunits. Consequently, abnormally hyperphosphorylated NF subunits, commonly found in several neurodegenerative diseases, are intimately associated with neural dysfunction and considered a hallmark of neurodegeneration. In addition, demyelinating diseases might be associated with hypophosphorylated NFs, compromised axonal transport and decreased axonal diameter, since the phosphorylation of NFs occurs in close proximity to myelin sheaths, which release signals needed to induce phosphorylation of NFs in mature axons [13].

In the next sections, we discuss the recent findings from our laboratory indicating that signaling mechanisms involved in the regulation of IF phosphorylation/dephosphorylation are important targets of neurotoxins, metabolites accumulating in neurodegenerative diseases as well as thyroid hormones, emphasizing the relevance of cytoskeletal homeostasis on the brain function/dysfunction. To assess the effects of the neurotoxicants on the phosphorylation level of IF proteins, we developed an approach to measure the *in vitro* incorporation of radioactive phosphate (^{32}P-orthophosphate) into these proteins [14]. In order to shed light onto the signaling cascades targeted by them, we used pharmacological and immunological approaches, specific enzyme inhibitors, channel blockers, or glutamate antagonists as well as monoclonal antibodies directed to signaling cascades or specific phosphorylation sites. We conclude that misregulated cell signal transduction interferes with the phosphorylation/dephosphorylation of IFs disrupting the homeostasis of the cytoskeleton of astrocytes and neurons and this is associated with cell dysfunction and neurodegeneration in experimental models of neurotoxicity. **Figure 2** corresponds to a schematic representation of the consequences of misregulated NF phosphorylation for neuronal function.

http://dx.doi.org/10.5772/66926

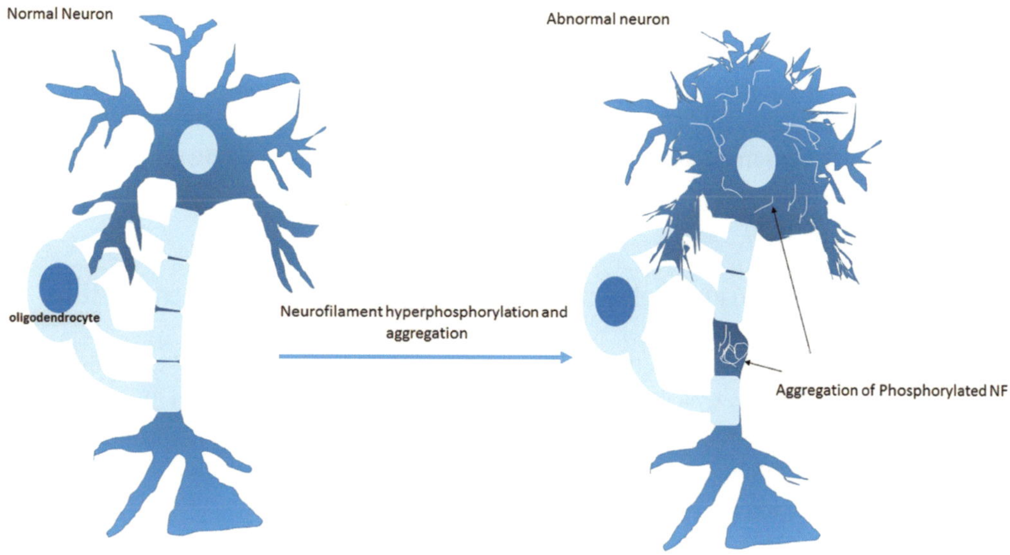

Figure 2. Schematic representation of disrupted neurofilament phosphorylation. The hyperphosphorylation of neurofilaments can change the cytoskeleton architecture and lead to neurofilament aggregation in perikarya and in axon accounting for cell damage.

3.2. Central roles of Ca^{2+} and glutamate receptors on the regulation of cytoskeletal dynamics in neural cells

Changes in the cytoplasmic free Ca^{2+} concentration constitute one of the main pathways by which information is transferred from extracellular signals received by animal cells to intracellular sites. However, an augmented Ca^{2+} influx through the NMDA receptor or voltage-dependent calcium channels (VDCCs) can be responsible for the activation of lethal metabolic pathways in neural cells. Overactivation of glutamate receptors produces neuronal membrane depolarization. This causes the influx of Ca^{2+} into the cytoplasm and subsequently triggers cascade events leading to excitotoxic neuronal death. Excitotoxicity is recognized as a major pathological process of neuronal death in neurodegenerative diseases involving the CNS. In this regard, compelling findings point to the cytoskeleton as an end point of excitotoxic mechanisms.

Different toxins and stress conditions are implicated in the misregulation of intracellular Ca^{2+}-dependent processes in cells and different cell types exhibit a diverse range of transient responses to their stimuli. Exposure of tissue slices to neurotoxicants or metabolites in toxic concentrations triggers the activation of ionotropic and metabotropic glutamate receptors as well as L-VDCC and the endoplasmic reticulum (ER) Ca^{2+} channels. These receptors and channels activate several intracellular-signaling complexes altering cell behavior in a spatiotemporally regulated manner. Metabolism of cyclic nucleotides, membrane phospholipids as well as endogenous enzymatic regulators are the key biochemical steps coordinating cell response to an extracellular stimulus [15].

Calcium is a critical regulator of cytoskeletal dynamics. Dysregulation of Ca^{2+} homeostasis is an important event in driving the disruption of assembly/disassembly equilibrium as well as the interaction of cytoskeletal proteins with regulatory proteins or cell organelles. In particular, IF proteins are directly regulated by Ca^{2+} levels, which crosslink signaling cascades and connect physiological or pathological extracellular signals with the IF cytoskeleton influencing multiple aspects of cell behavior. Consequently, abnormally elicited Ca^{2+} signals provoking misregulation of key phosphorylation cascades are able to disrupt cytoskeletal homeostasis and this is commonly associated with the cell damage.

4. Toxicity of diphenyl ditelluride on the cytoskeleton of neural cells

Many processes in the organic synthesis, vulcanization of rubber and in metal-oxidizing solutions to tarnish metals, such as silver, extensively use tellurium. Diphenyl ditelluride $(PhTe)_2$ is the simplest of the aromatic, diorganoyl ditelluride compounds used in organic synthesis. Indeed, developmental exposure to $(PhTe)_2$ is teratogenic and is associated with long-term behavioral and neurochemical changes in rats. Until recently, the general toxicity of $(PhTe)_2$ was considered to be exclusively related to the oxidation of thiol-containing proteins (for review, see [16]). However, compelling evidence from our laboratory points to an important role played by signaling mechanisms involved in regulating IF phosphorylation/dephosphorylation as target of $(PhTe)_2$ neurotoxicity. In addition, we evidence a remarkable role of Ca^{2+} mediating these actions secondary to glutamate receptors and L-VDCC activation.

The neurotoxicity of $(PhTe)_2$ is spatiotemporally regulated, consistent with the window of susceptibility of signaling cascades as well as the structural and functional heterogeneity of neurons in different brain regions. In this regard, exposure of cortical slices from 18- and 21-day-old rats to $(PhTe)_2$ shows unaltered phosphorylation of IF proteins, while IFs of acute cortical slices from younger pups (9 and 15 days old) are hypophosphorylated. Activated ionotropic glutamate receptors, L-VDCC and ryanodine channels result in PP1-mediated hypophosphorylation of GFAP and NF subunits pointing to the cortical cytoskeleton as a preferential target of the action of phosphatases in this window of vulnerability. Activation of PP1 is modulated by dopamine and cyclic AMP-regulated neuronal phosphoprotein 32 (DARPP-32), an important endogenous Ca^{2+}-mediated inhibitor of PP1 activity. Depending on the site of phosphorylation, DARPP-32 is able to produce opposing biochemical effects, that is, inhibition of PP1 activity or inhibition of protein kinase A (PKA) activity [17]. Decreased cAMP and PKA catalytic subunits support that $(PhTe)_2$ disrupts the cytoskeletal-associated phosphorylating/dephosphorylating system of neurons and astrocytes through PKA-mediated inactivation of DARPP-32, promoting PP1 release and hypophosphorylation of IF proteins of those neural cells [18]. Regarding neurons, hypophosphorylation of IF proteins could be associated with cell dysfunction since decreased phosphorylation of KSP repeats in the carboxyl-terminal domains of NF-M and NF-H correlates with impaired axonal transport and increased NF-packing density.

In contrast with younger rats, hippocampal slices of 21-day-old rats acutely exposed to $(PhTe)_2$ result in hyperphosphorylated IFs. Hippocampal IF hyperphosphorylation is partially dependent on L-VDCC, NMDA and ER Ca^{2+} channels. The signal evoked by $(PhTe)_2$ is also transduced through metabotropic glutamate receptors on the plasma membrane, leading to the activation of phospholipase C (PLC) that produces the intracellular messengers inositol

1,4,5-trisphosphate (IP3) and diacylglycerol (DAG). IP3 binds to specific receptors on the ER changing the conformation of IP3 receptors and opening the channel. Released Ca^{2+} and DAG directly activate PKCaMII and PCK, resulting in the hyperphosphorylation of some of the critical amino acid residues in the carboxyl-terminal tail domain of NF-L known to interfere with filament assembly. In addition, the activation of Erk1/2 and p38MAPK results in hyperphosphorylation of KSP repeats of NFM. Interestingly, PKCaMII and PKC are upstream of MAPK activation implying in a significant cross-talk among signaling pathways elicited by $(PhTe)_2$ that connect the glutamate metabotropic cascade with the activation of Ca^{2+} channels. The final molecular result is the extensive phosphorylation of amino- and carboxyl-terminal sites on IF proteins and deregulated cytoskeletal homeostasis [19].

4.1. Diphenyl ditelluride disrupts the cytoskeleton and provokes neurodegeneration in acutely injected young rats

The *in vivo* exposure to $(PhTe)_2$, in which the neurotoxicant reaches the brain via systemic circulation, also results in different susceptibilities of the IF proteins from neural cells. This can be evidenced in cerebral cortex and hippocampus of 15-day-old rats acutely injected with a toxic dose of $(PhTe)_2$ (0.3 μmol/kg body weight) [20]. Cortical hyperphosphorylation of neuronal and glial IF proteins is an early and persistent event up to 6 days after injection, accompanied by increased levels of GFAP and NF-L. Upregulated gene expression as well as GFAP and vimentin hyperphosphorylation could be a response to injury and take part in the program of reactive astrogliosis, as further demonstrated in striatum [21] and cerebellum [22] of $(PhTe)_2$-injected rats. In addition, hippocampal IFs are not responsive to the insult until weaning. A strong evidence supports an important role of astrocytes in a more severe cortical than hippocampal damage following the *in vivo* $(PhTe)_2$ insult. This supports a direct action of the neurotoxicant on intracellular signaling pathways and highlights the relevance of the interplay between glial and neuronal cells to adapt the cellular metabolic response to the insult even when the brain connections are only partially preserved, as shown in acute brain slices.

Of importance, neurodegeneration is part of the deleterious *in vivo* effects of $(PhTe)_2$ toxicity, as demonstrated in the striatum [23] and cerebellum [22] of $(PhTe)_2$-injected rats. Neurodegeneration is associated with alterations in Ca^{2+} homeostasis and glutamatergic neurotransmission, upstream of inhibited Akt and activated caspase 3. We therefore propose that excitotoxicity is a main mechanism of neurodegeneration caused by this compound in the developing rat brain. On the other hand, most of the actions of $(PhTe)_2$ disrupting the homeostasis of the cytoskeleton in neural cells are mediated by high Ca^{2+} levels. Moreover, a link among disrupted IF homeostasis, activated astrocytes and neuronal apoptosis in $(PhTe)_2$-injected rats has been demonstrated by immunohistochemical approaches. In addition, MAPK pathway might be a link between altered IF equilibrium and neural cell damage, since MAPK is implicated in IF hyperphosphorylation and neurodegeneration as well in the brain structures attained by $(PhTe)_2$ toxicity. Further supporting the cytoskeleton as an end point of neurotoxicity, hyperphosphorylated NFs can inhibit their proteolytic breakdown by calpain, a Ca^{2+}-activated protease. In addition, abnormally phosphorylated NFs accumulate in the perikarya and the phospho-NF aggregates can thus become cytotoxic by the enduring impairment of axonal transport of NFs (see **Figure 2**). The increased time the NF spent in the cell body is thought to result in further aberrant phosphorylation and may prevent them from entering the axon, resulting in a deleterious feedback loop [24].

In summary, we propose that complex and integrated actions mediate the $(PhTe)_2$ toxicity directed to the cytoskeleton of neural cells. These molecular mechanisms induce spatiotemporal responses of the cells because of the different windows of susceptibility of the developmental brain. Nonetheless, the Ca^{2+} -initiated events highlight a role for this neurotoxicant as a disruptor of the cytoskeleton.

5. Cytoskeleton as a target of amino acids and their metabolites

Misregulated cytoskeletal homeostasis is among the molecular mechanisms underlying the neural cell dysfunction in brain tissue exposed to high levels of amino acids and/or their metabolites. In humans, several neurological impairments are associated with enzymatic deficiencies or defects in proteins involved in cellular metabolism of neural cells, causing accumulation of metabolic intermediates associated with neuronal damage. We discuss some aspects of the molecular mechanisms underlying the disruption of cytoskeletal homeostasis in response to branched-chain keto acids (BCKAs) derived from leucine, isoleucine and valine. We also addressed the effects of homocysteine and quinolinic acid (QUIN), a metabolite of tryptophan metabolism, directed to the cytoskeleton.

5.1. Branched chain α-keto acids and the cytoskeleton of neural cells

The branched-chain ketoacids, α-ketoisocaproic acid (KIC), α-keto-β-methylvaleric acid (KMV) and α-keto-isovaleric acid (KIV) are produced from the respective branched-chain amino acids (BCAAs) leucine, isoleucine and valine, in the reaction catalyzed by the branched-chain α-keto acid dehydrogenase (BCKAD) complex. A deficiency of the BCKAD complex is an inherited metabolic disease known as maple syrup urine disease (MSUD) which lead to the accumulation of BCAAS and BCKAs in tissues and body fluids resulting in dramatic cerebral symptoms [25].

Curiously, cortical slices of young rats exposed to high levels of the BCAAs individually preserve the homeostasis of the cytoskeleton. On the other hand, their respective keto acids provide an interesting example of the fine-tune regulation of the cytoskeleton, since KIC [26] and KMV [27] were differently deleterious to the homeostasis of the cytoskeleton. KIC and KMV alter the dynamics of IF proteins of astrocytes and neurons through different transduction mechanisms dependent on excessive intracellular Ca^{2+} influx, while KIV appears not to be involved in the disruption of the IF cytoskeleton [28].

The effect of KIC is outlined by hypophosphorylation of GFAP, NF-M and NF-L in very young rats (up to 12 days of age) changing to hyperphosphorylation of the same proteins later in development (17 days of age). Nonetheless, both responses of the cytoskeletal-associated phosphorylating system are regulated by Ca^{2+} currents through the NMDA and L-VDCC, as well as by the intracellular Ca^{2+}storage release from the ER, leading to a differential activation of protein phosphatases or kinases [28]. These paradoxical findings provide an interesting

http://dx.doi.org/10.5772/66926

insight into the differential susceptibility of cortical IF cytoskeleton to the exposure to pathological levels of this metabolite. The different vulnerabilities of the cytoskeleton of cortical cells during development might be ascribed to the temporal maturation mediated by a multitude of developmental processes and signaling pathways. It is conceivable that they are associated with the pathological role of the developmentally regulated glutamate receptors in neural cells since the expression patterns of glutamate receptor subunit genes change during the ontogeny of the brain. Distinct regional and temporal patterns of the expression of types and subtypes of the glutamate ionotropic receptors during ontogeny may possibly explain the different signaling pathways targeting the cytoskeleton of cortical neural cells during development.

Interestingly, KMV disturbs the IF-associated cytoskeletal phosphorylation only in 12-day-old rats without changing the phosphorylation level of these proteins in younger or older animals, showing a specific window of vulnerability of cytoskeleton to KMV insult in the cerebral cortex of developing brain. Strikingly, this effect was dependent on intracellular Ca^{2+} concentrations; however, in this case ɣ-amino butyric acid A and B ($GABA_A$ and $GABA_B$, respectively) rather than glutamate receptors were involved in this action. This is in agreement with $GABA_A$ and $GABA_B$ receptors mediating the induction and maintenance of Ca^{2+} levels [27].

Overall, we propose that BCKAs in supra-physiological concentrations disrupt the cytoskeleton of rat brain through misregulation of the phosphorylating system associated with the IF cytoskeleton. We evidenced developmentally regulated mechanisms in which Ca^{2+}-mediated excitotoxicity plays a critical role in destabilizing the cytoskeleton that may ultimately disrupt normal cell function and viability. Although evidence from animal models should be taken with caution, we can propose that the disrupted cytoskeleton is part of the physiopathology of MSUD.

5.2. Hyperhomocysteinemia and the cytoskeleton of neural cells

Homocysteine (Hcy) is a sulfur-containing amino acid generated during methionine metabolism. Genetic mutations impairing Hcy metabolism cause accumulation of this amino acid attaining high levels in blood, leading to severe hyperhomocysteinemia and brain damage. Otherwise, along with genetic factors, mild-moderate hyperhomocysteinemia is associated with nutritional imbalance and hormonal factors. Mild hyperhomocysteinemia, which markedly enhance the vulnerability of neuronal cells to excitotoxicity and oxidative imbalance, is also common in older people, constituting an independent risk factor for stroke and cognitive impairment [29].

Various existing experimental evidences from our group link hyperhomocysteinemia and cytoskeletal misregulation, supporting that disrupted cytoskeleton could be an end point of neural dysfunction in this neurometabolic disorder. Experiments with brain slices acutely exposed to mild Hcy levels (100 μM) showed greater vulnerability of hippocampal cytoskeleton as compared with cortical one. Moreover, a window of vulnerability of the cytoskeleton of hippocampal cells is evidenced, since misregulated phosphorylation is detected only at postnatal day 17 [30], reflecting an altered activity of the endogenous phosphorylating system associated with the IFs in this brain structure. As expected, NMDA receptors, L-VDCC and extracellular Ca^{2+}influx result in PKC and PKCaMII activation. The prevention of Hcy action through the inhibition of PKC and MEK, a step that is upstream of MAPK cascade (Raf-1/MEK/MAPK), is consistent with an effect at the level of the monomeric GTPase Raf-1, supporting a role for PKC phosphorylating and activating Raf-1 in the Hcy-induced modulation of the cytoskeleton.

In contrast with hypophosphorylation found in hippocampal slices, the chemically induced chronic hyperhomocysteinemia differently alters the signaling mechanisms directed to the cytoskeleton, producing PP1-, PP2A- and PP2B-mediated hypophosphorylation of NF subunits and GFAP in hippocampal slices of 17-day-old rats without affecting the cerebral cortex [31] through glutamate and Ca^{2+}-mediated mechanisms. Further evidence that homocysteine targets the cytoskeleton came from cytoskeletal reorganization in primary astrocytes and neurons exposed to homocysteine [32]. Dramatically altered actin cytoskeleton in primary astrocytes exposed to 100 μM Hcy is consistent with the role of actin as a main determinant of cell morphology. Concomitant disrupted GFAP meshwork underlies the remodeled actin cytoskeleton and altered cell morphology. These findings provide further evidence of the cross-talk among the different cytoskeletal subsystems and the roles played by the toxic levels of Hcy.

Therefore, taking into account our experimental evidence it is conceivable that disturbed cell signaling is an important determinant of the disrupted homeostasis of the cytoskeleton as a whole, with widespread consequences on cell function that could be associated with human hyperhomocysteinemia.

6. Cytoskeleton is a target of quinolinic acid neurotoxicity

Quinolinic acid is a neuroactive metabolite of the kynurenine pathway normally found in nanomolar concentrations in human brain and cerebrospinal fluid (CSF). QUIN is antagonist of NMDA receptor and it has a high *in vivo* potency as an excitotoxin supporting involvement in the pathogenesis of a variety of human neurological diseases. The neurotoxicity of QUIN results from complex mechanisms including presynaptic receptors, energetic dysfunction, oxidative stress, transcription factors and behavior [33]. We experimentally demonstrate that the disruption of the cytoskeleton, in particular, misregulation of the phosphorylation system associated with the IFs, is a target of QUIN toxicity in injected rat striatum, tissue slices and primary astrocytes and neurons in culture.

6.1. Effects of intrastriatally injected quinolinic acid on the cytoskeleton of neural cells

Acute intrastriatal injection of QUIN (150 nmol/0.5 μL) in adolescent rats (30 days old) provokes NF-L and GFAP hyperphosphorylation 30 min after infusion, evidencing the susceptibility of the cytoskeleton of both neurons and astrocytes in the early events of QUIN toxicity. Hyperphosphorylated NF-LSer55 destabilizes the NF structure and this might represent an early step in the pathophysiological cascade of deleterious events exerted by QA in rat striatum. Experimental insights to get light on the molecular mechanisms underlying this effect point to NMDA-mediated Ca^{2+} events and oxidative stress upstream of activated second messenger-dependent protein kinases PKA, PKC and PKCaMII, but not MAPKs after QUIN infusion [34].

A link between misregulation of cell-signaling mechanisms, disruption of IF phosphorylation and cell damage as part of QUIN toxicity becomes more evident analyzing the long-lasting effect of the acute intrastriatal injection of QUIN in adolescent rats on the dynamics of the phosphorylating system until 21 days after injection [35]. The acutely injected QUIN alters the homeostasis of IF phosphorylation in a selective manner, progressing from striatum to cerebral cortex and hippocampus. Twenty-four hours after QUIN injection, the IFs are hyperphosphorylated in the striatum. This effect progresses to cerebral cortex causing hypophosphorylation at day 14 and appears in the hippocampus as hyperphosphorylation at day 21 after QUIN infusion, PKA and PKCaMII mediating this effect. However, MAPKs (Erk1/2, JNK and p38MAPK) are hyperphosphorylated/activated only in the hippocampus, suggesting different signaling mechanisms in these two brain structures during the first weeks after QUIN infusion. Also, PP1 and PP2B-mediated hypophosphorylation of the IF proteins in the cerebral cortex 14 days after QUIN injection reinforces the selective signaling mechanisms in different brain structures. Increased GFAP immunocontent in the striatum and cerebral cortex 24 h and 14 days after QUIN injection, respectively, suggests reactive astrocytes in these brain regions. Yet, we observe biochemical and histopathological alterations in the striatum, cortex and hippocampus, as well as altered behavioral tests in response to the long-lasting exposure to QUIN through glutamate and Ca^{2+}-mediated mechanisms. Thus, it is tempting to propose that the long-lasting deleterious effect of intrastriatal QUIN injection could be due to the fact that QUIN interferes with the highly regulated signaling mechanisms targeting the cytoskeleton in the immature brain [36].

6.2. Insight into the molecular basis of quinolinic acid action toward the cytoskeleton

Studies in acute brain slices further support the role of glutamatergic signaling and Ca^{2+} overload disturbing the cytoskeletal equilibrium downstream of QUIN exposure. Moreover, this experimental approach brings light on the cell-specific mechanisms targeting the cytoskeleton in astrocytes and neurons when the cell connections are partially preserved. In astrocytes, the QUIN action is mainly due to increased Ca^{2+} influx through NMDA and L-VDCC. In neuronal cells, QUIN acts through the activation of metabotropic glutamate receptors and influx of Ca^{2+} through NMDA receptors and L-VDCC, as well as Ca^{2+} release from intracellular stores. These mechanisms then set off a cascade of events including the activation of PKA, PKCaMII and PKC, which phosphorylate head domain sites on GFAP and NFL. Moreover, Cdk5 is activated downstream of mGluR5, phosphorylating the KSP repeats on NFM and NFH. Metabotropic glutamate receptors type 1 (mGluR1) is upstream of PLC, which, in turn, produce DAG and IP3 promoting hyperphosphorylation of KSP repeats on the tail domain of NFM and NFH [37].

6.3. The cytoskeleton of astrocytes and neurons responds differently to quinolinic acid toxicity

The susceptibility of the cytoskeleton to toxic levels of QUIN is also detectable in isolated astrocytes and neurons growth in primary cultures [38]. In astrocytes, Ca^{2+}-mediated glutamate mechanisms target the endogenous phosphorylating system, since metabotropic glutamate receptors and Ca^{2+} influx through NMDA receptors are upstream of PKA, PKCaMII and PKC activation, provoking GFAP hyperphosphorylation. Interestingly, the misregulated phosphorylation system leads to a reversible and dramatically altered actin cytoskeleton with concomitant change of morphology to fusiform and/or flattened cells with retracted cytoplasm and disruption of the GFAP meshwork [39] supporting the dynamic behavior of the cytoskeleton.

Interestingly, neurons show greater vulnerability to QUIN than astrocytes (10×). Neurons exposed to QUIN presented PKA- and PKC-mediated hyperphosphorylation of NF subunits. These effects are also downstream of ionotropic and metabotropic glutamate signaling and Ca^{2+} influx through NMDA receptors and L-VDCC. The misregulated signaling pathways disrupt the neuronal cytoskeleton, evaluated by altered neurite/neuron ratios and neurite outgrowth. It is important to consider that microtubules play a central role in cell polarity [40]. In particular, microtubules are the main determinants of neuronal polarity and regulation of microtubule dynamics includes tubulin posttranslational modifications [40] and phosphorylation of microtubule-associated proteins (MAPs), whose binding to microtubules is essential for neurite formation [41]. As an example, activated GSK-3β leads to increased phosphorylation of some MAPs, destabilizing microtubules with consequence for neurite stabilization [42]. Therefore, the neurite destabilization could derive from both NFs and microtubules disruption.

Interestingly, we found a protective role of astrocyte-conditioned medium on the disrupted neuronal cytoskeleton and morphometric alterations, suggesting that QUIN-induced trophic factors secreted by astrocytes are able to modulate signaling mechanisms targeting the neuronal cytoskeleton. More interestingly, co-cultured astrocytes and neurons preserve their cytoskeletal organization and cell morphology together with unaltered activity of the phosphorylating system associated with the cytoskeleton. In other words, co-cultured astrocytes and neurons tightly and actively interact with one another reciprocally protecting themselves against QUIN injury [38]. This evidence raise the question about the role played by the activated microglia eliciting signals essential to destabilize the astrocytic and neuronal cytoskeleton but this hypothesis remains to be clarified.

All together, we conclude that among the multiple mechanisms through which accumulated QUIN is able to induce cell damage, our experimental evidence points to Ca^{2+}-mediated mechanisms directed to the cytoskeletal disruption as an end point of QUIN toxicity. Both *in vivo* and *ex vivo* approaches clearly demonstrate a wide spectrum of misregulated signaling mechanisms downstream of QUIN action directly affecting the cytoskeleton and disrupting cell homeostasis. We also provide evidence that impaired physiological equilibrium of the signaling cascades directed to the cytoskeleton underlies QUIN cytotoxicity and is associated with neurodegeneration. The *in vitro* results showing disorganized cytoskeleton and altered cell morphology further support the cytoskeleton as a hallmark of stress condition that could be implicated in the human brain disorders associated with high QUIN levels.

7. Cytoskeleton of neural cells is a target of thyroid hormones

Thyroid hormones are essential for the development and function of central nervous system. In brain, these hormones are essential for myelination [43, 44], neuritogenesis [45], synaptic plasticity [46–48], IF phosphorylation [49–54], cell differentiation and maturation [55]. Considering the role of these hormones on brain development, thyroid diseases might account for brain injury as well as alteration in mood and cognition [56].

http://dx.doi.org/10.5772/66926

The classical mechanism of thyroid hormone action involves the modulation of nuclear receptors by 3,5,3′-triiodo-L-thyronine (T_3). The nuclear receptors are ligand-dependent transcription factors, which are involved in the genomic-dependent effects of thyroid hormones. However, there are numerous physiological effects of these hormones that cannot be mediated by the genome-like mechanism, due the short time frame in which the response occurs. Nongenomic actions of thyroid hormones are defined as events that (i) do not primarily involve the cell nucleus, (ii) are rapid in onset (minutes or a few hours) relative to transcription and translation and (iii) do not require gene transcription and protein synthesis [57]). These events are triggered by rapid/nongenomic responses that are frequently associated with secondary messenger-signaling pathways.

7.1. Insight into the molecular basis of genomic and nongenomic action of thyroid hormones toward the cytoskeleton of neural cells

The first evidence of nongenomic actions of thyroid hormones targeting the cytoskeleton demonstrated the thyroxine-dependent modulation of actin polymerization in cultured astrocytes. Thyroxine (T_4) was involved in the conversion of soluble actin to a fibrous form through nongenomic mechanism [58].

While many of the T_3 actions are mediated by genomic-dependent mechanisms, T_4 and reverse 3,3′,5′-triiodothyronine (reverse T_3, rT_3) exert direct, The nongenomic effects in neural cells. Both T_4 and rT_3 hormones control actin polymerization in cultured astrocytes without affecting gene expression. The authors suggested that these events might contribute to thyroid hormone's influence on brain development. Subsequently, the same research group showed that both T_4 and rT_3, but not T_3, directly regulate the F-actin content of elongating neurites of cerebellar neurons. These results provide a molecular mechanism for the influence of thyroid hormones on brain development that is independent of regulated gene expression [59].

Trentin and Moura Neto [60] demonstrated that T_3 altered the organization of GFAP in cerebellar astrocytes in culture. GFAP filaments that normally spread in the cytoplasm of astrocytes became organized around the cell nucleus. In addition, Zamoner and coworkers [51] showed that both T_3 and T_4 induced GFAP phosphorylation and reorganization in glioma C6 cells through the inhibition of RhoA GTPase. The modulation of GFAP was accompanied by increased proliferation of glioma cells. Taking together, these results suggest that thyroid hormones may be important regulators of astrocyte growth and differentiation.

Despite the evidence that nongenomic actions of thyroid hormones initiated at the plasma membrane via integrin $\alpha V\beta 3$ [57, 61], the complexity of the processes underlying the differential mechanisms of action to thyroid hormones suggests the existence of multiple binding sites for these hormones. In this context, it has been previously demonstrated that both T_3 and T_4 may modulate the GABAergic system and induce PKA- and PKCaMII-mediated hyperphosphorylation of vimentin, GFAP, NF-M and NF-L in cerebral cortex from very young rats (up to 10 days of age) [50]. However, only T_4 caused hyperphosphorylation of the same proteins later in development (15 days of age) through GABA-independent mechanisms [49]. These paradoxical findings provide an interesting insight into the differential susceptibility of cortical IF cytoskeleton to thyroid hormone exposure.

Calcium-dependent mechanisms play a central role on the thyroid hormone-induced modulation of the phosphorylating system associated with IFs. Zamoner and colleagues [49] demonstrated that the nongenomic mechanisms underlying the effects T_4 targeting the IF-associated phosphorylating system in cerebral cortex from 15-day-old rats are dependent on extracellular Ca^{2+} influx through VDCC, as well as Ca^{2+} release from ER stores.

Taking into account that in rat the myelination peak is coincident with postnatal day 15 and that this is a period of intense synaptogenesis, the NF hyperphosphorylation induced by T_4 in cerebral cortex from 15-day-old rats appears to be correlated to synaptogenesis and myelination (for review, see [53]).

In summary, we could suggest that nongenomic actions of T_4 targeting the cytoskeleton of glial cells and neurons might account for neuronal cell migration, myelination, synaptogenesis and synaptic plasticity. Moreover, the modulation of NF phosphorylation by thyroid hormone may control axonal caliber.

7.2. Hypothyroidism and the cytoskeleton of neural cell

The effects of thyroid hormones in central nervous system during development include the modulation of the cytoskeleton dynamics. Hypothyroidism in the developing rat brain is associated with oxidative stress and aberrant intraneuronal accumulation of NFs in the perikaryon of Purkinje neurons (see **Figure 2**). The authors suggested that the neuron alterations observed in the developing hypothyroid brain are comparable to those seen in neurodegenerative diseases [62]. Corroborating these findings, it has been shown that the effects of hypothyroidism on neuronal cytoskeleton involve the developmental modulation of specific isoforms of protein expression, which induce stoichiometric imbalance between the NF triplet [52]. In addition, thyroid hormone deficiency induces a delay and a partial arrest of astrocyte differentiation, supported by the decreased expression of GFAP both in cortical [52] and in hippocampal astrocytes [54], which was accompanied by downregulation of the astrocyte glutamate transporters. These findings are associated with the extracellular signal-regulated kinase (ERK)1/2 and c-jun terminal kinase (JNK) activation. NF hyperphosphorylation might account for the aberrant intraneuronal accumulation of these cytoskeletal structures previously described [62].

Our research group demonstrated the hyperphosphorylation of tail KSP repeats on NF-H in hypothyroid cortical and hippocampal neurons [52, 54]. The carboxyl-terminal phosphorylation of NF-H progressively restricts association of NFs with kinesin, the axonal anterograde motor protein and stimulates its interaction with dynein, the axonal retrograde motor protein [63]. This event could represent one of the mechanisms by which carboxyl-terminal phosphorylation would slow NF axonal transport.

Taking into account our experimental evidence, we propose that the consequences of congenital hypothyroidism to neural cells involve IF hyperphosphorylation, misregulation of glutamate-glutamine cycle, oxidative stress and glutamate excitotoxicity. These events suggest a compromised astroglial defense system that is probably playing a role in the physiopathology of the neurological dysfunction of hypothyroidism (**Figure 3**).

http://dx.doi.org/10.5772/66926

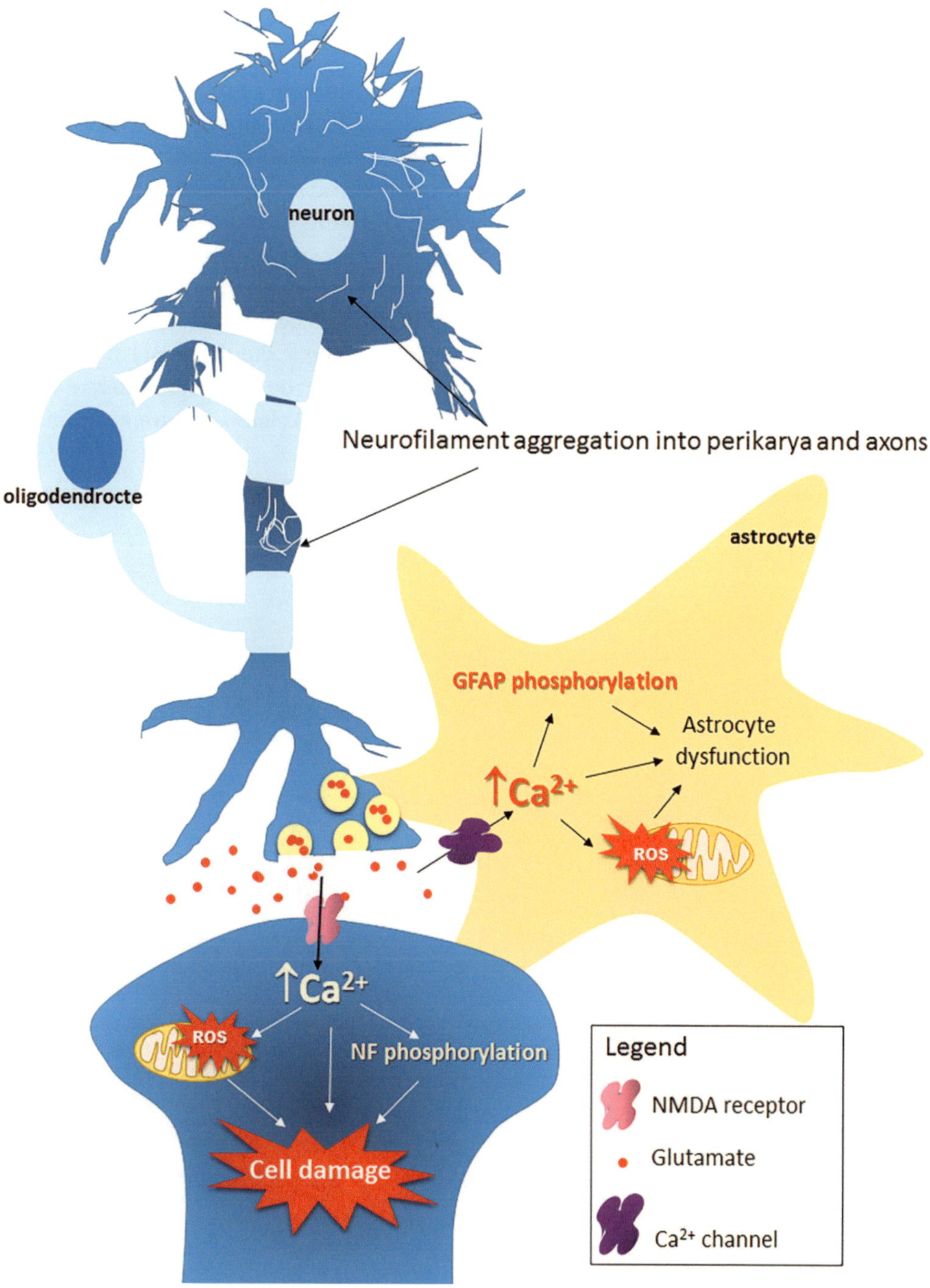

Figure 3. Role of glutamate excitotoxicity on intermediate-filament dynamics and cell damage. Congenital hypothyroidism leads to glutamate excitotoxicity, calcium overload, and oxidative stress. These events are related to intermediate-filament (GFAP and NF) hyperphosphorylation and neural cell damage.

8. General conclusion

Studies of our group on the endogenous phosphorylating system associated with the IF proteins of neural cells point to a critical role of disrupted cytoskeleton in response to a variety of signals both in physiological and in pathological conditions. Our findings highlight the IFs as a preferential target of the signal transduction pathways. Importantly, a large body of evidence shows a link among misregulation of cell-signaling mechanisms, disruption of IF phosphorylation and cell damage in response to different stress signals. While the exact signaling pathways regulating NF phosphorylation remains elusive, there is increasing evidence that known signal transduction cascades are involved. These actions can be initiated by the activation of NMDA-, L-VDCC, or G protein-coupled receptors and the signal is transduced downstream of Ca^{2+} mobilization or monomeric GTPase activation through different kinase/phosphatase pathways, regulating the dynamics of the cytoskeleton. **Figure 4** summarizes

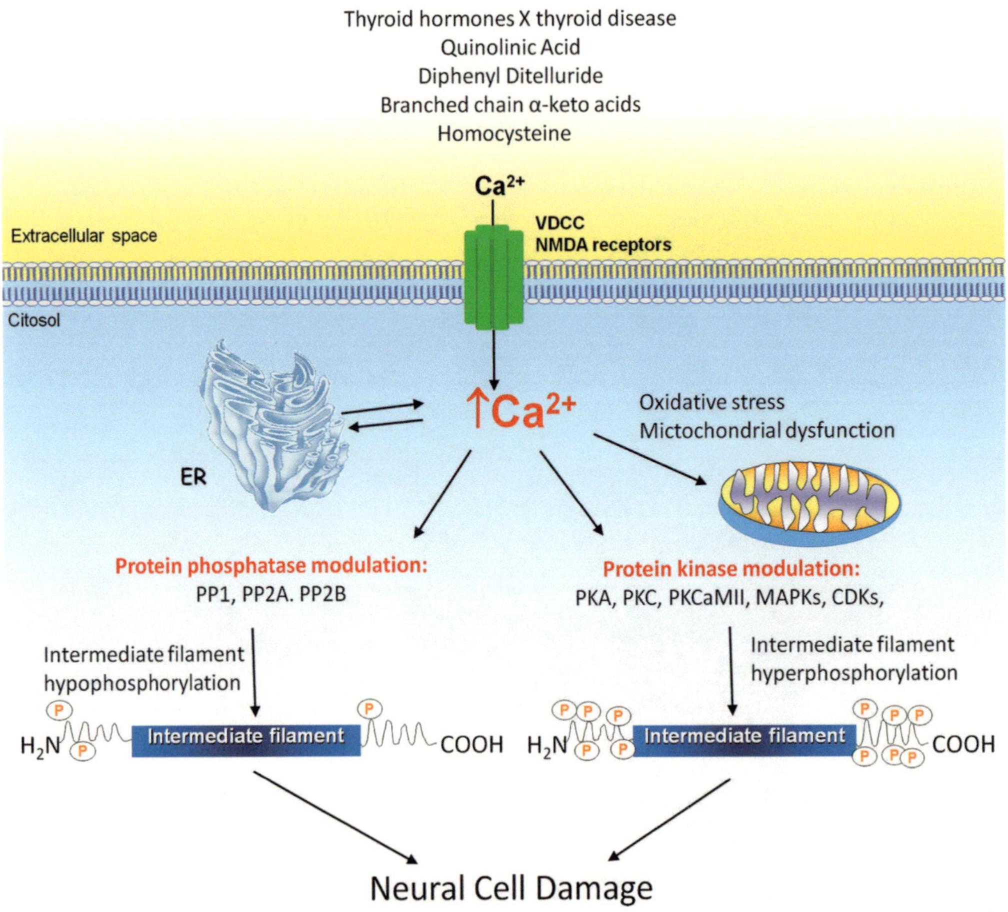

Figure 4. Summary of calcium-associated mechanisms triggered by thyroid hormones, quinolinic acid, diphenyl ditelluride, branched-chain keto acids, and homocysteine targeting intermediate-filament phosphorylation in neural cells. Calcium influx through the NMDA receptor or voltage-dependent calcium channels (VDCC) can be responsible for the activation of lethal metabolic pathways in neural cells. Augmented intracellular Ca^{2+} levels might be associated with the modulation of diverse cell-signaling pathways and exhibit a diverse range of responses to their stimuli.

the calcium-associated mechanisms triggered by thyroid hormones, quinolinic acid, $(PhTe)_2$, BCKAs and homocysteine targeting IF phosphorylation in neural cells.

Despite the focus on the misregulation of IF dynamics in response to signaling mechanisms downstream of metabolites and neurotoxicants, we should consider that cytoskeleton is a complex meshwork of interconnecting filaments [1]. In this regard, the morphological alterations demonstrated in primary cells in culture mainly reflect the reorganization of the meshwork of filaments. Taking into account our findings, we propose that misregulation of kinase/phosphatase cascades downstream of stressors could disrupt the cytoskeleton as a whole and this might be an important determinant of neural dysfunction associated with the action of neurotoxicants and in neurometabolic conditions.

Author details

Ariane Zamoner[1] and Regina Pessoa-Pureur[2]*

*Address all correspondence to: rpureur@ufrgs.br

1 Department of Biochemistry, Center of Biological Sciences, Federal University of Santa Catarina, Florianópolis, Brazil

2 Department of Biochemistry, Institute of Basic Sciences of Health, Federal University of Rio Grande do Sul, Porto Alegre, Brazil

References

[1] Huber F, Boire A, Lopez MP, Koenderink GH. Cytoskeletal crosstalk: when three different personalities team up. Curr Opin Cell Biol. 2015;32:39–47.

[2] Herrmann H, Bar H, Kreplak L, Strelkov SV, Aebi U. Intermediate filaments: from cell architecture to nanomechanics. Nat Rev Mol Cell Biol. 2007;8(7):562–73.

[3] Herrmann H, Aebi U. Intermediate filaments: molecular structure, assembly mechanism and integration into functionally distinct intracellular Scaffolds. Annu Rev Biochem. 2004;73:749–89.

[4] Laser-Azogui A, Kornreich M, Malka-Gibor E, Beck R. Neurofilament assembly and function during neuronal development. Curr Opin Cell Biol. 2015;32:92–101.

[5] Gentil BJ, Tibshirani M, Durham HD. Neurofilament dynamics and involvement in neurological disorders. Cell Tissue Res. 2015;360(3):609–20.

[6] Seifert G, Schilling K, Steinhauser C. Astrocyte dysfunction in neurological disorders: a molecular perspective. Nat Rev Neurosci. 2006;7(3):194–206.

[7] Hol EM, Pekny M. Glial fibrillary acidic protein (GFAP) and the astrocyte intermediate filament system in diseases of the central nervous system. Curr Opin Cell Biol. 2015;32:121–30.

[8] Der Perng M, Su M, Wen SF, Li R, Gibbon T, Prescott AR, et al. The Alexander disease-causing glial fibrillary acidic protein mutant, R416W, accumulates into Rosenthal fibers by a pathway that involves filament aggregation and the association of alpha B-crystallin and HSP27. Am J Hum Genet. 2006;79(2):197–213.

[9] Ubersax JA, Ferrell JE, Jr. Mechanisms of specificity in protein phosphorylation. Nat Rev Mol Cell Biol. 2007;8(7):530–41.

[10] Omary MB, Ku NO, Tao GZ, Toivola DM, Liao J. "Heads and tails" of intermediate filament phosphorylation: multiple sites and functional insights. Trends Biochem Sci. 2006;31(7):383–94.

[11] Sihag RK, Inagaki M, Yamaguchi T, Shea TB, Pant HC. Role of phosphorylation on the structural dynamics and function of types III and IV intermediate filaments. Exp Cell Res. 2007;313(10):2098–109.

[12] Holmgren A, Bouhy D, Timmerman V. Neurofilament phosphorylation and their proline-directed kinases in health and disease. J Peripher Nerv Syst. 2012;17(4):365–76.

[13] McLean NA, Popescu BF, Gordon T, Zochodne DW, Verge VM. Delayed nerve stimulation promotes axon-protective neurofilament phosphorylation, accelerates immune cell clearance and enhances remyelination in vivo in focally demyelinated nerves. PLoS One. 2014;9(10):e110174.

[14] Funchal C, de Almeida LM, Oliveira Loureiro S, Vivian L, de Lima Pelaez P, Dall Bello Pessutto F, et al. In vitro phosphorylation of cytoskeletal proteins from cerebral cortex of rats. Brain Res Brain Res Protoc. 2003;11(2):111–8.

[15] Mehta A, Prabhakar M, Kumar P, Deshmukh R, Sharma PL. Excitotoxicity: bridge to various triggers in neurodegenerative disorders. Eur J Pharmacol. 2013;698(1–3):6–18.

[16] Pessoa-Pureur R, Heimfarth L, Rocha JB. Signaling mechanisms and disrupted cytoskeleton in the diphenyl ditelluride neurotoxicity. Oxid Med Cell Longev. 2014;2014: 458601.

[17] Hakansson K, Lindskog M, Pozzi L, Usiello A, Fisone G. DARPP-32 and modulation of cAMP signaling: involvement in motor control and levodopa-induced dyskinesia. Parkinsonism Relat Disord. 2004;10(5):281–6.

[18] Heimfarth L, Loureiro SO, Reis KP, de Lima BO, Zamboni F, Lacerda S, et al. Diphenyl ditelluride induces hypophosphorylation of intermediate filaments through modulation of DARPP-32-dependent pathways in cerebral cortex of young rats. Arch Toxicol. 2012;86(2):217–30.

[19] Heimfarth L, Loureiro SO, Reis KP, de Lima BO, Zamboni F, Gandolfi T, et al. Cross-talk among intracellular signaling pathways mediates the diphenyl ditelluride actions on the hippocampal cytoskeleton of young rats. Chem Res Toxicol. 2011;24(10):1754–64.

[20] Heimfarth L, Loureiro SO, Zamoner A, Pelaez PDL, Nogueira CW, Da Rocha JBT, et al. Effects of in vivo treatment with diphenyl ditelluride on the phosphorylation of

cytoskeletal proteins in cerebral cortex and hippocampus of rats. Neurotoxicology. 2008;29(1):40–7.

[21] Heimfarth L, Reis KP, Loureiro SO, de Lima BO, da Rocha JB, Pessoa-Pureur R. Exposure of young rats to diphenyl ditelluride during lactation affects the homeostasis of the cytoskeleton in neural cells from striatum and cerebellum. Neurotoxicology. 2012;33(5):1106–16.

[22] Heimfarth L, Loureiro SO, Dutra MF, Petenuzzo L, de Lima BO, Fernandes CG, et al. Disrupted cytoskeletal homeostasis, astrogliosis and apoptotic cell death in the cerebellum of preweaning rats injected with diphenyl ditelluride. Neurotoxicology. 2013;34:175–88.

[23] Heimfarth L, Loureiro SO, Dutra MF andrade C, Pettenuzzo L, Guma FT, et al. In vivo treatment with diphenyl ditelluride induces neurodegeneration in striatum of young rats: implications of MAPK and Akt pathways. Toxicol Appl Pharmacol. 2012;264(2):143–52.

[24] Perrot R, Berges R, Bocquet A, Eyer J. Review of the multiple aspects of neurofilament functions and their possible contribution to neurodegeneration. Mol Neurobiol. 2008;38(1):27–65.

[25] Manara R, Del Rizzo M, Burlina AP, Bordugo A, Citton V, Rodriguez-Pombo P, et al. Wernicke-like encephalopathy during classic maple syrup urine disease decompensation. J Inherit Metab Dis. 2012;35(3):413–7.

[26] Funchal C, Zamoner A, dos Santos AQ, Loureiro SO, Wajner M, Pessoa-Pureur R. Alpha-ketoisocaproic acid increases phosphorylation of intermediate filament proteins from rat cerebral cortex by mechanisms involving Ca2+ and cAMP. Neurochem Res. 2005;30(9):1139–46.

[27] Funchal C, Dall Bello Pessutto F, de Almeida LM, de Lima Pelaez P, Loureiro SO, Vivian L, et al. Alpha-keto-beta-methylvaleric acid increases the in vitro phosphorylation of intermediate filaments in cerebral cortex of young rats through the gabaergic system. J Neurol Sci. 2004;217(1):17–24.

[28] Funchal C, de Lima Pelaez P, Loureiro SO, Vivian L, Dall Bello Pessutto F, de Almeida LM, et al. alpha-Ketoisocaproic acid regulates phosphorylation of intermediate filaments in postnatal rat cortical slices through ionotropic glutamatergic receptors. Brain Res Dev Brain Res. 2002;139(2):267–76.

[29] Hainsworth AH, Yeo NE, Weekman EM, Wilcock DM. Homocysteine, hyperhomocysteinemia and vascular contributions to cognitive impairment and dementia (VCID). Biochim Biophys Acta. 2016;1862(5):1008–17.

[30] Loureiro SO, Heimfarth L, Pelaez Pde L, Vanzin CS, Viana L, Wyse AT, et al. Homocysteine activates calcium-mediated cell signaling mechanisms targeting the cytoskeleton in rat hippocampus. Int J Dev Neurosci. 2008;26(5):447–55.

[31] Loureiro SO, Heimfarth L, Pelaez Pde L, Lacerda BA, Vidal LF, Soska A, et al. Hyperhomocysteinemia selectively alters expression and stoichiometry of intermediate

filament and induces glutamate- and calcium-mediated mechanisms in rat brain during development. Int J Dev Neurosci. 2010;28(1):21–30.

[32] Loureiro SO, Romao L, Alves T, Fonseca A, Heimfarth L, Moura Neto V, et al. Homocysteine induces cytoskeletal remodeling and production of reactive oxygen species in cultured cortical astrocytes. Brain Res. 2010;1355:151–64.

[33] Lugo-Huitron R, Ugalde Muniz P, Pineda B, Pedraza-Chaverri J, Rios C, Perez-de la Cruz V. Quinolinic acid: an endogenous neurotoxin with multiple targets. Oxid Med Cell Longev. 2013;2013:104024.

[34] Pierozan P, Zamoner A, Soska AK, Silvestrin RB, Loureiro SO, Heimfarth L, et al. Acute intrastriatal administration of quinolinic acid provokes hyperphosphorylation of cytoskeletal intermediate filament proteins in astrocytes and neurons of rats. Exp Neurol. 2010;224(1):188–96.

[35] Pierozan P, Goncalves Fernandes C, Ferreira F, Pessoa-Pureur R. Acute intrastriatal injection of quinolinic acid provokes long-lasting misregulation of the cytoskeleton in the striatum, cerebral cortex and hippocampus of young rats. Brain Res. 2014;1577:1–10.

[36] Pierozan P, Fernandes CG, Dutra MF, Pandolfo P, Ferreira F, de Lima BO, et al. Biochemical, histopathological and behavioral alterations caused by intrastriatal administration of quinolic acid to young rats. FEBS J. 2014;281(8):2061–73.

[37] Pierozan P, Zamoner A, Soska AK, de Lima BO, Reis KP, Zamboni F, et al. Signaling mechanisms downstream of quinolinic acid targeting the cytoskeleton of rat striatal neurons and astrocytes. Exp Neurol. 2012;233(1):391–9.

[38] Pierozan P, Ferreira F, de Lima BO, Pessoa-Pureur R. Quinolinic acid induces disrupts cytoskeletal homeostasis in striatal neurons. Protective role of astrocyte-neuron interaction. J Neurosci Res. 2015;93(2):268–84.

[39] Pierozan P, Ferreira F, Ortiz de Lima B, Goncalves Fernandes C, Totarelli Monteforte P, de Castro Medaglia N, et al. The phosphorylation status and cytoskeletal remodeling of striatal astrocytes treated with quinolinic acid. Exp Cell Res. 2014;322(2):313–23.

[40] Etienne-Manneville S. From signaling pathways to microtubule dynamics: the key players. Curr Opin Cell Biol. 2010;22(1):104–11.

[41] Caceres A, Kosik KS. Inhibition of neurite polarity by tau antisense oligonucleotides in primary cerebellar neurons. Nature. 1990;343(6257):461–3.

[42] Witte H, Bradke F. The role of the cytoskeleton during neuronal polarization. Curr Opin Neurobiol. 2008;18(5):479–87.

[43] Younes-Rapozo V, Berendonk J, Savignon T, Manhaes AC, Barradas PC. Thyroid hormone deficiency changes the distribution of oligodendrocyte/myelin markers during oligodendroglial differentiation in vitro. Int J Dev Neurosci. 2006;24(7):445–53.

[44] Fernandez M, Paradisi M, Del Vecchio G, Giardino L, Calza L. Thyroid hormone induces glial lineage of primary neurospheres derived from non-pathological and pathological

rat brain: implications for remyelination-enhancing therapies. Int J Dev Neurosci. 2009;27(8):769–78.

[45] Martinez R, Gomes FC. Neuritogenesis induced by thyroid hormone-treated astrocytes is mediated by epidermal growth factor/mitogen-activated protein kinase-phosphatidylinositol 3-kinase pathways and involves modulation of extracellular matrix proteins. J Biol Chem. 2002;277(51):49311–8.

[46] Fernandez-Lamo I, Montero-Pedrazuela A, Delgado-Garcia JM, Guadano-Ferraz A, Gruart A. Effects of thyroid hormone replacement on associative learning and hippocampal synaptic plasticity in adult hypothyroid rats. Eur J Neurosci. 2009;30(4):679–92.

[47] Vallortigara J, Alfos S, Micheau J, Higueret P, Enderlin V. T3 administration in adult hypothyroid mice modulates expression of proteins involved in striatal synaptic plasticity and improves motor behavior. Neurobiol Dis. 2008;31(3):378–85.

[48] Vallortigara J, Chassande O, Higueret P, Enderlin V. Thyroid hormone receptor alpha plays an essential role in the normalisation of adult-onset hypothyroidism-related hypoexpression of synaptic plasticity target genes in striatum. J Neuroendocrinol. 2009;21(1):49–56.

[49] Zamoner A, Heimfarth L, Loureiro SO, Royer C, Mena Barreto Silva FR, Pessoa-Pureur R. Nongenomic actions of thyroxine modulate intermediate filament phosphorylation in cerebral cortex of rats. Neuroscience. 2008;156(3):640–52.

[50] Zamoner A, Funchal C, Heimfarth L, Silva F, Pessoa-Pureur R. Short-term effects of thyroid hormones on cytoskeletal proteins are mediated by GABAergic mechanisms in slices of cerebral cortex from young rats. Cell Mol Neurobiol. 2006;26(2):209–24.

[51] Zamoner A, Funchal C, Jacques-Silva MC, Gottfried C, Mena Barreto Silva FR, Pessoa-Pureur R. Thyroid hormones reorganize the cytoskeleton of glial cells through Gfap phosphorylation and Rhoa-dependent mechanisms. Cell Mol Neurobiol. 2007;27(7):845–65.

[52] Zamoner A, Heimfarth L, Pessoa-Pureur R. Congenital hypothyroidism is associated with intermediate filament misregulation, glutamate transporters down-regulation and MAPK activation in developing rat brain. Neurotoxicology. 2008;29(6):1092–9.

[53] Zamoner A, Pessoa-Pureur R. Nongenomic actions of thyroid hormones: every why has a wherefore. Immunology, Endocrine & Metabolic Agents in Medicinal Chemistry. 2011;11(3):165–78.

[54] Cattani D, Goulart PB, de Liz Oliveira Cavalli VL, Winkelmann-Duarte E, dos Santos AQ, Pierozan P, et al. Congenital hypothyroidism alters the oxidative status, enzyme activities and morphological parameters in the hippocampus of developing rats. Mol Cell Endocrinol. 2013;375(1–2):14–26.

[55] Trentin AG, Alvarez-Silva M, Moura Neto V. Thyroid hormone induces cerebellar astrocytes and C6 glioma cells to secrete mitogenic growth factors. Am J Physiol Endocrinol Metab. 2001;281(5):E1088–94.

[56] Samuels MH. Cognitive function in untreated hypothyroidism and hyperthyroidism. Curr Opin Endocrinol Diabetes Obes. 2008;15(5):429–33.

[57] Davis PJ, Goglia F, Leonard JL. Nongenomic actions of thyroid hormone. Nat Rev Endocrinol. 2016;12(2):111–21.

[58] Siegrist-Kaiser CA, Juge-Aubry C, Tranter MP, Ekenbarger DM, Leonard JL. Thyroxine-dependent modulation of actin polymerization in cultured astrocytes. A novel, extranuclear action of thyroid hormone. J Biol Chem. 1990;265(9):5296–302.

[59] Farwell AP, Dubord-Tomasetti SA, Pietrzykowski AZ, Stachelek SJ, Leonard JL. Regulation of cerebellar neuronal migration and neurite outgrowth by thyroxine and 3,3′,5′-triiodothyronine. Brain Res Dev Brain Res. 2005;154(1):121–35.

[60] Trentin AG, Moura Neto V. T3 affects cerebellar astrocyte proliferation, GFAP and fibronectin organization. Neuroreport. 1995;6(2):293–6.

[61] Bergh JJ, Lin HY, Lansing L, Mohamed SN, Davis FB, Mousa S, et al. Integrin alphaV-beta3 contains a cell surface receptor site for thyroid hormone that is linked to activation of mitogen-activated protein kinase and induction of angiogenesis. Endocrinology. 2005;146(7):2864–71.

[62] Rahaman SO, Ghosh S, Mohanakumar KP, Das S, Sarkar PK. Hypothyroidism in the developing rat brain is associated with marked oxidative stress and aberrant intraneuronal accumulation of neurofilaments. Neurosci Res. 2001;40(3):273–9.

[63] Motil J, Chan WK, Dubey M, Chaudhury P, Pimenta A, Chylinski TM, et al. Dynein mediates retrograde neurofilament transport within axons and anterograde delivery of NFs from perikarya into axons: regulation by multiple phosphorylation events. Cell Motil Cytoskeleton. 2006;63(5):266–86.

Chapter 13

Acting on Actin During Bacterial Infection

Elsa Anes

Additional information is available at the end of the chapter

http://dx.doi.org/10.5772/66861

Abstract

Bacterial resistance to antibiotics is becoming a major threat to public health. It is imperative to find new therapeutic interventions to fight pathogens. Thus, deciphering host-pathogen interactions may allow defining targets for new strategies for effective treatments of infectious diseases. This chapter focuses on the bacterial manipulation of the host cell actin cytoskeleton. We discuss three infectious processes. The first is pathogen establishment of infection/invasion, explaining cellular uptake pathways that rely on actin, such as phagocytosis and macropinocytosis. The second process focus on the establishment of a replication niche, a process that subverts cytoskeletal functions associated with membrane trafficking namely phagosome maturation and cellular innate immune responses. Finally, pathogen dissemination is an emerging field that microfilaments have shown to participate: pathogen motility through the cytoplasm and from cell-to-cell or on the outer surface of the plasma membrane mimicking a receptor tyrosine kinase signaling pathway that helps the projection of pathogens to neighboring cells. It also establishes a connection with the innate immunity related with induction of cell signaling to inflammation, inflammasome activation, and programmed cell death. These studies revealed several potential targets related to actin cytoskeleton manipulation to design new therapeutic strategies for bacterial infections.

Keywords: actin, Rho GTPases, bacterial pathogens, phagocytosis, macropinocytosis, virulence mechanisms, innate immunity

1. Introduction

The cell cytoskeleton is composed of three distinct protein families each of which is assembled from monomers to form polymer networks namely from actin, tubulin, or intermediate-filament proteins. Host and pathogens have developed intrinsic interactions with the cytoskeletal system, playing a central role in several stages of their life cycles. Deciphering the complexity of these interactions is revealing new insights about the mechanisms of bacterial pathogenicity but also on defining new host targets for alternative therapies to available antibiotics.

Indeed, clarifying these bacterial mechanisms of host subversion has led to many discoveries about host cell biology, including the identification of new cytoskeletal proteins, regulatory pathways, and mechanisms of cytoskeletal function. Microorganisms exploit actin, microtubules, and intermediate filaments in diverse ways, however, it is mainly the actin cytoskeleton that appears to play a critical role in infection and is the topic of this chapter.

In host cells, actin is involved in the polymerization of stable filaments to assure the cell architecture; at the cell surface originates dynamic movements mediated via assembly and disassembly of microfilaments contributing to contour changes as well cellular locomotion,

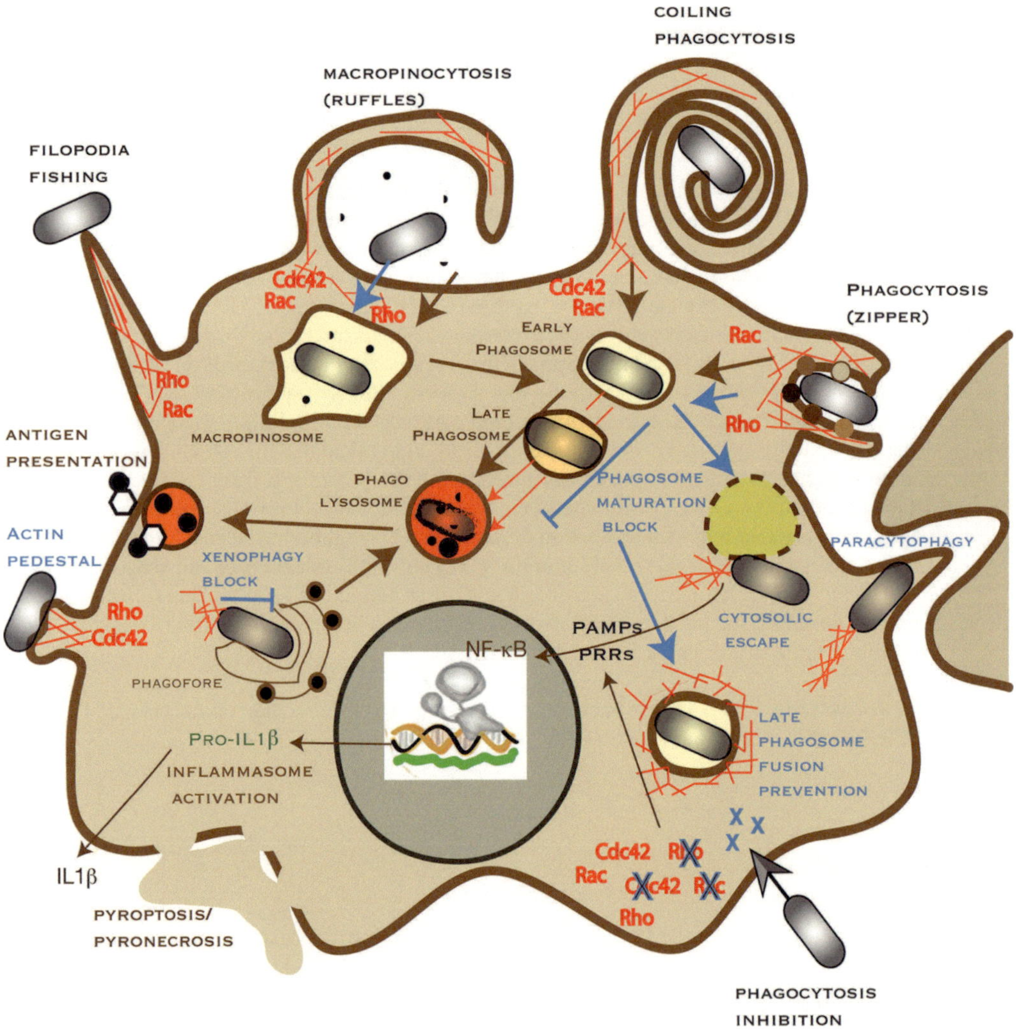

Figure 1. Schematic diagram of host cell actin rearrangements during bacterial infection. In red: actin filaments and actin polymerization promoting Rho GTPases. In brown: cell responses to bacterial infection. In blue: bacteria hijacking mechanisms of the host actin cytoskeleton.

cell-to-cell adhesion, and signaling. In the cytoplasm, the actin skeleton provides tracks and tails to direct vesicle trafficking. Thus, the importance of the actin cytoskeleton for eukaryotic host physiology from cell movement, cell-to-cell adherence, endocytosis, vesicle trafficking, and cell signaling, among others, has provided pathogenic bacteria with a plethora of opportunistic chances to be exploited.

The roles of the actin cytoskeleton in host-pathogen interactions can be summarized according to groups of pathogens and how they interact with this system. Some promote attachment to the plasma membrane, forming specialized actin structures (pedestals), allowing strong adherence to host epithelial surfaces. Others induce actin polymerization to enter into nonprofessional phagocytic cells; while others prevent polymerization to avoid uptake by professional phagocytic cells. A few pathogens use the actin cytoskeleton to allow other specialized internalization processes to occur in phagocytic cells as an alternative or in addition to phagocytosis. Intracellular pathogens manipulate the cytoskeleton to prevent membrane trafficking or fusion events leading to the establishment of a niche inside a vacuole often avoiding delivery into the degradative environment of the lysosome. Finally, some pathogens escape from the phagosome vacuole to the cytosol and use the actin machinery to move within cells and to spread directly from the cytoplasm of one cell into the cytoplasm of an adjacent cell. Recently, actin dynamics during infection was related to innate immune responses that rely on activation of cytosolic pattern recognition receptors (cytosolic PRRs) for inflammasome or autophagy assembly and programmed cell death.

This chapter provides a comprehensive summary of various strategies used by both extracellular and intracellular bacteria to hijack the host actin cytoskeleton (**Figure 1**).

2. Acting on actin during pathogen establishment of infection/invasion

Pathogens often have to overcome epithelial barriers to gain entry into the host cells. The first of which is the epithelial mucosae and a few pathogens, along their evolution, have developed strategies to overcome these barriers by means of active invasion mechanisms. Therefore some intracellular pathogens have evolved strategies to induce or modulate their uptake into these nonprofessional phagocytic cells. Alternatively, as a barrier circumventing mechanism, they may use the cells of the immune system (professional phagocytic cells such as macrophages, neutrophils, and dendritic cells) that patrols those epithelia. Here pathogens may or not play an active role in host cell internalization. Usually professional phagocytes recognize pattern signatures of pathogens (e.g., lipopolysaccharides: LPS), or opsonized bacteria (e.g., complement C3 or IgGs), by means of surface receptors. Likewise phagocytes play an active role in bacteria internalization. As part of the immune system these cells are equipped with a series of insult mechanisms designed to clear pathogens (as the proteolysis at low pH in the phagolysosome). Likewise, extracellular pathogens modulate the host cell plasma membrane for attachment and inhibition of phagocytosis in order to survive. In contrast, intracellular pathogens developed strategies to circumvent the bactericidal mechanisms of immune cells via establishing a protective vacuolar niche.

Several actin dependent mechanisms exist for allowing the establishment of infection: (1) Conventional phagocytosis meaning the entry into professional phagocytes by bilateral membrane pseudopodia formation that tightly encloses the bacteria. Phagocytosis always involves close contact between particle and plasma membrane by multivalence receptor-ligand interactions following morphological changes assembling a zipper mechanism. The host plays a central role for the internalization event while no action is required from the pathogen; (2) induced phagocytosis, a process of active induction of internalization into nonprofessional phagocytes such as epithelial cells, by pathogen manipulation of the host cell contractile system; both the host and the pathogen have active roles in the event. Mechanistically the process occurs by strong interactions between bacterial ligands with cell receptors as in conventional phagocytosis; (3) macropinocytosis: here there may be no direct contact between ligand-pathogen and cell-receptors. Literally, macropinocytosis means—cell drinking—and always involves extensive signaling (e.g., via EGF receptor, a type of tyrosine kinase receptor) that induces pseudopodia unilateral formation surrounding large amount of extracellular volume. So particles including bacteria go in passively along with extracellular fluid. Conventional macropinocytosis may occurs in several types of cells including professional and nonprofessional phagocytes leading to the formation of a large vacuole, the macropinosome; (4) induced macropinocytosis involves pathogen manipulation of the host cell cytoskeleton through growth factor induced signaling or directly using secretion systems that injects virulence factors into the cytosol. While referred classically as trigger phagocytosis, according to the type of morphological changes (with multiple ruffles at the cell surface), there is no direct connection between pathogen and plasma membrane. Finally, (5) an unconventional form of phagocytosis may be used for the establishment of infection via actin cytoskeleton. This is termed as coiling phagocytosis and involves single folds of the phagocyte plasma membrane wrapping around microbes in multiple turns **(Figure 1)**.

2.1. Phagocytosis of bacteria and inhibition of phagocytosis by pathogens

Phagocytosis is a universal phenomenon involving the recognition and binding of a particle (over 0.5 μm in diameter), in a multivalence receptor-dependent manner, to its internalization and degradation within the phagocytic cell [1]. Mechanistically the process of particle internalization from the plasma membrane is clathrin independent and requires actin polymerization [2]. Phagocytosis of one particle does not signal or permit the indiscriminate phagocytosis of other particles bound to the cell surface. In fact particle ingestion is not automatically triggered by initial particle binding, but requires the sequential recruitment of cell surface receptors into interactions with the remainder of the particle surface. The forming phagosome conforms to the shape of the particle as a close-fitting sleeve of plasma membrane, held in place by interactions between surface receptors and the particle surface, much as teeth hold a zipper together [3]. Phagocytosis can be broadly categorized into three steps: particle binding (along with receptor-cell signaling), internalization (i.e., phagosome formation and invagination) and phagosome maturation (i.e., biogenesis of the degradative compartment: the phagolysosome).

The phases prior to the establishment of interactions between bacterial ligands and phagocytic receptors may involve pathogen fishing by cell structures—this process is also dependent of filamentous actin (F-actin), filopodia extensions **(Figure 1)**. Filopodia serves differently in

http://dx.doi.org/10.5772/66861

pathogens and immune cells: pathogens will use it to approach cell membranes for invasion while macrophages will take advantage of these structures for fishing surrounding molecules in order to patrol the environment for possible invaders [4].

Phagocytosis was first discovered in the lower eukaryote amoebae that use it for feeding. In higher organisms, phagocytosis is fundamental for host defence against invading pathogens and contributes to the immune and inflammatory responses [5] including turnover and remodeling of tissues and disposal of dead cells. All cells may to some extent perform phagocytosis [6]. However in mammals, phagocytosis is the hallmark of specialized cells including macrophages, dendritic cells, and polymorphonuclear neutrophils—these cells are collectively referred to as professional phagocytes [6]. In certain circumstances, other cell types, such as fibroblasts engulfing apoptotic cells and bladder epithelial cells consuming erythrocytes, are able to perform conventional phagocytosis as efficiently as professional phagocytes [6].

Professional phagocytes express a series of cell surface receptors which recognize a variety of microbial ligands. Receptors on the surface of the phagocytic cell orchestrate a set of signaling events that are required for particle internalization. However, most pathogens possess many different ligands on their surface. Their phagocytic uptake occurs via multiligand interactions, which induce the engagement of many receptors at the same time.

Two major categories of receptors involved in pathogen recognition are opsonic receptors and nonopsonic receptors (pattern-recognition receptors: PRRs) [1]. Receptors for opsonins such as IgG antibodies and the complement fragment C3bi engage FcγRs and complement receptors (CR), respectively. PRRs include toll-like receptors (TLRs) and other receptor families as C-type lectins receptors that recognize sugar residues as mannose or fucose and lipopolysaccharides (LPS). TLRs often function as coreceptors in phagocytosis by their discrimination of a broad range of microbial products, including LPS and peptidoglycan. The role of TLRs in accelerating and modulating phagosome maturation is still a matter of debate [7].

Bacteria opsonized by complement C3b, by IgG or having lipoarabinomannans at the cell wall surface will be recognized by complement receptors such as CR1 and CR3/4, Fc receptors or Man-6P receptors respectively, each triggering phagocytosis without stimulating a strong superoxide burst. The entry via these phagocytic receptors leads to the maturation of the forming phagosome into a very degradative lysosomal compartment that will destroy microbes [8]. All these receptors will be downregulated during phagocyte activation either through bacterial proinflammatory components as in the case of LPS or cytokines as IFNγ [8].

Activated macrophages will in turn reprogram their expression profile in order to increase the ability to kill pathogens via oxidative bursts and decrease protein digestion extension from amino-acids to small peptides, for antigen presentation [9].

Phagocytosis uses the actin cytoskeleton to construct a cup and close the cup by contractile activities [10]. Latter along phagosome maturation the actin cytoskeleton is also utilized for vesicle trafficking and fusion along the endocytic pathway [11]. The induced polymerization of filamentous actin (F-actin) from globular actin (G-actin) beneath the site of attachment of the particle is the driving force behind ingestion and proceeds from signal transduction downstream of the phagocytic receptors [1]. The precise signaling cascades linking activated receptors to actin polymerization are not fully understood yet it is well known that Rho GTPase family plays critical roles in controlling these cytoskeletal rearrangements [1]. These, RhoA, Rac1, and cell division cycle 42 (Cdc42) act as molecular switches in controlling actin dynamics by regulating the actin-related protein 2/3 (Arp2/3) complex [12]. Arp2/3 requires activation by nucleation-promoting factors, such as the Wiskott-Aldrich syndrome protein (WASP) family. Nucleation-promoting factors exist in an autoinhibited conformation until activated by Cdc42 and Rac1, as well as by phosphoinositide (PI) signaling (discussed latter in this chapter). Effectors such as Cdc42 and the phosphoinositide 4,5-bisphosphate PI(4,5)P2 (PIP2) synergize to activate WASP homolog N-WASP which triggers actin polymerization via Arp2/3 [13]. As the newly formed actin branch grows, the plasma membrane is forced out, extending the membrane as pseudopodia **(Figure 1)**.

Various extracellular and intracellular cues including those from pathogens stimulate Rho GTPases, leading to actin-mediated membrane manipulation. RhoA, Rac1, and Cdc42 have all been shown to accumulate at the nascent phagosome cup. These proteins are preferred targets for bacterial toxins that in turn modulate the organization of the actin skeleton allowing invasion into nonprofessional phagocytic cells and preventing phagocytosis into professional phagocytes. These toxins modify the activity of Rho GTPases through covalent modification or regulation of the nucleotide state. Toxins such as *Clostridium difficile* toxin A and B modify Rho leading to inactivation of its function. This bacterium and the toxin it produces are a global health problem especially affecting the elderly who need to be prescribed prolonged doses of antibiotics. In fact extracellular bacteria, such as *Clostridium* spp., release toxins that glycosylate Rho GTPases in order to disorganize actin to reduce immune cell migration and phagocytosis and also to break down epithelial cell barriers [14].

Another group of toxins regulates the nucleotide state and thus the function of various Rho GTPases by acting as GTPase-activating proteins (GAPs). *Yersinia* spp. an enteropathogenic group of bacteria have secretion systems that inject a type of these Rho GAP toxins, Yop virulence factors leading to actin filamentation blocking and consequently to inhibition of phagocytosis in all host cells to where a contact is established with either professional or non-professional phagocytic cells [15].

Pseudomonas has the capacity to inactivate all Rho GTPases [16]. *Pseudomonas aeruginosa* is a Gram-negative opportunistic pathogen that causes life-threatening infections in cystic fibrosis patients, individuals with burn wounds, and the immuno-compromised. *P. aeruginosa* pathogenicity involves cell-associated and secreted virulence factors as ExoS one of four type III cytotoxins injected into the cytosol. *In vivo* the Rho GAP activity of ExoS stimulates the reorganization of the actin cytoskeleton by inhibition of Rac and Cdc42 and stimulates actin stress fiber formation by inhibiting of Rho [16]. The consequences are the prevention of phagocytosis. Moreover, the perturbation of F-to G-actin content together with cytosolic stress is sensed by the PRR pyrin triggering caspase 1 and inflammasome assembly leading to inflammation and cell death by pyroptosis.

Many intracellular bacterial pathogens have evolved to survive and even proliferate within immune phagocytic cells. Depending on the route of entry, the fate of intracellular bacteria varies significantly. Some opsonized bacteria as *Brucella,* the agent of brucellosis, for example, are destroyed efficiently within macrophages while the nonopsonised survive [17]. An essential feature of the pathogenicity of *Salmonella* is its capacity to cross a number of barriers requiring invasion of a large variety of phagocytic and nonphagocytic cells (reviewed in Ref. [18]). Virulent *Salmonella enterica* serovar Thyphimurium infection of macrophages triggers cell lysis while opsonized noninvasive mutants do not thus reinforce the idea that distinct overcomes depend on the internalization route [19]. The cytotoxicity of serovar Typhimurium is related to the capacity of this organism to invade cells. Mutants lacking invasion proteins encoded by the salmonella pathogenicity island 1 genome region (SPI-1) failed to induce cell lysis in murine macrophages [20]. This is an important step of salmonella infection allowing the pathogen escaping to macrophages to reach the basolateral membrane of the gut cells for invasion.

The uptake of *Mycobacterium* spp. by phagocytes has been intensively studied since these cell types, especially macrophages, are the preferred targets of this successful pathogen. An important class of *Mycobacterium* pathogens includes tuberculosis bacilli. This intracellular facultative pathogen controls the bacterial load during macrophage internalization by interfering with actin polymerization at the phagocytic cup [21]. This is a necessary step in virulence for preventing apoptosis and therefore to prevent pathogen intracellular killing [22]. For this, during early phases of *Mycobacterium* infection, the microRNA 142-3p is overexpressed in response to phagocytosis and interferes with the expression of N-WASP and consequently with the Arp2/3 complex required for actin nucleation at the cell membrane [21]. Therefore, a low bacterial load is accomplished intracellularly, preventing the apoptosis of the infected cells. In addition, recently, miR-142-3p was shown to directly regulate protein kinase Cα (PKCα), a key gene involved in phagocytosis [23].

The heterodimeric host surface receptor complement-receptor 3 (CR-3), mediates uptake of opsonized and nonopsonized mycobacteria. Interestingly, CR-3 is targeted by other intracellular pathogens, such as *Coxiella burnetii*, the Q-fever agent, in order to avoid phagocytosis. This strategy is based on ensuring a spatial location of CR-3 outside the pseudopod extensions [24].

Lipid modification by receptor signaling creates the potential for radiating signals that can affect large areas of the plasma membrane. Phospholipid kinases, lipid phosphatases, and hydrolases are activated during phagocytosis. Classes of phospholipids typically found on the inner face of biomembranes include phosphatidylinositol (PI). The generation of phosphoinositides derived from PI via phosphorylation events will generate classes of important lipids enrolled in cell signaling and phagocytosis as example of phosphatidylinositol (4)-phosphate (PI(4)P=PIP), PI(5)P, PI(4,5)P2 (PIP2), PI (3,4)P2, and PI(3,4,5)P3 (PIP3). As mentioned previously in this chapter, these phosphoinositides, especially PIP2 and PIP3, are capable of binding and increasing the activity of proteins that modify membrane chemistry and the actin cytoskeleton. As an example, PIP2 increases the activity of WASP, a protein that stimulates actin polymerization via Arp2/3.

This class of PIs in addition to their relevance in particle internalization is important during the phase of phagosome maturation into a degradative compartment, the phagolysosome. In phagosomal membranes PIP2 activates the actin nucleators of the Ezrin, Moesin, and Radixin family inducing polymerization of F-actin and therefore phagosome maturation [11]. This will be addressed later in this chapter in the context of the manipulation of the actin cytoskeleton by pathogens in order to establish an intracellular niche.

2.2. Induced phagocytosis by invasive pathogens

Classically, the manipulation of the actin cytoskeleton by invasive pathogens was classified into two general mechanisms according to the type of morphological changes that occur in the host cell—the zipper and trigger phagocytosis [3]. Entry of uropathogenic *Escherichia coli, Yersinia, Helicobacter, Listeria,* and *Neisseria* into epithelial cells is reminiscent of the classical model of zipper phagocytosis. The trigger model will be addressed as macropinocytosis in the next section of this chapter as it is not in fact a phagocytosis event. Moreover, the zipper mechanism may also be triggered actively by pathogens.

Adherence to nonprofessional phagocytic cells, epithelium by a pathogen is necessary to avoid mechanical clearance and is the first step of colonization by for example enteropathogens. Thus bacterial pathogens exhibit a large variety of cell surface adhesins, including fimbriae (pili) and afimbrial adhesins some of which participate in the internalization step. Likewise, in this type of entry, a bacterial adhesin binds to a host cell surface receptor involved in cell-to-cell adhesion and/or activates regulatory proteins that modulate cytoskeleton dynamics. Moreover, adherence and internalization into epithelial cells looks to be a strategy used by pathogens to escape destruction by immune cells as described below.

Most type I pili expressed by pathogenic *E. coli* bind to host mannose-containing glycoproteins some expressed in gut epithelial cells including M cells (microfold cells of Payer's Patches) [25]. Others such as FimH from uropathogenic *E. coli* can bind to β1 and $\alpha 3$ integrins and thereby promote bacterial internalization following a process that to date has only been described in urinary bladder epithelial cells. Uropathogenic *E. coli* (UPEC) cause the majority of community-onset urinary tract infections (UTI). Early in acute cystitis, UPEC gains access to an intracellular niche that protects a population of replicating bacteria from arriving phagocytes [26]. Transition bacillary forms of UPEC (1–2 μm in length) are readily engulfed, while filamentous UPEC resist phagocytosis, even when in direct contact with neutrophils and macrophages. Despite these strong host defenses, a subpopulation of UPEC is able to persist for months in a quiescent reservoir state which may serve as a seed for recurrent infections [27].

Yersinia spp. such as *Yersinia enterocolitica* and *Yersinia pseudotuberculosis* invades gut mucosae at the ileum terminal end and multiplies in the underlying lymphoid tissue. Invasin and YadA (Yersinia adhesion A) are crucial for yersinia adherence via β1 integrins and matrix components, respectively. β1 integrins exist on the basolateral face of enterocytes and on the apical surface of the epithelia derived M cells. The coalescence of integrins following bacteria invasin linkage will lead to yersinia internalization by a "zipper mechanism". Binding of invasin to β1 integrin activates focal adhesion tyrosine kinase and triggers a complex cascade implicating Rac1-Arp2/3 pathways but also phosphoinositide-3-kinase (PI3K) leading to the closure of the phagocytic cup. In contrast, YadA binds diverse extracellular matrix components, such as collagen, laminin, and fibronectin, thus indirectly mediating integrin binding [28]. Yersinia species also hijack host cell phosphoinositide metabolism for their uptake. Rac-1 recruits, and Arf6 activates the type I phosphatidylinositol-4-phosphate-5-kinase (PtdIns(4)P(5)Ka), which forms PIP2 at the entry site, and this lipid may regulate phagocytic cup formation by coordinating membrane traffic and controlling F-actin polymerization [29].

Helicobacter pylori is another example of pathogen that adheres to mucosa via β1integrins and invades nonphagocytic cells. Efficient infection of cultured epithelial cells seems to be restricted to certain *H. pylori* strains. This pathogen uses a type IV secretion system (T4SS) targeting β1 integrins to translocate the virulence factor CagA into the cytosol. The adhesin CagL present in the T4SS pilus surface bridge activates the integrin on the basolateral membrane of gastric epithelial cells. In all cases, however, invasion of *H. pylori* seems to involve a typical zipper-like entry process. Both PI3-K and PKC are required for bacterial uptake and induction of cytoskeletal rearrangements [30]. Curiously preinfection of cultured gastric cells with yersinia expressing Yop virulence factors that interfere with the same signaling events impaired phagocytosis of *H. pylori* [30]. Internalized *H. pylori* was shown to be located in tight phagosomes and in close association with condensed actin filaments and localized tyrosine phosphorylation signals. Similar to UPEC in bladder epithelial cells, invasion of epithelial cells by *H. pylori* may constitute one of the evasion strategies used by this pathogen to circumvent the host immune response and persist in stomach.

Curiously the vaccinal strain for tuberculosis *Mycobacterium bovis* BCG has been used as the more effective treatment for bladder cancer [31]. The bacillus induces phagocytosis in tumor cells via their surface fibronectin attachment protein (FAP) to β1integrins. After phagocytosis a strong cytotoxic effect is displayed via T-helper CD8 stimulation leading to antitumor activity.

Listeria monocytogenes is a food-borne Gram-positive bacterium that makes use of two surface proteins, Internalin A (InlA) and B (InlB), to engage, in a species-specific manner, to host adhesion molecules E-cadherin and hepatocyte growth factor receptor Met respectively, to induce its internalization [32]. Only InlA is critical for invasion of the gut epithelial cells. The specific engagement of E-cadherin initiates activation of the adherens junction machinery inducing the recruitment of β-catenin, Rho GAP protein ARHGAP10, α-catenins to the site of the entry. Internalization is then further mediated by Rac- and Arp2/3-dependent actin polymerization. In contrast to this, InlB is essential for *Listeria* uptake by most nonphagocytic cell types, such as hepatocytes, endothelial cells, fibroblasts, and certain epithelial cell lines. Additionally, it is known that ActA, a *Listeria* protein required for actin-tail formation and intracellular cytosolic movement, can also mediate *Listeria* uptake by epithelial cells [32]. Recently a new phagocytic process was characterized that allows human endothelial cells to internalize listeria independent of all known pathogenic bacterial surface proteins. Here bacteria adhesion is mediated by Rho kinase and the control of the internalization step is coordinated by formins (as FHOD1 and FMNL3) a class of actin nucleation proteins. The overall control of the event is mediated by cytoskeletal proteins usually enrolled in cell shape and locomotion including Rho, focal adhesions, and PI kinases [33].

Neisseria gonorrhoeae, is an exclusive human pathogen that primarily infects the urogenital epithelia, causing the sexually transmitted disease gonorrhoea. Entry of *N. gonorrhoeae* into human epithelial cells is multifactorial. Initial attachment is mediated by pili (a T4SS), followed by tight adherence via the phase-variable colony opacity (Opa) proteins. These are a family of 11 outer membrane proteins variably expressed at the surface of the bacterium. However, only OpaA confers invasion into epithelia [34]. This entry is mediated by heparan sulfate proteoglycan (HSPG) receptors of the syndecan family expressed on the target cell surface. Pilus engagement has also been demonstrated to play a role in host cell cytoskeletal rearrangements inducing microvilli formation at the cell surface to surround the bacteria for a zipper mechanism of internalization [35].

In endothelial cells, the T4SS-pilus-mediated adhesion of *Neisseria meningitidis* induces the formation of membrane protrusions similar to microvilli leading to bacterial uptake. These protrusions result from a Rho- and Cdc42-dependent cortical actin polymerization, and from the activation of the ErbB2 tyrosine-kinase receptor and the Src kinase, leading to tyrosine phosphorylation of cortactin, an activator of Arp2/3 [36]. Adhesion of *N. meningitidis* to endothelial cells promotes the local formation of membrane protrusions reminiscent of epithelial microvilli structures that surround bacteria and provoke their internalization within intracellular vacuoles.

2.3. Macropinocytosis, induced macropinocytosis, and coiling phagocytosis

Unique molecular properties associated with the process of macropinocytosis are beginning to be elucidated. Because of their size and the fact that they may be formed without activation by ligands, the large vacuoles (macropinosomes) formed during this pinocytosis event can contain extracellular fluid and pathogens. At the mechanistic level, phagocytosis and macropinocytosis present many similarities including the involvement of phosphoinositol phosphate signaling and actin cytoskeleton reorganization. During macropinocytosis it is not observed a direct connection between bacteria/cargo and multiple receptors but it was demonstrated the relevance of tyrosine kinase receptors involved in responses to growth factors as the epidermal growth factor and platelet-derived growth factor. The consequence of intensive actin remodeling results in ruffling protrusions at the cell surface, or in unilateral large pseudopodia formation leading to the formation of large macropinosomes. Activated receptor tyrosine kinases, as well as the Src family kinases, are clearly observed on newly formed macropinosomes. Therefore in concert with the morphological definition provided by Lewis in 1931 based on ruffling formation, and elevation in response to growth factor stimulation can be used to define macropinocytosis [37].

Macropinocytosis has been observed in professional phagocytes as well in epithelial cells. Immature dendritic cells and activated macrophages display high levels of constitutive macropinocytosis [38]. The consequent internalization of large volumes of extracellular solute that accompanies macropinocytosis facilitates their capacity to continuously survey the extracellular space for foreign material. In fact, this increased levels of macropinocytosis upon encounter with the antigen/pathogen enhances both antigen capture and antigen presentation by dendritic cells as well as the complete clearance of pathogens after macrophage activation by inflammatory stimulus [38].

In epithelial cells, an induced form of macropinocytosis was observed after infection with pathogens such as *Shigella, Salmonella,* enterophatogenic *E.coli* (EPEC), and *Mycobacterium tuberculosis*. Therefore, individual pathogens have developed a range of strategies to modulate the host's normal macropinocytic pathways both to invade the host cells and to manipulate the lipid and protein composition of the encapsulating macropinosome to promote cell uptake and then survival. A few virulence factors secreted by pathogens are able to induce ruffling similar to the growth factors named above. The closure of ruffles back to themselves will entrap pathogens into a large vacuole (micropinosome) incorrectly named in distinct publications as "spacious phagosome".

Invasive enteropathogens, such as *Shigella flexneri* and *S. enterica* serovar Typhimurium, use the trigger mechanism of invasion in epithelial cells to induce membrane ruffles and macropinocytosis. This is a phenomenon dependent on a type III secretion system encoded by both bacteria. The T3SS effectors activate host Cdc42 and Rac1 albeit via distinct cellular relays. In *Salmonella*, SopE acts as a guanyl-nucleotide-exchange factor for Rho [39]. This induced Rho GTPase perturbation is recognized in the cytosol by PRRs (NOD1 sensor) inducing a proinflammatory response and innate immune responses. SigD/SopB is another protein secreted by the SPI-1 T3SS of *Salmonella* to invade nonphagocytic cells. The phosphatidyl-inositol phosphatase activity of SigD/SopB induces rapid disappearance of PIP2 from invaginating regions of the cytoplasmic membrane leading indirectly to Rho activation and macropinocytosis. Once inside the host cell, *Salmonella* induces the recovery of normal cytoskeleton dynamics via SptP, a SPI-1 effector with Cdc42 and Rac1 GAP activity that returns these proteins to their nonactivated state.

In comparison, the effectors IpaC, IpgB1, and VirA of *Shigella* bind to initiate a focal adhesion structure required for internalization via a process that recruit Rho isoforms [40]. Consequently, the injection of the effectors IpaC, IpgB1, and VirA by *S. flexneri* induces Rac1/Cdc42-dependent actin polymerization. Finally, the translocated effector IpaA binds vinculin and enhances its association to actin filaments, thus mediating the localized depolymerization of actin, which is required to close the phagocytic cup [40].

S. flexneri invasion has been classically described as a macropinocytosis-like process, however the role of macropinosomes in intracellular bacterial survival remains elusive. There is evidence that bacterial entry and membrane ruffling are associated with different bacterial effectors and host responses during *S. flexneri* invasion. Rho isoforms are recruited differentially to either entering bacteria or membrane ruffles, and entry has been proposed to occur initially via effector mediated contact of *S. flexneri* to specific receptors suggesting entry is akin to receptor mediated phagocytosis. In fact, the host surface molecules β1-integrins and CD44 (hyaluronic acid receptor) are needed for *Shigella* entry [40].

Recently, the mechanism of *Shigella* invasion of epithelial cells was observed using advanced large volume correlative light electron microscopy (CLEM) indicating a combination of induced phagocytosis and macropinocytosis [41]. Here, the macropinocytic event instead of being the major effector for internalization was in fact shown to be required for release of the bacteria from the phagosome and cytosolic escape later in phagocytosis. Macropinocytic vesicles formed at the invasion site are functionally involved in vacuolar rupture. This unique and surprising pathogenic strategy stands in stark contrast to other invasive pathogens that induce direct lysis of their surrounding vacuole via the action of destabilizing bacterial proteins.

S. enterica is an invasive, T3SS-employing pathogen and shares many common host entry characteristics with *S. flexneri*. It was hypothesized that salmonella containing vacuole and macropinosomes may be distinct, as they are sorted into different intracellular routes [42].

These evidence suggest that pathogen induced enhanced uptake of extracellular fluid in *S. enterica serovar* Typhimurium-infected epithelial cells is an event related to the invasion mechanisms used by this pathogen but not the major mechanism for bacteria internalization as referred in most published data.

Surface-adherent pathogens, such as enteropathogenic or enterohaemorrhagic *E. coli* (EPEC or EHEC, respectively), use their T3SS to secrete a transmembrane receptor into the host membrane to stimulate actin polymerization and generate cellular extensions called pedestals. EPEC uses the T3SS apparatus to inject the intimin receptor (Tir). Tir acts as a cell receptor of host kinases activating N-WASP and the actin nucleator Arp2/3 resulting in actin polymerization and pedestal formation at the site of the attachment. While stabilizing bacteria connection to epithelial cells the actin pedestal formation promotes T3SS mediated injection of additional effector proteins able to subvert other host pathways. Where bacteria are attached, microvilli are lost; the epithelial cells form cup-like pedestals upon which the bacteria rest. The underlying cytoskeleton of the epithelial cell is disorganized, with a proliferation of filamentous actin. Although EPEC have traditionally been considered to be noninvasive, accumulating evidence casts doubt on this assumption. From the earliest published electron micrographs of EPEC infection, bacteria have been observed within epithelial cells at the sites of attaching [43]. The virulence factor dependent on Tir signaling EspG contributes to the ability of EPEC pathogens to establish infection through a modulation of the host cytoskeleton involving transient microtubule destruction and actin polymerization in a manner akin to the *S. flexneri* VirA protein [28, 44].

Patients with inflammatory bowel disease exhibited an increased number of mucosae-associated *E. coli* with invasive properties. The adherent-invasive *E. coli* (AIEC) uses M cells to reach macrophages of Payer's Patches where they survive and replicate inside large macropinosomes that share features of phagolysosomes. To survive, these bacteria, inside the vacuoles, adapted to the harsh acidic environment that is the key signal to activate virulence genes. In fact infected macrophages with AIEC secrete large amounts of tumor necrosis factor alpha leading to local granuloma formation. Those macrophages will subsequently aggregate and fuse releasing bacteria that then will reach the basolateral domain of gut epithelial cells for invasion. Epithelial cell invasion is a key virulence factor only for EIEC, which may lead to a dysentery-like illness similar to that caused by *S. flexneri* [45].

Alveolar macrophages constitute the main defense against *M. tuberculosis* infection. However, tuberculosis bacilli resist phagocytic cell bactericidal mechanisms and replicate within them. Although *M. tuberculosis* survives within phagocytic cells, this bacterium may also bind and invade alveolar epithelial cells [46] and endothelial lymphatic cells [47]. Infection of epithelial cells was concomitant with large lamellipodia projections (ruffles) similar to macropinocytosis. Likewise, *Mycobacterium* can induce formation of macropinosomes however; this does not depend on a bacterial secretion system, as the culture media in the absence of pathogen was sufficient to induce this process. Since nonviable bacteria fail to induce macropinocytosis in opposition to live bacteria, the most prominent candidate to induce ruffling is pointed as being secretory products actively produced by life bacilli. There are no requirements for bacteria to attach directly to the plasma membrane. In endothelial cells, scanning electron microscopy (SEM) micrographs show that mycobacteria were internalized by characteristic phagocytosis-like and macropinocytosis events [47]. However the mycobacterial determinants leading to actin reorganization and pathogen active internalization are not clarified. It is very likely that the invasion and survival in epithelial and endothelial cells contributes to the one-third of the human population latently infected with this microorganism.

Coiling phagocytosis is an actin dependent endocytic event, morphologically accompanied by a typical pseudopodia that looks like whorls or wrapps around the bacteria in several turns (**Figure 1**). A definition of the phenomena is complex as it presents similarities to macropinocytosis and conventional phagocytosis: for the first due to the large pseudopodia; for the second due to cargo specific entrapment. In coiling phagocytosis, the single pseudopodia do not trap fluid droplets but enclose microbes; however, the multiple pseudopod whorls have largely self-apposed surfaces instead of those that are microbe-apposed surfaces. *Legionella pneumophila* and *Borrelia burgdorferi* the agents of Legionellosis and Lyme disease, respectively, use this form of endocytosis for establishment of the infection within macrophages. It was demonstrated that coiling phagocytosis is an active and selective process of the phagocytes, initially triggered by heat- and aldehyde-insensitive moieties of the microbial surface [48], suggesting that coiling and conventional phagocytosis are very closely related, most likely starting from the same phagocytosis-promoting receptor(s). The lack of difference between viable and killed microbes indicates that coiling phagocytosis is actively driven by the phagocytes and not by the microbes. This distinguishes coiling phagocytosis from nonclassical uptake mechanisms such as the induced phagocytosis or macropinocytosis. In this respect, the identification of granulocyte macrophage colony-stimulating factor (GM-CSF) and phorbol esters such as PMA as coiling-promoting substances may be a clue as to the regulatory mechanisms involved in coiling phagocytosis [48]. On the side of the phagocytes, coiling phagocytosis obviously is clearly a regulated mechanism, because the monocytes used it selectively for certain spirochetes, which is inconsistent with simply an accidental trapping of pericellular microbes.

In summary, deciphering the players that induce or prevent phagocytosis in one infection context may be used as strategies to clear pathogens in other context. It is an interesting observation that preinfection of cultured gastric cells with yersinia expressing Yop virulence factors that interfere with the same signaling events, impaired phagocytosis of *H. pylori*. This may be a potential starting strategy to fight gastric cancer due to this pathogen.

Define what receptors stimulate to induce a more bactericidal response of infected cells, how to control bacterial load that is internalized to induce apoptosis, as is the case of microRNAs that control WASP in tuberculosis context; how to neutralize factors that prevent Rho family of GTPases to modify actin in order to induce phagocytosis of extracellular pathogens, these are a few targets to explore deeply. Other relevant area to act is how to neutralize bacterial adhesins, secretion systems or their access to surface receptors as integrins to prevent epithelia invasion. It is imperative to decipher what are the virulence factors that mimics or induce growth factors that leads to induced macropinocytosis. In addition, it is important to find how to neutralize secretion systems that reorganize the actin cytoskeleton for macropinosome formation and therefore for pathogen invasion of epithelial and endothelial cells, important reservoirs of latent infections.

3. Acting on actin for the establishment of an intracellular niche

In addition to particle binding and internalization, phagocytosis includes the process of phagosome maturation leading to pathogen destruction in the acidic hydrolytic environment of the phagolysosome. These events are important innate immune mechanisms. Indeed a consequence of phagosome maturation is the activation of the antigen presentation machinery. Macropinocytosis culminates in the appearance of a large vacuole that, indeed follows the fate of the phagosome. Some pathogens have evolved to establish sustained infection in professional phagocytes preventing phagosome maturation as is the case of *M. tuberculosis* and *S. enterica*. Other's diverts the endocytic pathway into a distinct vacuole more similar to the secretory pathway (e.g., *Legionella pneumophila* associates with the endoplasmic reticulum). By doing this, pathogens establish an intracellular niche were they survive, escape the immune bactericidal responses and have access to nutrients. Finally, a group of pathogens are able to escape the endocytic pathway by lysing the vacuole and move to the cytosol (e.g., *Mycobacterium marinum* within macrophages; *M. tuberculosis* within endothelial cells; *Shigella*, listeria within epithelial cells) **(Figure 1)**.

The material in endosomes or phagosomes that is destined for lysosome degradation by endocytosis or phagocytosis reaches this compartment by fusing with the organelle. Critical for this is the membrane composition of the correct repertoire of lipids, membrane-bound proteins, and also proteins that shuttle on and off membranes. The manipulation of the phagosomal membrane by pathogens may block the ability of fusion with lysosomes leading to a vacuole that may be trafficked apart from the endocytic route. In alternative, the vacuole may be arrested from maturation along the endocytic pathway by pathogen membrane manipulation leading to continuous transient fusion events with upper compartments.

Phagosome maturation is known to be influenced by the lipid species present on the outer and most likely inner membrane, and published studies have focused mostly on kinases that generates PIP, and PIP2, which binds actin nucleation proteins [49]. Additionally, the ability to nucleate actin leading to F-actin polymerization from phagosomal membranes was associated to the formation and availability of actin tracks for organelles to move towards the actin-nucleating source, increasing vesicle trafficking, fusion events, and phagolysosome biogenesis **(Figure 1)** [50]. Identifying key roles for PIP and PIP2 opened the door for the analysis of several other lipids that interconnected with these phosphoinositides in the actin assembly process, as well as sphingolipids and fatty acids favouring phagosome maturation [11, 51]. Examples of F-Actin stimulatory factors includes the eicosanoide omega 6 arachadonic acid, ceramide and sphingosine-1-phosphate.

Several groups have explored the role of actin cytoskeleton during *Mycobacterium* late phases of phagocytosis. Pioneering work by de Chastellier and co-workers shows that *Mycobacterium* avium a pathogen common in AIDS patients, disrupt the macrophage actin filament network highlighting here the target for the bacterium that allows sustained intracellular survival. It was demonstrated that in contrast to nonpathogenic mycobacteria, pathogenic *M. tuberculosis* prevents actin polymerization on phagosomal membranes [11, 52]. Therefore, the enrichment of *M. tuberculosis* phagosomal membranes with classes of lipids that leads to PIP2 was shown to induce F-actin tracks from the vacuole membrane. This is concomitant with an increase of fusion events, phagolysosome biogenesis and, consequently *M. tuberculosis* intracellular killing [11]. Drug-induced manipulation of the pathogen actin nucleation-induced blockade represents interesting alternative therapies for tuberculosis.

Another pathogen that blocks phagosome maturation is *Salmonella*. Several hours after bacterial uptake into different host cell types, *Salmonella* induces the formation of an F-actin meshwork around the *Salmonella*-containing vacuole (SCV), which is a modified phagocytic compartment. SCV integrity is closely linked to a surrounding meshwork of actin that in contrast to what happens during mycobacteria infection, acts as a barrier that prevents membrane contact and, therefore vacuole fusion with other endocytic organelles [53]. This process does not require the Inv/Spa type III secretion system or cognate effector proteins, which induce actin polymerization during bacterial invasion. A second T3SS, the salmonella pathogenicity island 2 (SPI2), translocate effectors from the phagosomal membrane to the cytosol. The consequence of this event is the induced polymerization of actin around the SCV that will allow salmonella intravacuolar survival. The spv virulence locus will express the SpvB protein and ADP-ribosyl transferase that will promote actin depolymerisation in latter stages of infection. Treatment with actin-depolymerizing agents significantly inhibited intramacrophage replication of salmonella. Furthermore, after this treatment, bacteria were released into the host cell cytosol, whereas SPI-2 mutant bacteria remained within vacuoles [53]. In conclusion, while during *M. tuberculosis* infection actin assembly is prevented or F-actin is disrupted to allow the establishment of an intracellular niche, in the case of salmonella infection the generation of an F-actin induced mesh is required to maintain and position a vacuole that sustains bacterial growth.

4. Acting on actin for pathogen dissemination: actin-based motility of pathogens and innate immunity

Early after host invasion some pathogens escape lysosomal destruction and antigen presentation by escaping into the cytosol. Thereafter, actin polymerization is manipulated by several cytosolic pathogens such as *L. monocytogenes*, *S. flexneri*, *Burkholderia pseudomallei*, *Rickettsia* spp., and *M. marinum*. These generate and use actin tails to move within and between cells.

When intracellular moving bacteria reaches the plasma membrane, they push out long protrusions that are taken up by neighboring cells, facilitating the infection to spread from epithelial cell to cell in the absence of immune surveillance. At the cell-to-cell cytoplasmic membranes sites, the cytosolic actin-based moving pathogens induce the formation of surface protrusions that force the internalization from the infected cell into noninfected neighbor cells. The process of engulfment is called paracytophagy and involves internalization of a double membrane containing pathogen: the inner from the donor cell and the outer from the recipient cell (**Figure 1**) [54, 55]. At this point the pathogen may escape again to cytosol to start a new infection process.

In the case of enterophatogenic *E. coli* EPEC it was found that some actin pedestal of the attached EPECs also translocate along the cell surface, reaching speeds of 0.007 μm/s allowing bacteria to spread between attached cells [34] (**Figure 1**). While this model shares similarities with the *Listeria* or *Shigella* systems, the main difference is the presence of a membrane between the pathogen and the cell cytoskeleton (**Figure 1**: as in the case of filopodia fishing compared to paracytophagy). The actin polymerization system Arp2/3 complex has been manipulated by several pathogens differently. Some mimics the Wiskott-Aldrich syndrome protein (WASP) family [56], while other's recruit WASP directly to activate Arp2/3 [57]. Examples of the first include the actA protein of listeria and RickA of riquetsia. For the second examples exist as is the case of IcsA of *S. flexneri* and nondetermined factors of *M. marinum* but dependent on the ESAT-6 secretion system 1 [57]. *M. marinum* is a water-borne bacterium that naturally infects fish and amphibians and is an opportunistic pathogen for humans causing tuberculosis while *Rickettsia conorii* belongs to the spotted fever group of *Rickettsia* species transmitted by ticks [55].

The actin-based motility of *B. pseudomallei* the causative agent of melioidosis occurs by a mechanism distinct to that used by other intracytoplasmic pathogens. In fact, the actin tails induced by this pathogen contains Arp2/3 components but it is not clear in the enrollment of the intracellular motility of *B. pseudomallei* [58]. The overexpression of Scar1 a cellular actin nucleating promoting factor that in the context of *S. flexneri*, *L. monocytogenes* and *R. conorii*, blocks actin tail formation and motility, during *B. pseudomallei* infection as no effect on actin-based motility [58].

The predominance of a membrane surrounding vacuole during the infection of most intracellular pathogens looks to be related to immune protection from the defensive mechanisms that exist in the cytosol. The arrival of a pathogen or their PAMPs to the cytosol could "wake up" several patrol mechanisms that include cytosolic PRRs. The sensing by cytosolic innate receptors leads to an inflammatory response by secretion of proinflammatory cytokines and chemokines or a interferon type I response that overall leads to antimicrobial response; the stress in the cytosol induce inflammasome assembly [59].

Therefore, the arrival of the pathogens in the cytosol establishes a bridge to the innate immune response by contact of the pathogen-associated molecular patterns (PAMPS) with PRRs, such as NLRPs (Nod like, similar to Toll like receptors- TLRs on cell membranes). Additionally, and by causing cytosol stress, PAMPS will activate (via PRRs) the inflammasome, a complex structure of proteins similar to the apoptosome [60]. Inflammasome assembly will lead to pro-Interleukin1β (pro-IL-1β) and pro-IL-18 inflammatory cytokine activation via caspase 1 and to the programmed cell death dependent on caspase 1, as it is pyroptosis and pyronecrosis [22]. This is a natural immune response in gut and respiratory epithelial cells but not in endothelial vascular and lymphatic cells that lakes these cytosolic receptors and constitutes important host niches for intracellular pathogen survival [33, 47].

Rickettsiae possess a tropism to endothelial cells, a tissue that usually serves as barrier to intravascuolar blood from surrounding tissues. This tropism leads to the endothelial cell injury associated with complications of the disease. RickA (mentioned previously in this chapter) is a protein present in the pathogenic species *R. conorii*, but absent in *Rickettsia thyphi* [56]. This absence is responsible for an erratic actin-based motility of *R. thyphi* leading to the hypothesis of existence of multiple actin-polymerization mechanisms in pathogenic rickettsia. A consequence of this erratic movement may be the delayed spread from cell to cell and continuous replication of thyphi species leading to bacterial overload and necrotic cell lysis [56]. For *R. conorii* paracytophagy cell-to-cell-spread is the common mechanism for pathogen dissemination [55].

Macrophages, in contrast to endothelial cells, possess NLRs and other PRRs families. During *M. tuberculosis* as well as for *M. marinum* infection phagolysosomal rupture and bacteria escape to the cytosol usually leads to necrotic cell death [61, 62]. The existence of a functional RD1 region expressing ESAT-6 is relevant for the activation of the inflammasome, the necrotic cell death and the secretion of proinflammatory cytokines IL-1β [21]. In endothelial cells, however, the tubercle bacilli survives [47].

The detection of cytosolic LPS, as a consequence of disruption of replication vacuoles harboring Gram-negative bacteria was shown to trigger the activation of murine caspase-11 that leads to the assembly of a noncanonical inflammasome [63]. Caspase-11 (Casp-4 in humans) is also crucial for clearance of bacteria that escape the vacuole, such as *Burkholderia*. In addition, detection of *sdhA* mutants of *Legionella* and *sifA* mutants of *Salmonella* activate caspase-11-dependent pyroptosis [63]. Detection of cytosolic pathogens thus leads to caspase-1- or caspase-11-mediated pyroptosis and restricts bacterial growth.

Another potent host defense mechanism that restricts intracellular pathogens is autophagy. Some intracellular bacteria cause the formation of ubiquitinated aggregates around either bacterial structures or replication vacuoles, and the autophagic machinery can recognize these. The process of bacterial clearance by selective autophagy is called xenophagy. *Listeria* moves within the host cytoplasm through actin-based motility, promoted by the bacterial ActA protein, which is important for avoiding recognition by autophagy [64]. In contrast to the ActA protein, the *Shigella* IcsA protein that also promotes actin-based motility from one pole of the bacterium binds to the autophagy protein Atg5 thus targeting the bacterium to a phagophore. *Shigella* uses two different mechanisms to escape the host autophagic response: first, it secretes IcsB, a protein that competitively binds to IcsA and prevents its recognition by Atg5 thus preventing LC3 recruitment and the process of autophagy [65].

All together these findings let us to postulate that important strategies to fight pathogens will pass by control their life cycle in the cytosol. Either addressing the linkage of actin tails to Arp2/3 or WASP proteins or neutralizing the bacteria actin nucleators to prevent motility and spread to neighbor cells; either to induce death of the infected cell by apoptosis, pyroptosis, or necrotic lysis; either by exposition of pathogen signatures that leads to xenophagy; altogether these are a few potential strategies to address in the future.

5. Concluding remarks

During evolution, higher eukaryotic organisms have developed epithelial barriers and phagocytic immune cells to resist and fight infections. The discovery of antibiotics in the early part of the last century led to predictions that bacterial infections would be kept under tight control via natural systems and treatment with drugs. But the capacity of bacteria to evade natural protective systems and rapidly develop resistance to antibiotics had led to the current situation of bacteria posing major health problems in both the developed and underdeveloped world. There is now a major requirement to find alternative treatments to fight bacterial pathogens. Over the years, various studies have elucidated the mechanisms by which bacterial PAMPs, adhesins, and secretion systems together with their translocated effectors target and alter the host actin dynamics. Targeting the host actin machinery is important for the survival and pathogenesis of several extracellular, vacuolar, and cytosolic bacteria. Studying the manipulation of host actin by pathogens has vastly improved our understanding of various basic cell biological processes in host cells while giving key insights into both bacterial pathogenesis and host innate immunity. Together this opens a new and exciting field of research with the objective of discovering new classes of antibiotics that directly or indirectly interfere with this actin-modulating mechanism.

Author details

Elsa Anes

Address all correspondence to: eanes@ff.ulisboa.pt

Research Institute for Medicines iMed-ULisboa, Faculty of Pharmacy, Universidade de Lisboa, Lisbon, Portugal

References

[1] Niedergang F, Chavrier P. Regulation of phagocytosis by Rho GTPases. Current Topics in Microbiology and Immunology 2005;291:43–60.

[2] Cannon GJ, Swanson JA. The macrophage capacity for phagocytosis. Journal of Cell Science 1992;101:907–13.

[3] Swanson JA, Baer SC. Phagocytosis by zippers and triggers. Trends in Cell Biology 1995;5:89–93.

[4] Bornschlögl T. How filopodia pull: what we know about the mechanics and dynamics of filopodia. Cytoskeleton 2013;70:590–603.

[5] Aderem A, Underhill DM. Mechanisms of phagocytosis in macrophages. Annual Review of Immunology 1999;17:593–623.

[6] Rabinovitch M. Professional and non-professional phagocytes: an introduction. Trends in Cell Biology 1995;5:85–7.

[7] Yates RM, Russell DG. Phagosome maturation proceeds independently of stimulation of toll-like receptors 2 and 4. Immunity 2005;23:409–17.

[8] Yates RM, Hermetter A, Taylor GA, Russell DG. Macrophage activation downregulates the degradative capacity of the phagosome. Traffic 2007;8:241–50.

[9] Russell DG, VanderVen BC, Glennie S, Mwandumba H, Heyderman RS. The macrophage marches on its phagosome: dynamic assays of phagosome function. Nature Reviews Immunology 2009;9:594–600.

[10] Mercanti V, Charette SJ, Bennett N, Ryckewaert J-J, Letourneur F, Cosson P. Selective membrane exclusion in phagocytic and macropinocytic cups. Journal of Cell Science 2006;119:4079–87.

[11] Anes E, Kuhnel M, Boss E, Moniz-Pereira J, Habermann A, Griffiths G. Selected lipids activate phagosome actin assembly and maturation resulting in killing of pathogenic mycobacteria. Nature Cell Biology 2003;5:793–802.

[12] Pollard TD, Cooper JA. Actin, a central player in cell shape and movement. Science 2009;326:1208–12.

[13] Welch MD. The world according to Arp: regulation of actin nucleation by the Arp2/3 complex. Trends in Cell Biology 1999;9:423–7.

[14] Jank T, Giesemann T, Aktories K. Rho-glucosylating Clostridium difficile toxins A and B: new insights into structure and function. Glycobiology 2007;17:15R–22R.

[15] Cornelis GR, Wolf-Watz H. The yersinia yop virulon. Molecular Microbiology 1997;23:861–7.

[16] Krall R, Sun J, Pederson KJ, Barbieri JT. In vivo rho GTPase-activating protein activity of Pseudomonas aeruginosa cytotoxin ExoS. Infection and Immunity 2002;70:360–7.

[17] Alonso A, Portillo FG-D. Hijacking of eukaryotic functions by intracellular bacterial pathogens. International Microbiology 2004;7:181–91.

[18] Darwin KH, Miller VL. Molecular basis of the interaction of *Salmonella* with the intestinal mucosa. Clinical Microbiology Reviews 1999;12:405–28.

[19] Santos RL, Tsolis RM, Bäumler AJ, Smith R III, Adams LG. Salmonella enterica Serovar typhimurium induces cell death in bovine monocyte-derived macrophages by early sipb-dependent and delayedsipb-independent mechanisms. Infection and Immunity 2001;69:2293–301.

[20] Chen LM, Kaniga K, Galán JE. Salmonella spp. are cytotoxic for cultured macrophages. Molecular Microbiology 1996;21:1101–15.

[21] Bettencourt P, Marion S, Pires D, Santos LF, Lastrucci C, Carmo N, et al. Actin-binding protein regulation by microRNAs as a novel microbial strategy to modulate phagocytosis by host cells: the case of N-Wasp and miR-142-3p. Frontiers in Cellular and Infection Microbiology 2013;3:19.

[22] Welin A, Eklund D, Stendahl O, Lerm M. Human macrophages infected with a high burden of ESAT-6-expressing M. tuberculosis undergo caspase-1-and cathepsin B-independent necrosis. PloS One 2011;6:e20302.

[23] Naqvi AR, Fordham JB, Nares S. miR-24, miR-30b, and miR-142-3p regulate phagocytosis in myeloid inflammatory cells. The Journal of Immunology 2015;194:1916–27.

[24] Meconi S, Jacomo V, Boquet P, Raoult D, Mege JL, Capo C. Coxiella burnetii induces reorganization of the actin cytoskeleton in human monocytes. Infection and Immunity 1998;66:5527–33.

[25] Miller H, Zhang J, Kuolee R, Patel GB, Chen W. Intestinal M cells: the fallible sentinels? World Journal of Gastroenterology 2007;13:1477–86.

[26] Olson PD, Hunstad DA. Subversion of host innate immunity by uropathogenic *Escherichia coli*. Pathogens 2016;5(1):2.

[27] Justice SS, Hung C, Theriot JA, Fletcher DA, Anderson GG, Footer MJ, et al. Differentiation and developmental pathways of uropathogenic *Escherichia coli* in urinary tract pathogenesis. Proceedings of the National Academy of Sciences of the United States of America 2004;101:1333–8.

[28] Reis RSD, Horn F. Enteropathogenic *Escherichia coli, Samonella, Shigella* and *Yersinia*: cellular aspects of host-bacteria interactions in enteric diseases. Gut Pathogens 2010;2:8.

[29] Wong K-W, Isberg RR. Arf6 and phosphoinositol-4-phosphate-5-kinase activities permit bypass of the Rac1 requirement for beta1 integrin-mediated bacterial uptake. The Journal of Experimental Medicine 2003;198:603–14.

[30] Kwok T, Backert S, Schwarz H, Berger J, Meyer TF. Specific entry of *Helicobacter pylori* into cultured gastric epithelial cells via a zipper-like mechanism. Infection and Immunity 2002;70:2108–20.

[31] Sinn HW, Elzey BD, Jensen RJ, Zhao X, Zhao W, Ratliff TL. The fibronectin attachment protein of bacillus Calmette-Guerin (BCG) mediates antitumor activity. Cancer Immunology, Immunotherapy 2008;57:573–9.

[32] Cossart P, Pizarro-Cerda J, Lecuit M. Invasion of mammalian cells by Listeria monocytogenes: functional mimicry to subvert cellular functions. Trends in Cell Biology 2003;13:23–31.

[33] Rengarajan M, Hayer A, Theriot JA. Endothelial cells use a formin-dependent phagocytosis-like process to internalize the bacterium listeria monocytogenes. PLoS Pathogens 2016;12:e1005603.

[34] Dramsi S, Cossart P. Intracellular pathogens and the actin cytoskeleton. Annual Review of Cell and Developmental 1998;14:137–66.

[35] Griffiss JM, Lammel CJ, Wang J, Dekker NP, Brooks GF. *Neisseria gonorrhoeae* coordinately uses pili and opa to activate hec-1-b cell microvilli, which causes engulfment of the gonococci. Infection and Immunity 1999;67:3469–80.

[36] Lambotin M, Hoffmann I, Laran-Chich M-P, Nassif X, Couraud PO, Bourdoulous S. Invasion of endothelial cells by Neisseria meningitidis requires cortactin recruitment by a phosphoinositide-3-kinase/Rac1 signalling pathway triggered by the lipo-oligosaccharide. Journal of Cell Science 2005;118:3805–16.

[37] Kerr MC, Teasdale RD. Defining macropinocytosis. Traffic 2009;10:364–71.

[38] BoseDasgupta S, Pieters J. Inflammatory stimuli reprogram macrophage phagocytosis to macropinocytosis for the rapid elimination of pathogens. PLoS Pathogens 2014;10:e1003879.

[39] Guiney DG, Lesnick M. Targeting of the actin cytoskeleton during infection by Salmonella strains. Clinical Immunology 2005;114:248–55.

[40] Schroeder GN, Hilbi H. Molecular pathogenesis of *Shigella* spp.: controlling host cell signaling, invasion, and death by type III secretion. Clinical Microbiology Reviews 2008;21:134–56.

[41] Weiner A, Mellouk N, Lopez-Montero N, Chang Y-Y, Souque C, Schmitt C, et al. Macropinosomes are key players in early *Shigella* invasion and vacuolar escape in epithelial cells. PLoS Pathogens 2016;12:e1005602.

[42] Garcia-del Portillo F, Finlay BB. Salmonella invasion of nonphagocytic cells induces formation of macropinosomes in the host cell. Infection and Immunity 1994;62:4641–5.

[43] Donnenberg MS, Calderwood SB, Donohue-Rolfe A, Keusch GT, Kaper JB. Construction and analysis of tnphoa mutants of enteropathogenic escherichia coli unable to invade HEp-2 cells. Infection and Immunity 1990;58:1565–71.

[44] Hardwidge PR, Deng W, Vallance BA, Rodriguez-Escudero I, Cid VJ, Molina M, et al. Modulation of host cytoskeleton function by the enteropathogenic *Escherichia coli* and *Citrobacter rodentium* effector protein EspG. Infection and Immunity 2005;73:2586–94.

[45] Boudeau J, Glasser A-L, Masseret E, Joly B, Darfeuille-Michaud A. Invasive ability of an *Escherichia coli* strain isolated from the ileal mucosa of a patient with Crohn's disease. Infection and Immunity 1999;67:4499–509.

[46] García-Pérez BE, Mondragón-Flores R. Internalization of mycobacterium tuberculosis by macropinocytosis in non-phagocytic cells. Microbial Pathogens 2003;35:49–55.

[47] Lerner TR, de Souza Carvalho-Wodarz C, Repnik U, Russell MRG, Borel S, Diedrich CR, et al. Lymphatic endothelial cells are a replicative niche for Mycobacterium tuberculosis. The Journal of Clinical Investigation 2016;126:1093–108.

[48] Rittig MG, Jagoda JC, Wilske B, Murgia R, Cinco M, Repp R, et al. Coiling phagocytosis discriminates between different spirochetes and is enhanced by phorbol myristate acetate and granulocyte-macrophage colony-stimulating factor. Infection and Immunity 1998;66:627–35.

[49] Griffiths G. On phagosome individuality and membrane signalling networks. Trends in Cell Biology 2004;14:343–51.

[50] Kjeken R, Egeberg M, Habermann A, Kuehnel M, Peyron P, Floetenmeyer M, et al. Fusion between phagosomes, early and late endosomes: a role for actin in fusion between late, but not early endocytic organelles. Molecular Biology of the Cell 2004;15:345–58.

[51] Kuhnel M, Mayorga LS, Dandekar T, Thakar J, Schwarz R, Anes E, et al. Modelling phagosomal lipid networks that regulate actin assembly. BMC Systems Biology 2008;2:107.

[52] Anes E, Peyron P, Staali L, Jordao L, Gutierrez MG, Kress H, et al. Dynamic life and death interactions between Mycobacterium smegmatis and J774 macrophages. Cellular Microbiology 2006;8:939–60.

[53] Méresse S, Unsworth KE, Habermann A, Griffiths G, Fang F, Martínez Lorenzo MJ, et al. Remodelling of the actin cytoskeleton is essential for replication of intravacuolar Salmonella. Cellular Microbiology 2001;3:567–77.

[54] Robbins JR, Barth AI, Marquis H, de Hostos EL, Nelson WJ, Theriot JA. Listeria monocytogenes exploits normal host cell processes to spread from cell to cell. Journal of Cell Biology 1999;146:1333–50.

[55] Schaible UE, Haas A. Intracellular Niches of Microbes: A Pathogens Guide through the Host Cell. Wiley-VCH Verlag GmbH&Co. KGaA, Weinheim; 2009. p. 738 DOI: 10.1002/9783527629176.

[56] Jeng RL, Goley ED, D'Alessio JA, Chaga OY, Svitkina TM, Borisy GG, et al. A Rickettsia WASP-like protein activates the Arp2/3 complex and mediates actin-based motility. Cellular Microbiology 2004;6:761–9.

[57] Stamm LM, Pak MA, Morisaki JH, Snapper SB, Rottner K, Lommel S, et al. Role of the WASP family proteins for Mycobacterium marinum actin tail formation. Proceedings of the National Academy of Sciences of the United States of America 2005;102:14837–42.

[58] Breitbach K, Rottner K, Klocke S, Rohde M, Jenzora A, Wehland J, et al. Actin-based motility of Burkholderia pseudomallei involves the Arp 2/3 complex, but not N-WASP and Ena/VASP proteins. Cellular Microbiology 2003;5:385–93.

[59] Asrat S, de Jesús DA, Hempstead AD, Ramabhadran V, Isberg RR. Bacterial pathogen manipulation of host membrane trafficking. Annual Review of Cell and Developmental Biology 2014;30:79–109.

[60] Stutz A, Golenbock DT, Latz E. Inflammasomes: too big to miss. The Journal of Clinical Investigation 2009;119:3502–11.

[61] Simeone R, Bobard A, Lippmann J, Bitter W, Majlessi L, Brosch R, et al. Phagosomal rupture by mycobacterium tuberculosis results in toxicity and host cell death. PLoS Pathogens 2012;8:e1002507.

[62] van der Wel N, Hava D, Houben D, Fluitsma D, van Zon M, Pierson J, et al. *M. tuberculosis* and *M. leprae* translocate from the phagolysosome to the cytosol in myeloid cells. Cell 2007;129:1287–98.

[63] Aachoui Y, Leaf IA, Hagar JA, Fontana MF, Campos CG, Zak DE, et al. Caspase-11 protects against bacteria that escape the vacuole. Science 2013;339:975–8.

[64] Yoshikawa Y, Ogawa M, Hain T, Yoshida M, Fukumatsu M, Kim M, et al. Listeria monocytogenes ActA-mediated escape from autophagic recognition. Nature Cell Biology 2009;11:1233–40.

[65] Ogawa M, Yoshikawa Y, Kobayashi T, Mimuro H, Fukumatsu M, Kiga K, et al. A Tecpr1-dependent selective autophagy pathway targets bacterial pathogens. Cell Host & Microbe 2011;9:376–89.

Chapter 14

Heterotrimeric G Proteins and the Regulation of Microtubule Assembly

Sukla Roychowdhury and Jorge A. Sierra-Fonseca

Additional information is available at the end of the chapter

http://dx.doi.org/10.5772/66929

Abstract

Microtubules (MTs), a major component of cell cytoskeleton, exhibit diverse cellular functions including cell motility, intracellular transport, cell division, and differentiation. These functions of MTs are critically dependent on their ability to polymerize and depolymerize. Although a significant progress has been made in identifying cellular factors that regulate microtubule assembly and dynamics, the role of signal transducing molecules in this process is not well understood. It has been demonstrated that heterotrimeric G proteins, which are components of G protein-coupled receptor (GPCR) signaling pathway, interact with microtubules and play important roles in regulating assembly/dynamics of this cytoskeletal filament. While α subunit of G proteins (Gα) inhibits microtubule assembly and accelerates microtubule dynamics, G$\beta\gamma$ promotes tubulin polymerization. In this chapter, we review the current status of G-protein modulation of microtubules and cellular and physiological aspects of this regulation. Molecular, biochemical, and cellular methodologies that have been used to advance this field of research are discussed. Emphasis has been given on G-protein-microtubule interaction in neuronal differentiation as significant progress has been made in this field. The outcome from this research reflects the importance of GPCRs in transducing extracellular signals to regulate a variety of microtubule-associated cellular events.

Keywords: cytoskeleton, G-proteins, microtubules, neuronal differentiation, G$\beta\gamma$, tubulin, G protein-coupled receptor, GTP-binding proteins

1. Introduction

The major component of microtubules (MTs) is the heterodimeric protein tubulin, consisting of α and β subunits, which are assembled into linear protofilaments. The protofilaments associate laterally to form the microtubule, a 25-nm-wide hollow cylindrical polymeric structure [1]. Due to the asymmetry of the αβ-tubulin heterodimer, MTs are polar structures with two distinct ends. These ends possess different polymerization rates: a slow-growing minus end with an exposed α-tubulin subunit, and a fast-growing plus end, at which the β-tubulin subunit is exposed [2, 3]. MT assembly occurs in two phases: nucleation, which is facilitated by a third tubulin isoform, γ-tubulin; and elongation, during which αβ-tubulin heterodimers are added to the plus end [1, 4]. Tubulin is a unique guanine nucleotide-binding protein containing one exchangeable binding site and one nonexchangeable binding site. GTP at both sites is needed for optimal assembly, and GTP at the exchangeable site is hydrolyzed after assembly [5, 6]. This hydrolysis creates an MT consisting largely of GDP-tubulin; however, a small region of GTP-bound tubulin, called a "GTP cap," remains at the end. This cap allows MTs to polymerize. The loss of the cap results in a transition from growth to shortening (called a "catastrophe"), whereas the reacquisition of the GTP cap results in a transition from shortening to growing (called a "rescue"). This behavior, known as dynamic instability, allows MTs to be remodeled rapidly in cells. An important consequence of dynamic instability is that it allows microtubules to search for specific target sites within the cell more effectively [7–9]. The MT assembly process is depicted in **Figure 1**.

MT assembly and stability can be affected by a wide variety of proteins. In this regard, microtubule-associated proteins (MAPs) play a very important role. Members of this group of proteins, such as MAP2 and tau, are known to promote MT assembly and stabilize MTs *in vivo* and *in vitro* [10–13]. The phosphorylation of MAPs is critical for their function, since

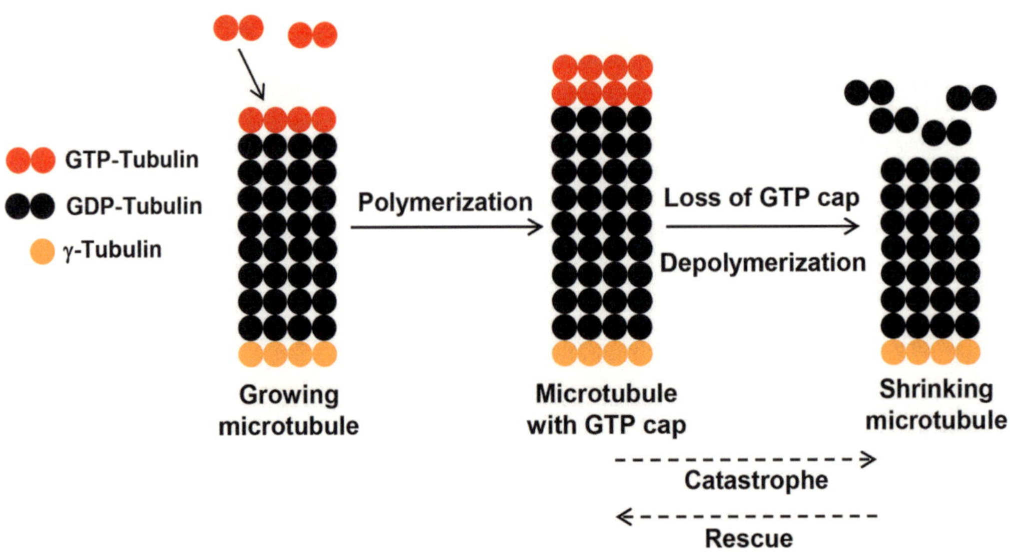

Figure 1. Polymerization/depolymerization of MTs. MTs is polymerized from tubulin heterodimer consisting of α and β subunits. A third tubulin isoform, γ-tubulin, serves as a template for nucleation, which allows proper MT assembly. As shown in the figure, MT assembly requires tubulin to be in GTP bound form. However, it is hydrolyzed to GDP when it is incorporated in MTs, except at the plus (+) end where tubulin remains at GTP bound form (a GTP cap). This cap allows MTs to polymerize. The loss of GTP cap (by hydrolysis) results in the transition to MT depolymerization (a "catastrophe"). GTP cap can be regained by binding to tubulin-GTP and MT polymerization is reestablished (a "rescue").

phosphorylated MAPs separate from MTs, causing MTs to become more susceptible to disassembly and destabilization [14, 15]. Destabilization of MTs can be promoted by a large number of proteins collectively termed "catastrophe promoters," as they favor the transition of MTs from elongation to shortening. Examples of these proteins include stathmin/Op18, a small heat-stable protein that is abundant in many types of cancer cells, katanin, and some kinesin-related motor proteins [16, 17]. Also, many drugs known to alter tubulin polymerization are considered valuable tools in studying the mechanisms of MT assembly. Some of these drugs, such as nocodazole, depolymerize MTs, whereas others, such as taxol, promote MT assembly [18–21]. Even though MTs are composed of α/β-tubulin heterodimers in all eukaryotic cells, MTs exhibit great functional diversity. One possible explanation is that both α- and β-tubulin undergo a series of posttranslational modifications that allow MTs to engage in a variety of cellular activities [22]. These modifications include tyrosination/detyrosination, acetylation, glutamylation, and phosphorylation [23]. Although much progress has been made in identifying and characterizing the cellular factors that regulate MT assembly and dynamics, the precise spatial and temporal control of the process is not clearly understood.

Over the past decades, an effort has been made to understand the regulation of MT assembly and dynamics by signal transducing G proteins, as reviewed in Refs. [24, 25]. G proteins are heterotrimer, consisting of guanine nucleotide-binding α plus $\beta\gamma$ subunits. The G-protein-signaling cascade begins with the agonist-induced activation of a G protein-coupled receptor (GPCR), which allows GTP to bind to the α subunit of the heterotrimer, and subsequently, the GTP-bound-activated Gα changes its association with G$\beta\gamma$ in a manner that permits both subunits to participate in the regulation of intracellular effector molecules [26]. The traditional pathway for GPCR signaling is shown in **Figure 2**. The GPCR family of proteins is highly diverse; more than 1000 gene-encoding GPCRs are found in the human genome [27, 28]. GPCRs participate in the regulation of a wide variety of physiological functions, including cell growth and differentiation, neurotransmission, immune system function, and hormonal signaling. Participation in such a multitude of processes makes GPCRs a very attractive drug target, and approximately 30% of commercially available drugs are designed to target GPCRs [29]. GPCRs consist of seven transmembrane domains, connected by three extracellular loops and three intracellular loops. The extracellular region is responsible for agonist binding (neurotransmitters, hormones, and odorants, among others), and the intracellular region is responsible for interacting with heterotrimeric G proteins [30]. In humans, there are 21 isoforms of Gα subunits, 6 Gβ isoforms, and 12 isoforms of Gγ [31]. G-protein heterotrimers are typically classified into four classes depending on the Gα subunit: Gαs (for stimulation of adenylyl cyclase), Gαi (for inhibition of adenylyl cyclase), Gαq (which regulates phospholipase), and Gα12/13, which is involved in the regulation of monomeric G proteins and other molecules, such as PKC [31, 32]. Typical effectors of Gα signaling include adenylyl cyclase, phospholipase C, phospholipase A_2, ion channels, and several kinases and transcription factors. Termination of the signal occurs when GTP bound to the α subunit is hydrolyzed by its intrinsic GTPase activity that causes its functional dissociation from the effector and reassociation with $\beta\gamma$ [26, 33–35]. Thus, G proteins act as molecular switches that can be turned "on" and "off" through the GTPase cycle. While the signal-transducing ability of heterotrimeric G proteins was once believed to depend fully on the α subunit, it has now become clear

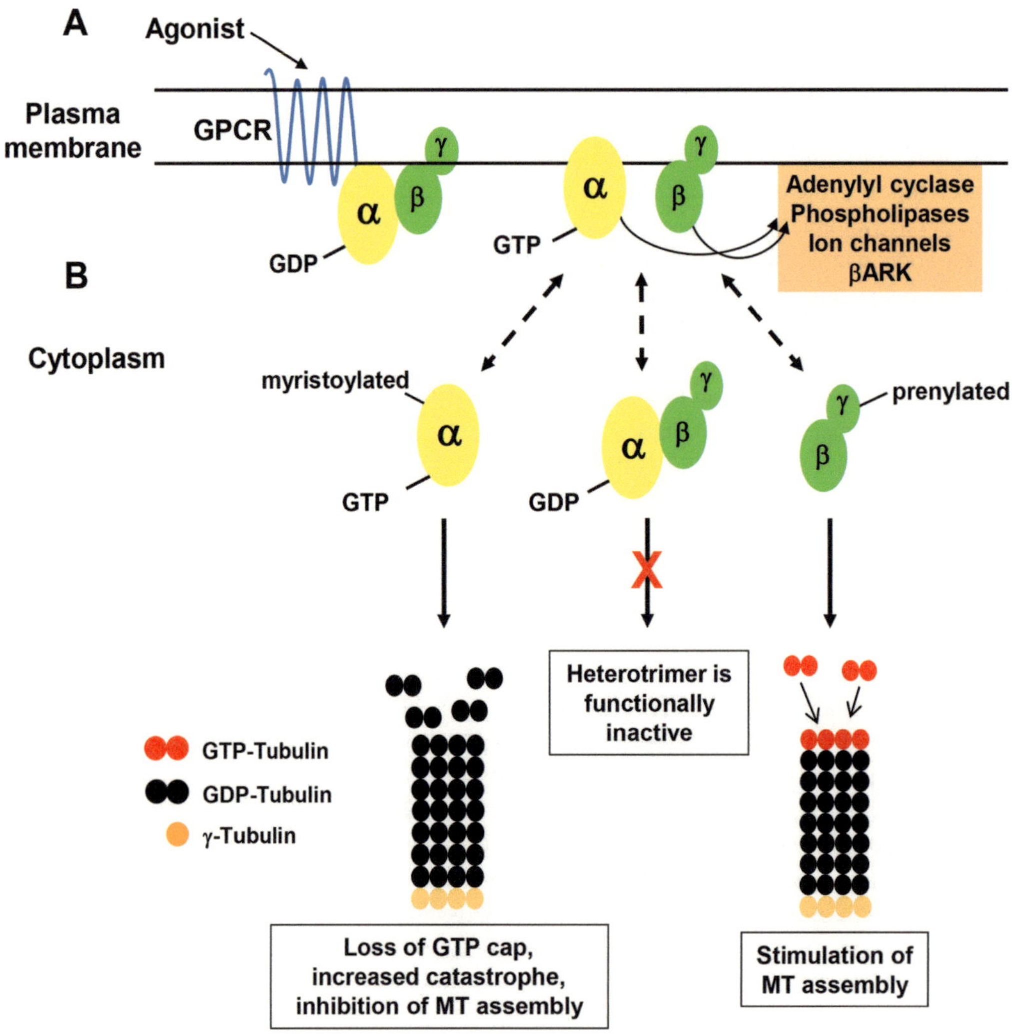

Figure 2. G-protein-mediated signaling and the regulation of MT assembly. (**A**) The traditional pathway for G-protein signaling cascade begins with the agonist-induced activation of a GPCR (G protein-coupled receptor), which allows GTP to bind to the α subunit of the heterotrimer and subsequently the GTP-bound-activated Gα changes its association with G$\beta\gamma$ in a manner that permits both subunits to participate in the regulation of intracellular effector molecules. (**B**) α and $\beta\gamma$ subunits of heterotrimeric G proteins interact with tubulin/MTs (in cytoplasm) and influence MT assembly and dynamics. Results generated from *in vitro* studies using purified proteins and cultured cells suggest that the Gα subunit inhibits MT assembly and promotes MT disassembly by interacting with tubulin-GTP and initiating GTP hydrolysis of tubulin, therefore causing MT depolymerization. The G$\beta\gamma$ subunit, on the other hand, promotes MT assembly. The G$\alpha\beta\gamma$ heterotrimer is functionally inactive (similar to that observed in traditional GPCR pathway) and does not interact functionally with Tubulin/MTs. Upon activation, Gα dissociates from G$\beta\gamma$ subunits, and both subunits then interact with tubulin/MTs and modulate assembly/dynamics. It is suggested that G-protein-MT interaction is an important step for G-protein-mediated cell activation.

that the $\beta\gamma$ subunit is capable of interacting with numerous effector molecules to influence a variety of signaling pathways [36, 37]. Among the effector molecules interacting with G$\beta\gamma$ are phospholipases, K^+ and Ca^{2+} channels, GPCR kinases, members of the MAP kinase signaling pathway, monomeric G proteins, regulators of G protein signaling (RGS), and phosphoinositide-3 kinase (PI3K) [37–42].

Although G proteins are likely to be membrane-bound when coupled to receptors, results from several laboratories in past decades demonstrate their association with several subcellular compartments including MTs. G protein-MT interactions have been shown to modulate the assembly, dynamics and functions of MTs (**Figure 2**). This chapter focuses on our current understanding of G protein regulation of MT assembly and cellular and physiological aspects of this regulation.

2. Heterotrimeric G proteins and the tubulin/MT system

Gα and MT assembly/dynamics. Direct interactions between tubulin and α subunits of Gs, Gi1 have been demonstrated [43] and these interactions were shown to activate GTPase activity of tubulin, inhibit microtubule assembly, and accelerate microtubule dynamics [44–47]. To elucidate the role of Gα in microtubule assembly *in vitro*, purified Gα subunits as well as tubulin were used in the reconstitution assay. G protein α subunits Gi1α, Gsα, and Goα were shown to activate the GTPase activity of tubulin and inhibit microtubule assembly. The assembly of tubulin-GTP (or tubulin-GppNHp) into microtubules was inhibited by Gi1α (80–90%) in the absence of exogenous GTP. The addition of exogenous GTP—but not the addition of hydrolysis-resistant GppNHp—overcame the inhibition of microtubule assembly by Gi1α [45], thus, it appears that GTP hydrolysis resulting from the association of tubulin and Gi1α plays a critical role in modulating microtubule assembly. Gα appears to bind to tubulin and activate the intrinsic GTPase of tubulin in a manner similar to what occurs during MT formation. However, unlike the formation of microtubules from tubulin dimers, Gα dissociates from the tubulin-Gα complex subsequent to GTP hydrolysis [45]. This finding is consistent with the possibility that Gα would accelerate MT dynamic instability. Analysis of the dynamics of individual microtubules by video microscopy has demonstrated that Gi1α increases the catastrophe frequency [45]. To determine the role of Gsα in MT dynamics *in vivo*, PC12 cells were transfected with Gsα-GFP [47]. Transfected cells were treated with cholera toxin to activate Gsα-GFP or forskolin to stimulate adenylate cyclase and to increase cAMP. Cholera-toxin activation of Gsα-GFP resulted in a displacement of Gsα-GFP from the plasma membrane. It was found that activated Gsα released from the plasma membrane was directly bound to cellular microtubules and then colocalized with microtubules. As a result, activated Gsα made MTs more dynamic, decreasing the pool of insoluble MTs, without changing the total cellular tubulin content [47].

Gβγ and MT assembly. The Gβγ subunit has the opposite effect on tubulin polymerization, as it was found that Gβγ promotes MT assembly *in vitro* [48]. Assembly was monitored by negative staining electron microscopy and measuring protein in polymers collected by centrifugation. The effect of different combinations of βγ on MT assembly was tested. Tubulin that was purified free of microtubule-associated proteins was incubated at 37°C in the presence of β1γ2 or β1γ1 (transducin βγ) for 45 min to 1 h. Microtubule assembly was stimulated markedly when β1γ2 was present at ~1:20 molar ratio with tubulin; in contrast, β1γ1 had no effect on microtubule assembly [48]. An electron microscopic analysis indicated the formation of very few microtubules either by tubulin alone or in the presence of β1γ1. In the presence of β1γ2, however, robust microtubule polymerization occurred. Protein estimation in the pellets also indicated a 71% increase in the presence of β1γ2. An SDS-PAGE of the samples further confirmed the increase in tubulin concentration in the pellet formed in the presence of β1γ2. No detectable change in pellet protein concentration (compared to controls) was observed in the presence of β1γ1. β immunoreactivity was detected exclusively in the microtubule fraction after assembly in the presence of β1γ2, suggesting a preferential association with microtubules rather than with soluble tubulin. A number of proteins, including the γ subunit of Gβγ, undergo a process of posttranslational modification termed "prenylation" and this modification is important for the biological functions of these proteins. For example, prenylation of γ subunits is required for the high-affinity interactions of Gβγ with α subunits or effector molecules [49–51]. Interestingly, it was found that a mutant β1γ2, β1γ2 (C68S), which does not undergo prenylation of γ subunit, did not stimulate the formation of MTs, suggesting that the functional interaction of Gβγ with MTs require the same specificity as other effector molecules of Gβγ [49–51].

To investigate the potential link between Gβγ and MT assembly *in vivo*, cultured PC12 and NIH3T3 cells were used. The role of Gβγ in MT assembly was demonstrated using nocodazole, a microtubule-depolymerizing drug [52]. Colchicine and the synthetic compound nocodazole are both antimitotic drugs and known to exert their effects by a similar mechanism, that is, by binding to tubulin dimers and inhibiting the subsequent addition of tubulin molecules to microtubules. However, the potential usefulness of nocodazole is due to its readily reversible and rapid activity [53, 54]. Nocodazole-induced depolymerization of microtubules drastically inhibited (~68%) the interaction between Gβγ and tubulin [52]. This result was further confirmed by the isolation of polymerized tubulin (MT) and soluble tubulin (ST) fractions from PC12 cells. Although Gβγ was found in both fractions, a tubulin-Gβγ interaction was found preferentially in MT fractions rather than ST fractions as demonstrated by coimmunoprecipitation analyses. This is consistent with *in vitro* studies, in which Gβγ was preferentially associated with MTs assembled from β1γ2 [48]. Removal of nocodazole from the cultured media allowed MTs to repolymerize to their fullest extent and tubulin-Gβγ interaction was restored completely in the MT fraction. These results clearly demonstrate that the association of Gβγ with MTs is important for MT assembly and/or stability. The interactions between Gβγ and tubulin/MTs were also assessed by immunofluorescence microscopy. Microtubules in PC12 cells are well defined and extend to the cell periphery. Gβγ was more concentrated in the perinuclear region where they were colocalized with microtubules. The network of microtubule structure collapsed and Gβγ labeling was dispersed, when cells were treated with nocodazole for 4 h. Microtubules reappeared after the removal of nocodazole, when cells were incubated in fresh media for 4 h. Gβγ labeling was also appeared in perinuclear region where they were colocalized with MTs [52]. In addition to interphase cells, Gβγ-tubulin association was also observed in mitotic spindle in PC12 cells.

Gβγ-γ-tubulin interactions. γ-Tubulin, a member of tubulin superfamily, is a centrosomal protein and its role in MT nucleation is well documented [55–58]. In addition to its binding of αβ-tubulin, Gβγ was also found to interact with γ-tubulin [33]. However, unlike αβ-tubulin,

http://dx.doi.org/10.5772/66929

the interaction between γ-tubulin and Gβγ is not inhibited by nocodazole, suggesting that the interaction between Gβγ and γ-tubulin is not dependent upon microtubules [33]. Both Gβγ and γ-tubulin were colocalized in the centrosomes of PC12 cells. Interestingly, γ-tubulin and Gβγ immuno-reactivity appears to be increased significantly in duplicated centrosomes at the onset of mitosis, and Gβγ was consistently found to colocalize with tubulin at mitotic spindle, particularly at the spindle pole areas [33]. Earlier studies in *Caenorhabditis elegans* and *Drosophila* have also demonstrated that Gβγ is involved in cell division by positioning the mitotic spindle and attaching microtubules to the cell cortex [21, 22]. In *C. elegans* embryos, Gβγ was shown to be important in the regulation of migration of the centrosome around the nucleus [21]. These studies collectively suggest an important role of Gβγ in centrosome functions, perhaps through its interactions with γ-tubulin. Although centrosome-associated γ-tubulin is known to be involved in MT nucleation, most γ-tubulin in cells are found in the cytoplasm as γ-tubulin ring complex (γTuRC) and it has been shown that γTuRC translocate to centrosome to mediate MT nucleation [59, 60]. Since Gβγ immunoreactivity also increases significantly in duplicated chromosomes at the onset of mitosis [52], it is possible that Gβγ may allow translocation of γ-tubulin to centrosomes. The γ-tubulin-Gβγ complex might then induce robust microtubule nucleation at the centrosome and formation of the mitotic spindle.

Gαβγ heterotrimer and MT assembly. Since G protein activation and subsequent dissociation of α and βγ subunits are necessary for G proteins to participate in signaling processes [26], it was determined if similar activation is required for modulation of microtubule assembly by G proteins. For that, Gαβγ heterotrimer was reconstituted from α and βγ subunits and its effect was tested on GTPase activation of tubulin and MT assembly. Myristoylated Gi1α and prenylated Gβ1γ2 were used to reconstitute the heterotrimer, since lipid modified G-protein subunits have been found to be more effective in interacting with tubulin and subsequent modulation of its functions [45, 48]. In addition, lipid-modified, G-protein subunits have been shown to reconstitute heterotrimers more effectively [61, 62]. Reconstituted heterotrimers have been shown to block Gi1α activation of tubulin GTPase and inhibit the ability of Gβ1γ2 to promote *in vitro* microtubule assembly [46], suggesting that G-protein activation is required for functional coupling between Gα/Gβγ and tubulin/MTs (**Figure 2**). The results also suggest that G protein-coupled receptors (GPCRs) may be involved in the regulation of MT assembly and dynamics *in vivo* by mobilizing G-protein subunits to bind to MTs. In doing so, GPCRs may control a variety of cellular activities. It appears that G-protein-MT interaction is an important step for G-protein-mediated cell activation.

3. G protein-microtubule interactions and cell division

Microtubules play a key role in cell division, participating in the exact organization and function of the spindle apparatus, a vehicle necessary for chromosomal segregation. Microtubules in the spindle are organized in such a way that the minus ends are near the spindle poles, while the plus ends extend toward the cell cortex or chromosomes [63]. Both α and βγ subunits of G proteins Gi and Go are consistently found to be associated with mitotic spindle. Genetic studies in *C. elegans*, *Drosophila*, and mammalian cells have revealed that G-protein subunits are involved in regulating mitotic spindle for centrosome/chromosome movements in cell division [64–69]. G-protein α subunits of Gi are involved in cell division by regulating microtubule-pulling forces during chromosomal segregation through a receptor-independent pathway. Unlike the classical G-protein cycle, in which GPCR promotes the GDP/GTP exchange in Gα converting Gα in active GTP-bound form, in nonreceptor pathways, the GDP-bound form of Giα is stabilized through its interaction with guanine-nucleotide dissociation inhibitor (GDI) to regulate microtubule-pulling forces for chromosome movements [66, 70, 71]. Members of the GDI family of proteins, characterized by the presence of 20–25 amino-acid repeats termed "GPR" or "GoLoco" motifs, are known to stabilize the GDP-bound form of Giα by inhibiting the release of nucleotide. Thus, it appears that Giα participates in spindle function through a mechanism that is distinct from the receptor-mediated pathway. In addition to Giα, G$\beta\gamma$ has been shown to play a role in spindle position and orientation during cell division [64, 68]. The association of Goα and β (or G$\beta\gamma$) with spindle microtubules suggests that G-protein subunits may play an important role in the regulation of the assembly and disassembly of mitotic spindles through their ability to modulate microtubule assembly. Recently, it has been shown that reconstituted kinetochores *in vitro* bind preferentially to GTP rather than to GDP microtubules, suggesting that a protein exists in kinetochores that can distinguish between GTP conformation of the microtubules and allow the kinetochores to remain at the microtubule ends to ensure correct chromosome segregation [72]. Since Gα appears to interact preferentially to GTP (rather than the GDP-form of tubulin) and has been detected in mitotic spindles, it may be a likely candidate for segregating chromosomes through its interaction with microtubules.

4. G protein-microtubule interactions and neuronal differentiation

The process by which MT structure is remodeled in neurons is a central question in cell biology and recent research indicates an important role of G protein subunits in this process. During neuronal differentiation, two distinct domains emerge from the cell body: a long, thin axon that transmits signals, and multiple shorter dendrites, which are specialized primarily for receiving signals. The axon terminal contains synapses, specialized structures where neurotransmitters are released to communicate with target neurons. Cytoskeletal structures embodied within neurite extensions and growth cone formations are essential for establishing appropriate synaptic connections and signal transmission. MTs form dense parallel arrays in axons and dendrites that are required for the growth and maintenance of such neurites. In the axon, MTs are bundled by tau, a microtubule-associated protein (MAP), with their plus end oriented toward the nerve terminal. MAP2, a group of high molecular weight MAPs, participates in MT bundling in the dendrites (**Figure 3**). Unlike MTs, actin filaments in neurons are enriched in growth cones and organized into long bundles that form filamentous protrusions, or filopodia, veil-like sheets of branched actin that form lamellipodia [1, 7, 73]. The interaction between these two cytoskeletal filaments is important for the advancement of growth cones and axon guidance [74, 75].

It is clear that cytoskeletal components can detect biochemical signals and respond in order to change the neuronal cell morphology. However, the precise signaling pathways that lead

http://dx.doi.org/10.5772/66929

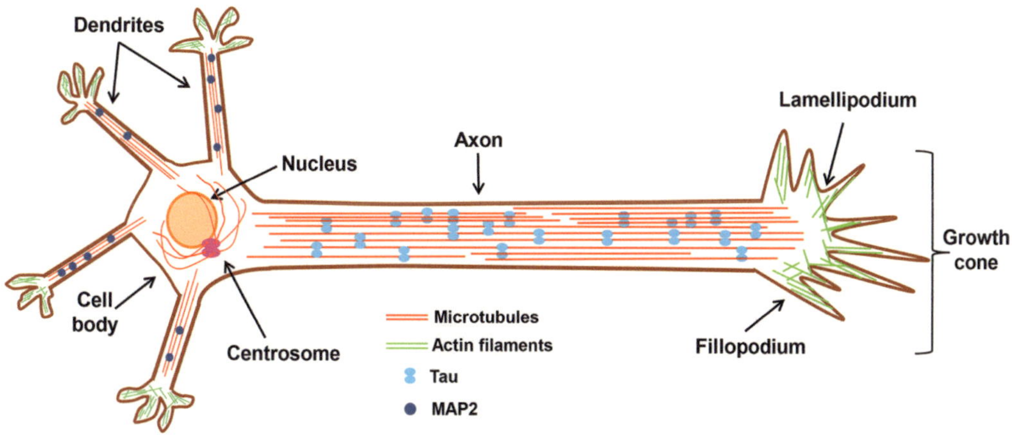

Figure 3. Neuronal cytoskeleton. The polarized and asymmetrical shape of neurons is achieved by means of a highly specialized cytoskeletal organization. In addition to cell body, MTs are found in the axon, dendrites, and the central domain of the growth cone. Tau, a microtubule-associated protein (MAP), participates in MT bundling in the axon, while MAP2 carries this function in the dendrites. Actin filaments are present in the growth cone and dendrites, where they form specialized structures such as lamellipodia and filopodia.

unique organization of MTs in neurons are not clearly understood [76]. PC12 cells have been used extensively for these studies as they respond to nerve growth factor (NGF) with growth arrest and exhibit a typical phenotype of neuronal cells that send out neurites [77]. NGF is a neurotrophic factor critical for the survival and maintenance of sensory and sympathetic neurons. The receptor commonly associated with this process is tyrosine kinase (TrkA) through which NGF exerts its effect [78]. PI3K appears to be the key molecule in this pathway and regulates localized assembly of MTs/actin filaments by downstream Akt/GSK3β pathways [79, 80]. The Rho and Ras families of small GTPases have also emerged as critical players in regulating the actin and MT cytoskeleton by modulating downstream effectors, including serine/threonine kinase, p21-activated kinase, ROCK, and mDia [81, 82]. GPCRs, as well as α and $\beta\gamma$ subunits of heterotrimeric G proteins, have also been shown to regulate neurite outgrowth [83–90]. These studies collectively suggested the role of α and $\beta\gamma$ subunits of G proteins in regulating neurite outgrowth. More recently, it has been demonstrated that both α and $\beta\gamma$ subunits of G proteins regulate neurite outgrowth and differentiation by interacting with MTs and by modulating MT assembly/dynamics [24].

Gsα and neuronal differentiation. NGF-induced neuronal differentiation of PC12 cells result in a translocation of Gsα, Gi1α, and Goα from cell bodies to cellular processes in which they appear to localize with microtubules [91]. Consistent with this, Gα in Neuro-2a cells, which spontaneously differentiate, showed a similar pattern of association of Gα with MTs [91]. The result has been further confirmed by transfecting PC12 cells with Gsα-GFP. Transfected cells were treated with cholera toxin to activate Gsα-GFP, or forskolin, to stimulate adenylyl cyclase and an increase in cAMP. Colocalization of Gsα along MTs was seen in cells treated with cholera toxin but not in those treated with forskolin, indicating that activation of Gsα induces Gsα translocation to the cytoplasm where it associates with MTs [47]. To understand the function of

Gsα/microtubule association in neuronal development and differentiation, real time trafficking of a Gsα-GFP fusion protein was used [92]. GFP-Gsα concentrated at the distal end of the neurites in NGF-differentiated living PC12 cells as well as in the cultured hippocampal neurons. Gsα appeared to translocate to the growing tip of neurites and to membrane ruffles of the newly formed extensions after NGF treatment, and it has been suggested that during neuronal differentiation, Gsα redistributes toward the areas of highly dynamic cytoskeletal activity. Neurite length as well as the number of neurites per cell was also increased in cells overexpressing Gsα-GFP in the presence of NGF. The effect was greatest in cells overexpressing constitutively active Gsα (GsQL). On the other hand, a dominant-negative Gαi-transducin chimera that interferes with Gsα binding to tubulin and activation of tubulin GTPase attenuated the neurite elongation and the neurite number both in PC12 cells and in primary hippocampal neurons. Thus, it appears that activated Gsα translocated from plasma membrane induced neurite outgrowth and development through interaction with tubulin/microtubules in the cytosol [92].

Gβγ and neuronal differentiation. The involvement of Gβγ in neuronal development and differentiation has been previously shown [68, 89]. Gβ1-deficient mice have been shown to have neural tube defects [94], and Gβ5-knockout mice have been shown to display abnormal behavior and develop multiple brain abnormalities [95]. It has also been shown that impaired Gβγ signaling promotes neurogenesis in the developing neocortex and increased neuronal differentiation of progenitor cells [68]. Although the mechanism by which Gβγ controls this process is not yet understood, the possibility that Gβγ may act on MTs has been suggested. Sachdev et al. [89] have also suggested that Gβγ-Tctex-1 complex plays a key role in regulating neurite outgrowth in primary hippocampal neurons, most likely by modulating MTs and actin filaments through activation of downstream signaling. These studies suggest a connection between Gβγ signaling and the modulation of MTs during neuronal differentiation and development.

More recently, using biochemical and immunofluorescence analysis, it has been demonstrated that Gβγ-MT interactions and modulation of MT assembly is critical for NGF-induced neuronal differentiation of PC12 cells [94]. To address this, PC12 cells were treated with NGF over the course of three days to allow for neuronal differentiation. Microtubules (MTs) and soluble tubulin (ST) fractions were extracted using a microtubule-stabilizing buffer. The interaction of Gβγ with MT and ST fractions was analyzed by coimmunoprecipitating tubulin-Gβγ complex using a Gβ-specific antibody (rabbit polyclonal anti-Gβ) or a mouse monoclonal anti-α tubulin antibody and determining tubulin and Gβγ immunoreactivity in the complex [94]. Gβγ-MT interaction was significantly increased (2–3 fold) in NGF-treated cells. We also found that MT assembly was stimulated significantly (from 45.3 ± 4.8 to 70.1 ± 3.6%) in NGF-differentiated PC12 cells. The association of Gβγ with MTs in NGF-differentiated cells was also assessed by immunofluorescence microscopy [93]. After NGF treatment, the majority of the cells displayed neurite formation. Gβγ was detected in the neurites and in cell bodies. The colocalization of Gβγ with MTs/tubulin was observed along the neuronal process and in the central portion of the growth cone, but not at the tip of the growth cones.

Overexpression of Gβγ in PC12 cells induced neurite outgrowth in the absence of NGF, further supporting the role of Gβγ in neuronal differentiation [93]. Since Gβ1γ2 promoted

MT assembly *in vitro*—and Gβ1γ1 had no effect [48], PC12 cells were transfected by either β1γ1 or β1γ2. YFP-tagged β1, γ2, or γ1 constructs were used for transfection. Cells were cotransfected with β1 and γ2, or β1 and γ1. Within 24 h of transfection, both β1γ1- and β1γ2-transfected PC12 cells were found to overexpress the proteins. At 48 h of transfection, YFP-β1γ2 transfected cells induced neurite formation (in the absence of NGF). Overexpressed protein (YFP-Gβ1γ2) was localized in the neurite processes, growth cones, and cell bodies. Moreover, overexpressed Gβγ exhibited a pattern of association with MTs similar to that observed in NGF-differentiated cells. The average neurite length of Gβ1γ2 (42.8 ± 2.1 μm) and Gβ1γ1 (33.5 ± 1.8 μm) is significantly higher than that of control cells (18.4 ± 0.6 μm), with Gβ1γ2 having the most potent effect on neurite outgrowth. Although the average neurite length in Gβγ-overexpressing cells (42.8 ± 2.1 μm) was slightly lower than that observed in NGF-differentiated PC12 cells (53.6 ± 1.8 μm), the result clearly indicates the effectiveness of overexpressed Gβγ in inducing neurite outgrowth in the absence of NGF.

Finally, the role of Gβγ in neuronal morphology, outgrowth and differentiation was further investigated using peptides and prenylation pathway inhibitors. For example, GRK2i, a Gβγ blocking peptide known to inhibit Gβγ-dependent effector functions, induced neurite damage as well as MTs and Gβγ aggregation. In addition, cellular aggregation was also frequently observed in the presence of GRK2i. The percentage of cell-bearing neurites was reduced significantly. On the other hand, synthetic peptide mSIRK, which is known to activate Gβγ signaling in cells by promoting the dissociation of Gβγ from α subunits, stimulated neurite formation. Since, γ-subunit of Gβγ is known to be posttranslationally modified by prenyl lipid, and prenylation deficient mutant of Gβγ (C68S) was shown to be functionally inactive, inhibitors of an enzyme of prenylation pathway (PMPMEase) was tested for their effects on MT assembly and neurite outgrowth. These inhibitors were found to alter MT organization and blocked neurite outgrowth. The results further demonstrate that βγ subunit of heterotrimeric G proteins play a critical role in neurite outgrowth and differentiation by interacting with MTs and regulating MT assembly and organization.

5. Conclusion

Heterotrimeric G proteins transduce signals from cell surface receptors (G protein-coupled receptors) to intracellular effector molecules that include adenylyl cyclase, phospholipases, and ion channels. New evidence suggests that the modulation of the MTs by G proteins is an emerging field of research and therefore an in-depth understanding of G-protein-MTs interaction is important for elucidation of the function, behavior, and morphology of mammalian cells. Key results of this unique interaction may have a broader impact on health and diseases including cancer, Alzheimer's, Parkinson's, depression, and addictive behavior. We foresee that the G-protein-MT dependent pathway could be exploited for developing novel drugs to combat such diseases in the future.

Acknowledgments

Research in author's laboratory described in this report was, in part, supported by G12MD007592 (NIHMHD) to the University of Texas at El Paso. An R01 subcontract to SR from the University of Illinois at Chicago (MH39595) supported some earlier research in author's laboratory. The biochemical, molecular, and confocal microscopy experiments were carried out at the Biomolecule Analysis Core Facility, Genome Analysis Core facility, and Cytometry/ screening/imaging Facility at the Border Biomedical Research Center (UTEP) supported by a grant (G12MD007592) from NIMHD (NIH). The authors like to thank Dr. Siddhartha Das for critically reading the manuscript and thoughtful suggestions.

Author details

Sukla Roychowdhury[1, 2*] and Jorge A. Sierra-Fonseca[1, 2]

*Address all correspondence to: sukla@utep.edu

1 Neuromodulation Disorders, Border Biomedical Research Center, University of Texas at El Paso, El Paso, Texas, USA

2 Department of Biological Sciences, University of Texas at El Paso, El Paso, Texas, USA

References

[1] Desai A, Mitchison TJ. Microtubule polymerization dynamics. Annu. Rev. Cell Dev. Biol. 1997;**13**:83–117. DOI: 10.1146/annurev.cellbio.13.1.83

[2] Jiang K, Akhmanova A. Microtubule tip-interacting proteins: a view from both ends. Curr. Opin. Cell Biol. 2011;**23**:94–101. DOI: 10.1016/j.ceb.2010.08.008

[3] Sakakibara A, Ando R, Sapir T, Tanaka T. Microtubule dynamics in neuronal morphogenesis. Open Biol. 2013;**3**:130061. DOI: 10.1098/rsob.130061

[4] Conde C, Cáceres A. Microtubule assembly, organization and dynamics in axons and dendrites. Nat. Rev. Neurosci. 2009;**10**:319–332. DOI: 10.1038/nrn2631

[5] David-Pfeuty T, Erickson HP, Pantaloni D. Guanosinetriphosphatase activity of tubulin associated with microtubule assembly. Proc. Natl. Acad. Sci. U. S. A. 1977;**74**:5372–5376. DOI: 10.1073/pnas.74.12.5372

[6] Nogales E, Wolf SG, Downing KH. Structure of the alpha beta tubulin dimer by electron crystallography. Nature. 1998;**391**:199–203. DOI: 10.1038/34465

[7] Mitchison T, Kirschner M. Dynamic instability of microtubule growth. Nature. 1984;**312**:237–242. DOI: 10.1038/312237a0

[8] Carlier MF, Didry D, Simon C, Pantaloni D. Mechanism of GTP hydrolysis in tubulin polymerization: characterization of the kinetic intermediate microtubule-GDP-Pi using phosphate analogues. Biochemistry. 1989;**28**:1783–1791. DOI: 10.1021/bi00430a054

[9] Gundersen GG, Gomes ER, Wen Y. Cortical control of microtubule stability and polarization. Curr. Opin. Cell Biol. 2004;**16**:106–112. DOI: 10.1016/j.ceb.2003.11.010

[10] Murphy DB, Borisy GG. Association of high-molecular weight proteins with microtubules and their role in microtubule assembly. Proc. Natl. Acad. Sci. U. S. A. 1975;**72**:2696–2700.

[11] Margolis RL, Rauch CT, Job D. Purification and assay of a 145-kDa protein (STOP145) with microtubule-stabilizing and motility behavior. Proc. Natl. Acad. Sci. U. S. A. 1986;**83**:639–643.

[12] Kowalski RJ, Williams RC Jr. Microtubule-associated protein 2 alters the dynamic properties of microtubule assembly and disassembly. J. Biol. Chem. 1993;**268**:9847–9855.

[13] Gamblin TC, Nachmanoff K, Halpain S, Williams Jr RC. Recombinant microtubule associated protein 2C reduces the dynamic instability of individual microtubules. Biochemistry. 1996;**35**:12575–12586. DOI: 10.1021/bi961135d

[14] Ebneth A, Drewes G, Mandelkow EM, Mandelkow E. Phosphorylation of MAP2c and MAP4 by MARK kinases leads to the destabilization of microtubules in cells. Cell Motil. Cytoskeleton. 1999;**44**:209–224. DOI:10.1002/(SICI)1097-0169(199911)44:3<209::AID-CM6>3.0.CO;2-4

[15] van der Vaart B, Akhmanova A, Straube A. Regulation of microtubule dynamic instability. Biochem. Soc. Trans. 2009;**37**:1007–1013. DOI: 10.1042/BST0371007

[16] Belmont L, Mitchison TJ. Identification of a protein that interacts with tubulin dimers and increases the catastrophe rate of microtubules. Cell. 1996;**84**:623–631. DOI: 10.1016/S0092-8674(00)81037-5

[17] Kline-Smith SL, Walczak CE. The microtubule-destabilizing kinesin XKCM1 regulates microtubule dynamic instability in cells. Mol. Biol. Cell. 2002;**13**:2718–31. DOI: 10.1091/mbc.E01-12-0143

[18] De Brabander M, Geuens G, Nuydens R, Willebrords R, Aerts F, De Mey J. Microtubule dynamics during the cell cycle: the effects of taxol and nocodazole on the microtubule system of Pt K2 cells at different stages of the mitotic cycle. Int. Rev. Cytol. 1986;**101**:215–74. DOI: 10.1016/S0074-7696(08)60250-8

[19] McGuire WP, Rowinsky EK, Rosenshein NB, Grumbine FC, Ettinger DS, Armstrong DK, Donehower RC. Taxol: a unique antineoplastic agent with significant activity in advanced ovarian epithelial neoplasms. Ann. Intern. Med. 1988;**111**:273–279. DOI: 10.7326/0003-4819-111-4-273

[20] Wilson L, Jordan MA. Microtubule dynamics: taking aim at a moving target. Chem. Biol. 1995;**2**:569–573. DOI: 10.1016/1074-5521(95)90119-1

[21] Vasquez RJ, Howell B, Yvon AM, Wadsworth P, Cassimeris L. Nanomolar concentrations of nocodazole alter microtubule dynamic instability in vivo and in vitro. Mol. Biol. Cell. 1997;**8**:973–985. DOI: 10.1091/mbc.8.6.973

[22] Hammond JW, Cai D, Verhey KJ. Tubulin modifications and their cellular functions. Curr. Opin. Cell Biol. 2008;**20**:71–76. DOI: 10.1016/j.ceb.2007.11.010

[23] Wloga D, Gaertig J. Post-translational modifications of microtubules. J. Cell Sci. 2010;**123**:3447–3455. DOI: 10.1242/jcs.063727

[24] Roychowdhury S, Rasenick MM. Submembranous microtubule cytoskeleton: regulation of microtubule assembly by heterotrimeric G proteins. FEBS J. 2008;**275**:4654–4663. DOI: 10.1111/j.1742-4658.2008.06614.x

[25] Dave RH, Saengsawang W, Yu JZ, Donati R, Rasenick MM. Heterotrimeric G-proteins interact directly with cytoskeletal components to modify microtubule-dependent cellular processes. Neurosignals. 2009;**17**:100–108. DOI: 10.1159/000186693

[26] Gilman AG. G proteins: transducers of receptor-generated signals. Ann. Rev. Biochem. 1987;**56**:615–649. DOI: 10.1146/annurev.bi.56.070187.003151

[27] Fredriksson R, Lagerström MC, Lundin LG, Schiöth HB. The G-protein-coupled receptors in the human genome form five main families. Phylogenetic analysis, paralogon groups, and fingerprints. Mol. Pharmacol. 2003;**63**:1256–1272. DOI: 10.1124/mol.63.6.1256

[28] Wang D, Li Y, Zhang Y, Liu Y, Shi G. High throughput screening (HTS) in identification new ligands and drugable targets of G protein-coupled receptors (GPCRs). Comb. Chem. High Throughput Screen. 2012;**15**:232–241. DOI: 10.2174/138620712799218626

[29] Salon JA, Lodowski DT, Palczewski K. The significance of G protein-coupled receptor crystallography for drug discovery. Pharmacol. Rev. 2011;**63**:901–937. DOI: 10.1124/pr.110.003350

[30] Latek D, Modzelewska A, Trzaskowski B, Palczewski K, Filipek S. G protein-coupled receptors—recent advances. Acta Biochim. Pol. 2012;**59**:515–529.

[31] Downes GB, Gautam N. The G protein subunit gene families. Genomics. 1999;**62**:544–552. DOI: 10.1006/geno.1999.5992

[32] Simon MI, Strathmann MP, Gautam N. Diversity of G proteins in signal transduction. Science. 1991;**252**:802–808. DOI: 10.1126/science.1902986

[33] Neves SR, Ram PT, Iyengar R. G protein pathways. Science. 2002;**296**:1636–1639. DOI: 10.1126/science.1071550

[34] Dohlman HG, Thorner J, Caron MJ, Lefkowitz RJ. Model systems for the study of seven transmembrane-segment receptors. Annu. Rev. Biochem. 1991;**60**:653–688. DOI: 10.1146/annurev.bi.60.070191.003253

[35] McCudden CR, Hains MD, Kimple RJ, Siderovski DP, Willard FS. G-protein signaling: back to the future. Cell. Mol. Life Sci. 2005;**62**:551–577. DOI: 10.1007/s00018-004-4462-3

[36] Sternweis P.C. The active role of beta gamma in signal transduction. Curr. Opin. Cell Biol. 1994;**6**:198–203. DOI: 10.1016/0955-0674(94)90136-8

[37] Smrcka AV. G protein beta gamma subunits: central mediators of G protein-coupled receptor signaling. Cell. Mol. Life Sci. 2008;**65**:2191–2214. DOI: 10.1007/s00018-008-8006-5

[38] Ueda N, Iñiguez-Lluhi JA, Lee E, Smrcka AV, Robishaw JD, Gilman AG. G protein beta gamma subunits. Simplified purification and properties of novel isoforms. J. Biol. Chem. 1994;**269**:4388–4395.

[39] Wickman KD, Iñiguez-Lluhl JA, Davenport PA, Taussig R, Krapivinsky GB, Linder ME, Gilman AG, Clapham DE. Recombinant G-protein beta gamma-subunits activate the muscarinic-gated atrial potassium channel. Nature. 1994;**368**:255–257. DOI: 10.1038/368255a0

[40] Faure M, Voyno-Yasenetskaya TA, Bourne HR. cAMP and beta gamma subunits of heterotrimeric G proteins stimulate the mitogen-activated protein kinase pathway in COS-7 cells. J. Biol. Chem. 1994;**269**:7851–7854.

[41] Ford CE, Skiba NP, Bae H, Daaka Y, Reuveny E, Shekter LR, Rosal R, Weng G, Yang CS, Iyengar R, Miller RJ, Jan LY, Lefkowitz RJ, Hamm HE. Molecular basis for interactions of G protein betagamma subunits with effectors. Science. 1998;**280**:1271–1274. DOI: 10.1126/science.280.5367.1271

[42] Shi CS, Lee SB, Sinnarajah S, Dessauer CW, Rhee SG, Kehrl JH. Regulator of G-protein signaling 3 (RGS3) inhibits Gbeta1gamma 2-induced inositol phosphate production, mitogen-activated protein kinase activation, and Akt activation. J. Biol. Chem. 2001;**276**:24293–24300. DOI: 10.1074/jbc.M100089200

[43] Wang N, Yan K, Rasenick MM. Tubulin binds specifically to the signal-transducing proteins, Gs alpha and Gi alpha 1. J. Biol. Chem. 1990;**265**:1239–1242.

[44] Wang N, Rasenick MM. Tubulin-G protein interactions involve microtubule polymerization domains. Biochemistry. 1991;**30**:10957–10965. DOI: 10.1021/bi00109a021

[45] Roychowdhury S, Panda D, Wilson L, Rasenick MM. G protein alpha subunits activate tubulin GTPase and modulate microtubule polymerization dynamics. J. Biol. Chem. 1999;**274**:13485–13490. DOI: 10.1074/jbc.274.19.13485

[46] Roychowdhury S, Martinez L, Salgado L, Das S, Rasenick MM. G protein activation is prerequisite for functional coupling between $G\alpha/G\beta\gamma$ and tubulin/microtubules. Biochem. Biophys. Res. Commun. 2006;**340**:441–448. DOI: 10.1016/j.bbrc.2005.12.026

[47] Yu JZ, Dave RH, Allen JA, Sarma T, Rasenick MM. Cytosolic G{alpha}s acts as an intracellular messenger to increase microtubule dynamics and promote neurite outgrowth. J. Biol. Chem. 2009;**284**:10462–10472. DOI: 10.1074/jbc.M809166200

[48] Roychowdhury S, Rasenick MM. G protein beta1gamma2 subunits promote microtubule assembly. J. Biol. Chem. 1997;**272**:31476–31581. DOI: 10.1074/jbc.272.50.31576

[49] Iñiguez-Lluhi JA, Simon MI, Robinshaw JD, Gilman AG. G protein beta gamma subunits synthesized in Sf9 cells. Functional characterization and the significance of the prenylation of gamma. J. Biol. Chem. 1992;**267**:23409–23417.

[50] Higgins JB, Casey PJ. In vitro processing of G protein gamma subunits. Requirements for assembly of an active beta gamma complex. J. Biol. Chem. 1994;**269**:9067–9073.

[51] Yasuda H, Lindorfer MA, Woodfork KA, Fletcher JE, Garrison JC. Role of the prenyl group on the G protein gamma subunit in coupling trimeric G proteins to A1 adenosine receptors. J. Biol. Chem. 1996;**271**:18588–18595. DOI: 10.1074/jbc.271.31.18588

[52] Montoya V, Gutierrez C, Najera O, Leony D, Varela A, Popova J, Rasenick M, Das S, Roychowdhury S. G protein βγ subunits interact with αβ and γ tubulin and play a role in microtubule assembly in PC12 cells. Cell Motil. Cytoskeleton. 2007;**64**:936–950. DOI: 10.1002/cm.20234

[53] Hoebeke J, Van Nigen G, De Brabander MJ. Interaction of nocodazole (R 91734), a new antitumoral drug, with rat brain tubulin. Biochem. Biophys. Res. Commun. 1979;**69**:319–324. DOI: 10.1016/0006-291X(76)90524-6

[54] DeBrabander MJ, Geuens G., Nuydens R., Willebrords R., and De Mey J. Microtubule assembly in livingcells after release from nocodazole block: the effect of metabolic inhibitors, taxol and PH. Cell Biol. Int. Rep.1981;**5**:913–920

[55] Joshi HC, Palacios MJ, McNamara L, Cleveland DW. Gamma-tubulin is a centrosomal protein required for cell cycle-dependent microtubule nucleation. Nature. 1992;**356**:80–83. DOI: 10.1038/356080a0

[56] Oakley BR. Gamma-tubulin: the microtubule organizer? Trends Cell Biol. 1992;**2**:1–5. DOI: 10.1016/0962-8924(92)90125-7

[57] Moritz M, Braunfeld MB, Sedat JW, Alberts B, Agard DA. Microtubule nucleation by gamma-tubulin-containing rings in the centrosome. Nature. 1995;**378**:638–640. DOI: 10.1038/378638a0

[58] Moudjou M, Bordes N, Paintrand M, Bornens M. Gamma-tubulin in mammalian cells: the centrosomal and the cytosolic forms. J. Cell Sci. 1996;**109**:875–887.

[59] Moritz M, Agard DA. Gamma-tubulin complexes and microtubule nucleation. Curr. Opin. Struct. Biol. 2001;**11**:174–181. DOI: 10.1016/S0959-440X(00)00187-1

[60] Job D, Valiron O, Oakley B. Microtubule nucleation. Curr. Opin. Cell Biol. 2003;**15**:111–117. DOI: 10.1016/S0955-0674(02)00003-0

[61] Wedegaertner PB, Wilson PT, Bourne HR. Lipid modifications of trimeric G proteins. J. Biol. Chem. 1995;**270**:503–506. DOI: 10.1074/jbc.270.2.503

[62] Mumby SM, Linder ME. Myristoylation of G-protein alpha subunits. Methods Enzymol. 1994;**237**:254–268. DOI: 10.1016/S0076-6879(94)37067-2

[63] Heidemann SR, McIntosh JR. Visualization of the structural polarity of microtubules. Nature. 1989;**286**:517–519. DOI: 10.1038/286517a0

[64] Gotta M, Ahringer J. Distinct roles for Galpha and Gbetagamma in regulating spindle position and orientation in Caenorhabditis elegans embryos. Nat. Cell Biol. 2002;**3**:297–301. DOI: 10.1038/35060092

[65] Schaefer M, Petronczki M, Dorner D, Forte M, Knoblich J. Heterotrimeric G proteins direct two modes of asymmetric cell division in the Drosophila nervous system. Cell. 2001;**107**:183–194. DOI: 10.1016/S0092-8674(01)00521-9

[66] Fuse N, Hisata K, Katzen AL, Matsuzaki F. Heterotrimeric G proteins regulate daughter cell size asymmetry in Drosophila neuroblast divisions. Curr. Biol. 2003;**13**:947–954. DOI: http://dx.doi.org/10.1016/S0960-9822(03)00334-8

[67] Du Q, Macara IG. Mammalian Pins is a conformational switch that links NuMA to heterotrimeric G proteins. Cells. 2004;**119**:503–516. DOI: 10.1016/j.cell.2004.10.028

[68] Sanada K, Tsai LH. G protein betagamma subunits and AGS3 control spindle orientation and asymmetric cell fate of cerebral cortical progenitors. Cell. 2005;**122**:119–131. DOI: 10.1016/j.cell.2005.05.009

[69] Siegrist SE, Doe CQ. Microtubule-induced Pins/Galphai cortical polarity in Drosophila neuroblasts. Cell. 2005;**123**:1323–1335. DOI: 10.1016/j.cell.2005.09.043

[70] De Vries L, Fischer T, Tronchère H, Brothers GM, Strockbine B, Siderovski DP, Farquhar MG. Activator of G protein signaling 3 is a guanine dissociation inhibitor for Galpha i subunits. Proc. Natl. Acad. Sci. U. S. A. 2000;**97**:14364–14369. DOI: 10.1073/pnas.97.26.14364

[71] Kimple RJ, Willard FS, Siderovski DP. The GoLoco Motif: heralding a new tango between G protein signaling and cell division. Mol. Interv. 2002;**2**:88–100. DOI: 10.1124/mi.2.2.88

[72] Severin FF, Sorger PK, Hyman AA. Kinetochores distinguish GTP from GDP forms of the microtubule lattice. Nature. 1997;**388**:888–891. DOI: 10.1038/42270

[73] Dehmelt L, Halpain S. Actin and microtubules in neurite initiation: are MAPs the missing link?. J. Neurobiol. 2004;**58**:18–33. DOI: 10.1002/neu.10284

[74] Witte H, Bradke F. The role of the cytoskeleton during neuronal polarization. Curr. Opin. Neurobiol. 2008;**18**:479–487. DOI: 10.1016/j.conb.2008.09.019

[75] Geraldo S, Gordon-Weeks PR. Cytoskeletal dynamics in growth-cone steering. J. Cell. Sci. 2009;**122**:3595–3604. DOI: 10.1242/jcs.042309

[76] Li R, Gundersen GG. Beyond polymer polarity: how the cytoskeleton builds a polarized cell. Nat. Rev. Mol. Cell Biol. 2008;**9**:860–873. DOI: 10.1038/nrm2522

[77] Greene LA, Tischler AS. Establishment of a noradrenergic clonal line of rat adrenal pheochromocytoma cells which respond to nerve growth factor. Proc. Natl. Acad. Sci. U. S. A. 1976;**73**:2424–2428. DOI: 10.1073/pnas.73.7.2424

[78] Patapoutian A, Reichardt LF. Trk receptors: mediators of neurotrophin action. Curr. Opin. Neurobiol. 2001;**11**:272–280. DOI: 10.1016/S0959-4388(00)00208-7

[79] Cantley LC. The phosphoinositide 3-kinase pathway. Science. 2002;**296**:1655–1657. DOI: 10.1126/science.296.5573.1655

[80] Zhou FQ, Zhou J, Dedhar S, Wu YH, Snider WD. NGF-induced axon growth is mediated by localized inactivation of GSK-3beta and functions of the microtubule plus end binding protein APC. Neuron. 2004;**42**:897–912. DOI: 10.1016/j.neuron.2004.05.011

[81] Govek EE, Newey SE, Van Aelst L. The role of the Rho GTPases in neuronal development. Genes. Dev. 2005;**19**:1–49. DOI: 10.1101/gad.1256405

[82] Hall A, Lalli G. Rho and Ras GTPases in axon growth, guidance, and branching. Cold Spring Harb. Perspect. Biol. 2010;**2**:a001818. DOI: 10.1101/cshperspect.a001818

[83] Reinoso BS, Undie AS, Levitt P. Dopamine receptors mediate differential morphological effects on cerebral cortical neurons in vitro. J. Neurosci. Res. 1996;**43**:439–453. DOI: 10.1002/(SICI)1097-4547(19960215)43:4<439::AID-JNR5>3.0.CO;2-G

[84] Kwon JH, Vogt Weisenhorn DM, Downen M, Roback L, Joshi H, Wainer BH. Beta-adrenergic and fibroblast growth factor receptors induce neuronal process outgrowth through different mechanisms. Eur. J. Neurosci. 1998;**10**:2776–2789. DOI: 10.1111/j.1460-9568.1998.00315.x

[85] Lotto B, Upton L, Price DJ, Gaspar P. Serotonin receptor activation enhances neurite outgrowth of thalamic neurones in rodents. Neurosci. Lett. 1999;**269**:87–90. DOI: 10.1016/S0304-3940(99)00422-X

[86] He JC, Gomes I, Nguyen T, Jayaram G, Ram PT, Devi LA, Iyengar R. The G alpha(o/i)-coupled cannabinoid receptor-mediated neurite outgrowth involves Rap regulation of Src and Stat3. J. Biol. Chem. 2005;**280**:33426–33434. DOI: 10.1074/jbc.M502812200

[87] Igarashi M, Strittmatter S, Vartanian T, Fishman MC. Mediation by G proteins of signals that cause collapse of growth cones. Science. 1993;**259**:77–84. DOI: 10.1126/science.8418498

[88] Wolfgang WJ, Clay C, Parker J, Delgado R, Labarca P, Kidokoro Y, Forte M. Signaling through Gs alpha is required for the growth and function of neuromuscular synapses in Drosophila. Dev. Biol. 2004;**268**:295–311. DOI: 10.1016/j.ydbio.2004.01.007

[89] Sachdev P, Menon S, Kastner DB, Chuang JZ, Yeh TY, Conde C, Caceres A, Sung CH, Sakmar TP. G protein beta gamma subunit interaction with the dynein light-chain component Tctex-1 regulates neurite outgrowth. EMBO J. 2007;**26**:2621–32. DOI: 10.1038/sj.emboj.7601716

[90] Wang K, Wong YH. G protein signaling controls the differentiation of multiple cell lineages. Biofactors. 2009;**35**:232–238. DOI: 10.1002/biof.39

[91] Sarma T, Voyno-Yasenetskaya T, Hope TJ, Rasenick MM. Heterotrimeric G-proteins associate with microtubules during differentiation in PC12 pheochromocytoma cells. FASEB J. 2003;**17**:848–859. DOI: 10.1096/fj.02-0730com

[92] Sarma T, Koutsouris A, Yu JZ, Krbanjevic A, Hope TJ, Rasenick MM. Activation of microtubule dynamics increases neuronal growth via the nerve growth factor (NGF)- and Gαs-mediated signaling pathways. J. Biol. Chem. 2015;**290**:10045–10056. DOI: 10.1074/jbc.M114.630632

[93] Sierra-Fonseca JA, Najera O, Martinez-Jurado J, Walker EM, Varela-Ramirez A, Khan AM, Miranda M, Lamango NS, Roychowdhury S. Nerve growth factor induces neurite outgrowth of PC12 cells by promoting Gβγ-microtubule interaction. BMC Neurosci. 2014;**15**:32. DOI: 10.1186/s12868-014-0132-4

[94] Okae H, Iwakura Y. Neural tube defects and impaired neural progenitor cell proliferation in Gbeta1-deficient mice. Dev. Dyn. 2010;**239**:1089–1101. DOI: 10.1002/dvdy.22256

[95] Zhang JH, Pandey M, Seigneur EM, Panicker LM, Koo L, Schwartz OM, Chen W, Chen CK, Simonds WF. Knockout of G protein β5 impairs brain development and causes multiple neurologic abnormalities in mice. J. Neurochem. 2011;**119**:544–554. DOI: 10.1111/j.1471-4159.2011.07457.x

Chapter 15

Novel Insights into the Role of the Cytoskeleton in Cancer

Xuan Zhang, Zenglin Pei, Chunxia Ji, Xiaoyan Zhang,
Jianqing Xu and Jin Wang

Additional information is available at the end of the chapter

http://dx.doi.org/10.5772/66860

Abstract

The cytoskeleton is a complex network of highly ordered intracellular filaments that plays a central role in controlling cell shape, division, functions, and interactions in human organs and tissues, but dysregulation of this network can contribute to numerous human diseases, including cancer. To clarify the various functions of the cytoskeleton and its role in cancer progression, in this chapter, we will discuss the microfilament (actin cytoskeleton), microtubule (β-tubulin), and intermediate filament (keratins, cytokeratins, vimentin, and lamins) cytoskeletal structures; analyze the physiological functions of the cytoskeleton and its regulation of cell differentiation; and investigate the roles of the cytoskeleton in cancer progression, the epithelial-mesenchymal transition process (EMT), and the mechanisms of multidrug resistance (MDR) in relation to the cytoskeleton. Importantly, the cytoskeleton, as a key regulator of the transcription and expression of many genes, is known to be involved in various physiological and pathological processes, which makes the cytoskeleton a novel and highly effective target for assessing the response to antitumor therapy in cancer.

Keywords: Cytoskeleton, Regulator, cell differentiation, drug resistance, EMT

1. Introduction

The cytoskeleton is a structure similar to a bird's nest; it can be tightly packed or sparse and is found in both prokaryotes and eukaryotes [1]. The main component of the cytoskeleton is protein, and the specific differences in structure never affect the type of proteins incorporated [2]. The cytoskeletons of prokaryotes display apparent plasticity in composition, without conservation of the core filament-forming proteins. However, the eukaryotic cytoskeleton has evolved in a variety of functions through the addition of accessory proteins and extensive gene duplication. The distribution of cytoskeletal filaments puts constraints on the likely prokaryotic line that made the leap into eukaryogenesis [3].

There are three main cytoskeletal structures in eukaryotes, microfilaments (MFs, ≈7 nm diameter), microtubules (MTs, ≈25 nm diameter), and intermediate filaments (IFs, ≈10 nm diameter) [1]. MFs are responsible for cell contraction and reinforcement of the cell surface and allow changes in cell morphology. Actin and tubulin are the main globular proteins that form MFs and MTs, respectively. Actin is a ubiquitous eukaryotic filament-forming protein. Actin filaments (also called microfilaments or F-actin) consist of two proto filament polymers wound together in a right-handed helix [3]. Eukaryotic actin is a member of a large and diverse superfamily of ATPases that includes Hsp70 chaperones and several classes of sugar/sugar alcohol kinases [4, 5] as well as eukaryotic actin-related proteins (ARPs) [6, 7]. The actin cytoskeleton is involved in actin-based cytoskeletal structures, including various functionally distinct structures of actin organization, and can be regulated by actin regulatory proteins. It is well known that the actin cortex consists of a dense mesh-like array of F-actin anchored to the cell membranes [8, 9]. This structure provides the core "skeleton" of the cell, functioning to define cell shape and provide resistance to mechanical stress. Reorganization of the actin cytoskeleton describes a process where cells actively alter the architecture of actin filaments to adjust cell shape in response to environmental requirements. Globular- (G-) actin is a highly conserved, polar protein with a molecular weight of 42 kDa that forms dimers and trimers in a process called actin nucleation; these structures then assemble into a double-stranded helical filament (F-actin) with a diameter of 7–9 nm (actin polymerization) [10–15]. Filopodia are thin, hair-like cellular protrusions that consist of parallel actin bundles cross-linked by interacting protein partners such as fascin, α-actinin, fimbrin, and formins [16]. Filopodia sense changes in the cellular microenvironment, such as growth factor concentrations, to guide cellular movement through the surrounding matrix [10–12]. Fascin is a highly conserved actin-bundling protein with three isoforms. While Fascin-1 is ubiquitously expressed during embryogenesis, its expression is later restricted to the endothelium, neuronal tissue, and testis [16]. Fascin-2 and Fascin-3 are expressed in the retinal epithelium and testis only [17]. Fascin is phosphorylated by protein kinase C (PKC), which regulates its actin bundling activity in accordance with the current microenvironmental conditions, which are communicated via surface integrins [16].

Microtubules are responsible for structural strength and cell shape. They allow organelles to move within cells. These structures act like rails on which kinesin and dynein can pull organelles. Most microtubules consist of 13 protofilaments that interact laterally to form a hollow tube and arise from the polymerization of heterodimers of a- and ß-tubulin, which are added to the plus-end of microtubules containing GTP in both subunits [3, 18].

As the major components of the cytoskeleton, intermediate filaments (IFs) are ubiquitous cytoskeletal elements that are encoded by about 70 genes in the human genome [19–21] and are divided into six groups based on their structure. These groups include the keratins, cytokeratins, mesoderm-specific intermediate filaments, neurofilaments and related proteins, lamins, and beaded filament proteins of the eye lens [21–24]. Although these IF proteins have very different amino acid sequences, the organization of the structural domain is similar [24]. The keratin group is defined as the group of intermediate filaments within epithelial cells, forming particles from 44 kDa to approximately 66 kDa that are characterized by high stability and chemical resistance [25]. As the major structural proteins of the nuclear lamina, the lamins are subdivided into A- and B-types, both of which belong to the type V intermediate filament protein family [26].

On the other hand, prokaryotes also have homologs of the eukaryotic microfilaments (actin), microtubules (tubulin), and intermediate filament proteins [27]. FtsZ was first found in the prokaryote as the cytoskeleton and forms filaments but not tubules [28]. The MreB and ParM proteins in prokaryotic cells function like actin in eukaryotic cells [29]. The third type of cytoskeleton in prokaryotes is crescentin, which is responsible for the shape of cells [30].

2. The physiological functions of the cytoskeleton and the regulation of cell differentiation

The cytoskeleton is the frame around or within the cell, and a system of intracellular filaments is crucial for cell shape, division, and function in all three domains of life [3, 4]. The classical functions of the cytoskeleton have been summarized as morphology determination, cell polarity formation, structural support, membrane constitution, cell motility, and receptor or channel localization in the plasma membrane [31–33].

The activity of actin, the main type of microfilament, is regulated by GTPases, which control the formation of actin filaments [34–36]. In patients with Alzheimer's disease (AD), actin filaments play a central role in maintaining and modifying synaptic connections [37]. As the key mediator between receptor activation during learning and a protein involved in regulating spine morphology [38, 39], actin not only plays a role in the nervous system but also has functions in the immune system [40]. For example, F-actin can mediate signaling in B cells and T cells [41, 42]. The dynamics of the actin cytoskeleton regulate adhesion and signal transduction in T cells/APCs [40]. SWAP-70 and HS1 are important downstream components of the TCR signaling pathway and are regulated by actin [43, 44].

The key function of intermediate filaments is to support the cell membrane, serving as a structural scaffold to maintain cell shape. Cell motility is significantly enhanced as a result of changes in intermediate filaments. Intermediate filaments are fixed to the membrane through transmembrane proteins such as cadherins, which are involved in the formation of cell-cell tight junctions and the distribution of traction forces that arise in the interspace between cells. Under stress stimulation, intermediate filaments are significantly upregulated to induce the rearrangement of the cytoskeleton [45, 46]. Keratins are proteins that form intermediate filaments of epithelial cell cytoskeleton and have an antiapoptotic function, regulate protein synthesis, and play a role in wound healing [25]. Epithelial cells obtain a specific pattern of keratin expression during differentiation and maturation; this pattern reflects the specificity of the tissue and the degree of maturation [25]. The different components of the cytoskeleton do not work alone, and microtubules, microfilaments, and intermediate filaments often interact with each other to complete cellular processes. They always participate in protein localization and cell signaling. A characteristic of differentiation is a change in cell shape that is dependent on the cytoskeleton. During mesenchymal stem cell differentiation, the actin cytoskeleton of mesenchymal stem cells changes during osteogenic and chondrogenic differentiation [47].

Previous studies have shown that adipocyte differentiation is associated with actin structure [48, 49]. Disruption of the actin cytoskeleton can regulate MKL1 and result in adipocyte differentiation [50]. Rearrangements in and the formation of processes by the cytoskeleton are associated with the synthesis of synaptopodin, which is a marker of differentiated podocytes [51]. Moreover, cell differentiation is regulated by activating or repressing some transcription factors and is linked to the cytoskeleton [52–57]. For example, Zoubiane demonstrated that microtubules were required for β-casein expression, which resulted in epithelial cell differentiation [55]; Ahmad discovered that the pattern of microtubules in HL-60 cells changed following differentiation, with α-tubulin appearing more regularly organized in the differentiated HL-60 cells [56]; Takiqawa also confirmed that the cytoskeleton, including microtubules and microfilaments, regulates the expression of a differentiated phenotype in chondrocytes [57].

3. The cytoskeleton and its role in cancer progression

The cytoskeleton is known to contribute to cancer. The cytoskeleton may induce cell proliferation and activate oncogenes, resulting in tumorigenesis [58]. In mammary carcinoma cells, the upregulation of WNT4 increased mesenchymal and cytoskeleton remodeling markers [59]. CKAP4 (cytoskeleton-associated protein 4) is a DKK1 binding protein expressed on the surface of cells, with DKK1/CKAP4 promoting pancreatic cancer and lung cancer [60]. DKK1 is considered a factor that can modulate the β-catenin pathway and stimulate cancer cells or noncancerous proliferation [61, 62]. Zyxin localizes to focal adhesion sites and stress fibers in response to mechanical cues and has been shown to control the assembly of the cytoskeleton, the generation of traction force, cell movement, and signal transduction. If zyxin is nonfunctional, the cytoskeleton may be disturbed, leading to cancer [59].

Many actin-bundling proteins are also linked to cancer progression and tumor chemoresistance [63]. Fascin proteins organize F-actin into parallel bundles and are required for the formation of actin-based cellular protrusions. The inhibition of fascin can interfere with the formation of filopodia and suppress the migration and invasion of tumors [64], making it possible to use fascin as a small molecular target to inhibit cancer metastasis.

Intermediate filaments interact with arcs and can inhibit the activity of arcs, which can transport intermediate filaments to cell nucleus. However, fewer intermediate filaments may alter the cell shape and lead to diseases such as tumors [65]. For example, the changes in nuclear architecture that are the pathological hallmarks of cancer cells are related to alterations in the lamins, with alterations in lamins A/C being verified in the colon [66], gut [67], lung [68], and prostate cancer [69].

4. The role of the cellular cytoskeleton in epithelial-mesenchymal transition (EMT)

Epithelial-mesenchymal transition (EMT) is a biological process resulting in the loss of polarity and cell junctions, and disturbances in the cytoskeleton [70]. The reorganization of the actin cytoskeleton is important for metastasis and the differentiation of epithelial cells to mesenchymal cells [71]. When cells undergo EMT, the number of β-actin fibers is reduced and the distribution of β-actin becomes diffused. RhoA induces action fiber formation and regulates the cytoskeleton. The Rho family also plays a role in tumor migration and EMT [72]. Cellular transformation was closely associated with changes in the distribution and amount of cytoplasmic actin isoforms [73]. Actin filaments are formed by the polymerization of G-actin, which induces the formation of a leading edge in cancer cells undergoing EMT through interactions with binding proteins and contractile proteins such as myosin II, which accelerates the movement of actin fibers on the substrate to the leading edge [74].

Tubulin plays an important role in EMT induction and contributes to TGF-β-induced membrane extensions or protrusions of human carcinoma cells undergoing EMT in three-dimensional collagen gels [75]. Acetylated α-tubulin can serve as a new marker of EMT and is expressed at a high level on normal epithelial cells, while the expression of acetylated α-tubulin decreases during TGF-β-induced EMT [76]. β-III tubulin can modulate snail expression during EMT in HT-29 and LS180 colon cancer cells [77]. When human mammary epithelial cells undergo EMT, the expression of Twist or Snail downregulates the tubulin tyrosine ligase enzyme, leading to the detyrosination of α-tubulin. The accumulation of detyrosinated Glu-tubulin is vital for the formation of microtentacles. These results provide new insight into tumor progression, as increasing α-tubulin detyrosination could promote EMT [78]. Because of their effect on tumor migration during EMT, the inhibition of microtubules can be a useful target for antitumor drugs [79, 80].

During the EMT process, intermediate filaments are significantly rearranged, typically switching from cytokeratin-rich to vimentin-rich networks [81]. Intermediate filaments can be expressed in different types of cells. For example, keratins are specifically expressed in epithelial cells; type III (mostly mesodermal) proteins are expressed in mesenchymal cells [82–84]. Epithelial cells express different keratins that are considered almost specific markers, whereas mesenchymal cells, endothelial cells, and hematopoietic cells express vimentin [82, 85, 86]. Vimentin is a type III intermediate filament that is significantly upregulated during EMT in epithelial cells; thus, vimentin can be used as a marker of EMT [87]. E-cadherin is one type of cell adhesion molecule that regulates EMT [88]. Reduced CK8 expression is regarded as an indicator of EMT, leading to more malignant forms of cancer [89]. Breast milk exosomes containing high levels of TGF-β2 can induce changes in both benign and malignant breast epithelial cells, consistent with the development and progression of breast cancer, which occurs due to alterations in cellular shape and the actin cytoskeleton and the loss of cell-cell junctions [90]. TGF-β-induced EMT was found to restrain cell invasion, which may be alleviated by the overexpression of hyperactivated Ras [91]. Endocytosis has emerged as a highly interconnected infrastructure of various cellular circuitries that is essential for the execution of different cellular programs, including those promoting a canonical EMT program and relying on the activation of Wnt, Notch, or TGF-β signaling [92]. On the other hand, miR-200 can inhibit EMT and the migration of cervical cancer cells through RhoE, which is an actin-binding protein [93]. Therefore, the cellular cytoskeleton plays a role in EMT by activating Wnt, Notch, or TGF-β signaling pathways, triggering the reprogramming of gene expression patterns via transcriptional changes and the altered expression of mRNA, including epithelial (E-cadherin, claudins, occludins, desmoplakin, mucin-1, and cytokeratin-8, -9, -18) and mesenchymal markers (fibronectin, FSP1, vitronectin, vimentin, smooth-muscle actin, and FGFR2

IIIb and IIIc species variants) (**Figure 1**). Further in-depth study is required to determine the features of the dynamic expression and arrangement of intracellular filaments during cancer invasion and migration.

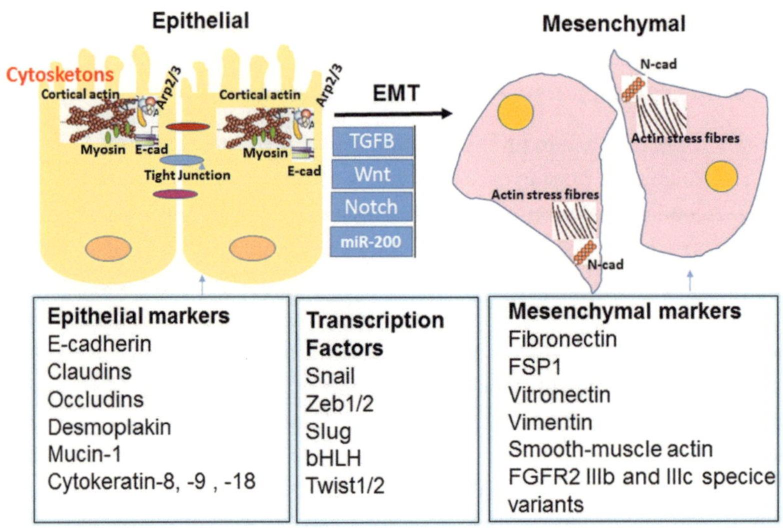

Figure 1. Cellular cytoskeletons in epithelial-mesenchymal transition process (EMT).

5. Mechanism of multidrug resistance (MDR) in relation to the cytoskeleton

Many patients develop drug resistance to anticancer agents, with the mechanisms including alterations in the ATP-binding cassette [94], microtubule dynamics, drug transport, and cell death [89], all of which involve tubulin and microtubules [95–97]. Microtubules have been considered a highly significant molecular target for anticancer agents, including microtubule-stabilizing agents. For example, paclitaxel binds to the β-tubulin subunit, accelerates the polymerization of tubulin, and stabilizes the resultant microtubules [98, 99]. Moreover, the paclitaxel-induced resistance of vimentin intermediate filaments to okadaic acid may occur through a microtubule-independent mechanism [100]. Townson also demonstrated that K8/18 filaments provided resistance to apoptosis in granulosa cell tumors by impairing FAS expression [101]. The organization of actin filaments associated with cellular differentiation may also influence the expression of P-glycoprotein (P-gp) through ezrin, radixin, and moesin in MDR osteosarcoma cells [92–104], which exhibit a significant increase in well-organized actin stress fibers [103], while inhibiting actin remodeling can suppress drug resistance in cancer [105].

The balance between polymerized and nonpolymerized tubulin will be a key determinant of the response to antimitotic-based chemotherapy. Cancer cells obtain mitotic drug resistance properties through β I-tubulin mutations [106], which is important for maintaining microtubule structure and sensitivity to microtubule-targeting agents. β-tubulin mutations confer resistance to epothilone and taxanes in ovarian cancer cells. Moreover, mutations in both a- and β-tubulin have been found to confer resistance to colchicine and vinblastine in Chinese hamster ovary (CHO) cells [107, 108]. The upregulation of β III-tubulin was further associated with resistance to paclitaxel and docetaxel [109–113]. On the other hand, as a negative regulator of β III-tubulin, HDAC3 can also mediate drug resistance [113].

As a major intermediate filament in the cells of epithelial-derived tumors, cytokeratin K8/18 expression is involved in cytokeratin-dependent drug resistance [114]. Hepatocyte cytokeratin plays a role in bile formation and resistance to bile acid challenge; however, the loss of K8 significantly increased liver injury in response to toxic stress in mice [115]. Caulin also demonstrated that normal and malignant epithelial cells deficient in K8/18 were approximately 100 times more sensitive to TNF-induced death [116], indicating that interaction between the damaging agent and cytokeratin might trigger signaling responses for cell survival.

In our previous study, we found that tissue remodeling proteins such as KRTHB3, KRT7, KRT8, KRT17, TPM4, CRYAB, SEPW1, LGALS3BP, and VATI were overexpressed in resistant KB-v1 cells, implying that the intracellular vesicular transport of many drugs is partly controlled by cytoskeletal filaments [117]. Research into the cytoskeleton is experiencing increased interest and rapid advancement, which will provide a greater mechanistic understanding of the molecular pathways and mechanisms contributing to drug resistance and will enable the development of more patient-tailored therapies.

Acknowledgements

This work was sponsored by a grant (16PJ1408800) to Dr. Jin Wang from the Shanghai Pujiang Program, a grant (15DZ2290200) from the Shanghai Science and Technology Commission, and a grant (SHDC12014104) from the Shanghai ShenKang Hospital Development Center, Shanghai, China.

Author details

Xuan Zhang, Zenglin Pei, Chunxia Ji, Xiaoyan Zhang, Jianqing Xu and Jin Wang*

*Address all correspondence to: wangjin@shaphc.org

Scientific Research Center, Shanghai Public Health Clinical Center, Shanghai, P.R. China

References

[1] Hardin J, Bertoni G, Kleinsmith LJ. Becker's World of the Cell (8th ed). New York: Pearson. 2015, 422–446.

[2] McKinley M, Valerie DO, Pennefather-O'Brien E, Harris R. Human Anatomy (4th Ed). New York: McGraw Hill Education. 2015, 29.

[3] Wickstead B, Gull K. The evolution of the cytoskeleton. J Cell Biol. 2011, 194 (4): 513–525.

[4] Flaherty KM, McKay DB, Kabsch W, Holmes KC. Similarity of the three-dimensional structures of actin and the ATPase fragment of a70-kDa heat shock cognate protein. Proc Natl Acad Sci USA. 1991, 88: 5041–5045.

[5] Bork P, Sander C, Valencia A. An ATPase domain common to prokaryotic cell cycle proteins, sugar kinases, actin, and hsp70 heat shock proteins. Proc Natl Acad Sci USA. 1992, 89: 7290–7294.

[6] Frankel S, Mooseker MS. The actin-related proteins. Curr Opin Cell Biol. 1996, 8: 30–37.

[7] Schafer DA, Schroer TA. Actin-related proteins. Annu Rev Cell Dev Biol. 1999, 15: 341–363.

[8] Bray D, White JG. Cortical flow in animal cells. Science. 1988, 239: 883–888.

[9] Charras GT, Hu CK, Coughlin M, Mitchison TJ. Reassembly of contractile actin cortex in cell blebs. J Cell Biol. 2006, 175: 477–490.

[10] Gerhardt H, Golding M, Fruttiger M, Ruhrberg C, Lundkvist A, Abramsson A, et al. VEGF guides angiogenic sprouting utilizing endothelial tip cell filopodia. J Cell Biol. 2003, 161(6): 1163–1177.

[11] Zheng JQ, Wan JJ, Poo MM. Essential role of filopodia in chemotropic turning of nerve growth cone induced by a glutamate gradient. J Neuro. 1996, 16(3): 1140–1149.

[12] Kalil K, Dent EW. Touch and go: guidance cues signal to the growth cone cytoskeleton. Curr Opin Neurobiol. 2005, 15(5): 521–526.

[13] Blanchoin L, Boujemaa-Paterski R, Sykes C, Plastino J. Actin dynamics, architecture, and mechanics in cell motility. Physiol Rev. 2014, 94(1): 235–263.

[14] Higgs HN. Discussing the morphology of actin filaments in lamellipodia. Trends Cell Biol. 2011, 21(1): 2–4.

[15] Holmes KC, Popp D, Gebhard W, Kabsch W. Atomic model of the actin filament. Nature. 1990, 347(6288): 44–49.

[16] Machesky LM, Li A. Invasive filopodia promotingmetastasis. Commun Integr Biol. 2010, 3(3): 263–270.

[17] Adams JC. Roles of fascin in cell adhesion and motility. Curr Opin Cell Biol. 2004, 16(5): 590–596.

[18] Howard J, Hyman AA. Dynamics and mechanics of the microtubule plus end. Nature. 2003, 422(6933): 753–758. doi:10.1038/nature01600.

[19] Nicholl ID, Quinlan RA. Chaperone activity of alpha-crystallins modulates intermediate filament assembly. EMBO J. 1994, 13(4): 945–953.

[20] Helfand BT, Chang L, Goldman RD. Intermediate filaments are dynamic and motile elements of cellular architecture. J Cell Sci. 2004, 117(Pt2): 133–141.

[21] Evans RM. Vimentin: the conundrum of the intermediate filament gene family. Bioessays. 1998, 20(1): 79–86.

[22] Fuchs, Weber K. Intermediate filaments: structure, dynamics, function, and disease. Annu Rev Biochem. 1994, 63: 345–382.

[23] Galou M, Gao J, Humbert J, Mericskay M, Li Z, Paulin D, et al. The importance of intermediate filaments in the adaptation of tissues to mechanical stress: evidence from gene knockout studies. Biol Cell. 1997, 89(2): 85–97.

[24] Szeverenyi I, Cassidy AJ, Chung CW, Lee BT, Common JE, Ogg SC, et al. The human intermediate filament database: comprehensive information on a gene family involved in many human diseases. Hum Mutat. 2008, 29(3): 351–360.

[25] Pastuszak M, Groszewski K, Pastuszak M, Dyrla P, Wojtuń S, Gil J. Cytokeratins in gastroenterology. Systematic review. Prz Gastroenterol. 2015, 10(2): 61–70.

[26] Dittmer TA, Misteli T. The lamin protein family. Genome Biol. 2011, 12: 222.

[27] Shih YL, Rothfield L. The bacterial cytoskeleton. Microbiol Mol Biol Rev. 2006, 70(3): 729–754.

[28] de Boer P, Crossley R, Rothfield L. The essential bacterial cell-division protein FtsZ is a GTPase. Nature. 1992, 359(6392): 254–256.

[29] Popp D, Narita A, Maeda K, Fujisawa T, Ghoshdastider U, Iwasa M, et al. Filament structure, organization, and dynamics in MreB sheets. J Biol Chem. 2010, 285 (21): 15858–15865.

[30] Ausmees N, Kuhn JR, Jacobs-Wagner C. The bacterial cytoskeleton: an intermediate filament-like function in cell shape. Cell. 2003, 115(6): 705–713.

[31] Jaeken L. A new list of functions of the cytoskeleton. IUBMB Life. 2007, 59(3): 127–133.

[32] Clegg JS. Intracellular water, metabolism and cell architecture. In Coherent Excitations in Biological Systems (Fröhlich H and Kremer F, eds). Springer Berlin Heidelberg, 1983, 162–177.

[33] Atema J. Microtubule theory of sensory transduction. J Theor Biol. 1973, 38: 181–190.

[34] Watanabe N, Kato T, Fujita A, Ishizaki T, Narumiya S. Cooperation between mDia1 and ROCK in Rho-induced actin reorganization. Nat Cell Biol. 1999, 1(3): 136–143.

[35] Aspenström P, Richnau N, Johansson AS. The diaphanous-related formin DAAM1 collaborates with the Rho GTPases RhoA and Cdc42, CIP4 and Src in regulating cell morphogenesis and actin dynamics. Exp Cell Res. 2006, 312(12): 2180–2194. doi: 10.1016/j.yexcr.2006.03.013.

[36] Matusek T, Gombos R, Szécsényi A, Sánchez-Soriano N, Czibula A, Pataki C, et al. Formin proteins of the DAAM subfamily play a role during axon growth. J Neurosci. 2008, 28(49): 13310–13319. doi: 10.1523/jneurosci.2727-08.2008.

[37] Stefen H, Chaichim C, Power J, Fath T. Regulation of the postsynaptic compartment of excitatory synapses by the actin cytoskeleton in health and its disruption in disease. Neural Plast. 2016, 2016: 2371970. doi: 10.1155/2016/2371970.

[38] Hotulainen P, Hoogenraad CC. Actin in dendritic spines: connecting dynamics to function. J Cell Biol. 2010, 189(4): 619–629.

[39] Lamprecht R. The role of actin cytoskeleton in memory formation in amygdala. Front Mol Neurosci. 2016; 9: 23.

[40] Comrie WA, Burkhardt JK. Action and traction: cytoskeletal control of receptor triggering at the immunological synapse. Front Immunol. 2016, 7: 68. doi: 10.3389/fimmu.2016.00068. eCollection 2016.

[41] Treanor B, Depoil D, Gonzalez-Granja A, Barral P, Weber M, Dushek O, et al. The membrane skeleton controls diffusion dynamics and signaling through the B cell receptor. Immunity. 2010, 32(2): 187–199.

[42] Campi G, Varma R, Dustin ML. T cell receptor microclusters as scaffolds for signaling. J Exp Med. 2005, 202(8): 1031–1036.

[43] Gomez TS, McCarney SD, Carrizosa E, Labno CM, Comiskey EO, Nolz JC, et al. HS1 functions as an essential actin-regulatory adaptor protein at the immune synapse. Immunity. 2006, 24(6): 741–752. 10.1016/j.immuni.2006.03.022

[44] Becart S, Altman A. SWAP-70-like adapter of T cells: a novel Lck-regulated guanine nucleotide exchange factor coordinating actin cytoskeleton reorganization and Ca2+ signaling in T cells. Immunol Rev. 2009, 232(1): 319–333.

[45] Toivola DM, Strnad P, Habtezion A, Omary MB. Intermediate filaments take the heat as stress proteins. Trends Cell Biol. 2010, 20: 79–91.

[46] Herrmann H, Strelkov SV, Burkhard P, Aebi U. Intermediate filaments: Primary determinants of cell architecture and plasticity. J Clin Invest. 2009, 119: 1772–1783

[47] Yourek G, Hussain MA, Mao JJ. Cytoskeletal changes of mesenchymal stem cells during differentiation. ASAIO J. 2007, 53(2): 219–228.

[48] Kanzaki M, Pessin JE. Insulin-stimulated GLUT4 translocation in adipocytes is dependent upon cortical actin remodeling. J Biol Chem. 2001, 276: 42436–42444.

[49] Noguchi M, Hosoda K, Fujikura J, Fujimoto M, Iwakura H, Tomita T, et al. Genetic and pharmacological inhibition of Rho-associated kinase II enhances adipogenesis. J Biol Chem. 2007, 282(40): 29574–29583.

[50] Nobusue H, Onishi N, Shimizu T, Sugihara E, Oki Y, Sumikawa Y, et al. Regulation of MKL1 via actin cytoskeleton dynamics drives adipocyte differentiation. Nat Commun. 2014, 5: 3368. doi: 10.1038/ncomms4368.

[51] Mundel P, Reiser J, Zúñiga Mejía Borja A, Pavenstädt H, Davidson GR, Kriz W, et al. Rearrangements of the cytoskeleton and cell contacts induce process formation during differentiation of conditionally immortalized mouse podocyte cell lines. Exp Cell Res. 1997, 236(1): 248–258.

[52] Braun T, Gautel M. Transcriptional mechanisms regulating skeletal muscle differentiation, growth and homeostasis. Nat Rev Mol Cell Biol. 2011, 12: 349–361.

[53] Farmer SR. Transcriptional control of adipocyte formation. Cell Metab. 2006, 4: 263–273.

[54] de Crombrugghe B, Lefebvre V, Nakashima K. Regulatory mechanisms in the pathways of cartilage and bone formation. Curr Opin Cell Biol. 2001, 13: 721–727.

[55] Zoubiane GS, Valentijn A, Lowe ET, Akhtar N, Bagley S, Gilmore AP, et al. A role for the cytoskeleton in prolactin-dependent mammary epithelial cell differentiation. J Cell Sci. 2004, 117: 271–280.

[56] Shariftabrizi A, Ahmadian S, Pazhang Y. Dynamics of γ-tubulin cytoskeleton in HL-60 leukemia cells undergoing differentiation and apoptosis by all-trans retinoic acid. Mol Med Rep. 2012, 5(2): 545–551.

[57] Takigawa M, Takano T, Shirai E, Suzuki F. Cytoskeleton and differentiation: effects of cytochalasin B and colchicine on expression of the differentiated phenotype of rabbit costal chondrocytes in culture. Cell Differ. 1984, 14(3): 197–204.

[58] Gaspar P, Holder MV, Aerne BL, Janody F, Tapon N. Zyxin antagonizes the FERM protein expanded to couple F-actin and Yorkie-dependent organ growth. Curr Biol. 2015, 25(6): 679–689. doi: 10.1016/j.cub.

[59] Vouyovitch C, Perry JK, Liu DX, Bezin L, Vilain E, Diaz JJ, et al. WNT4 mediates the autocrine effects of growth hormone in mammary carcinoma cells. Endocr Relat Cancer. 2016, pii: ERC-15-0528.

[60] Kimura H, Fumoto K, Shojima K, Nojima S, Osugi Y, Tomihara H, et al. CKAP4 is a Dickkopf1 receptor and is involved in tumor progression. J Clin Invest. 2016, 126(7): 2689–705.

[61] Yamabuki T, Takano A, Hayama S, Ishikawa N, Kato T, Miyamoto M, et al. Dikkopf-1 as a novel serologic and prognostic biomarker for lung and esophageal carcinomas. Cancer Res. 2007, 67(6): 2517–2525.

[62] Makino T, Yamasaki M, Takemasa I, Takeno A, Nakamura Y, Miyata H, et al. Dickkopf-1 expression as a marker for predicting clinical outcome in esophageal squamous cell carcinoma. Ann Surg Oncol. 2009, 16(7): 2058–2064.

[63] Stevenson RP, Veltman D, Machesky LM. Actin-bundling proteins in cancer progression at a glance. J Cell Sci. 2012, 125(Pt5): 1073–1079. doi: 10.1242/jcs.093799.

[64] Huang FK, Han S, Xing B, Huang J, Liu B, Bordeleau F, et al. Targeted inhibition of fascin function blocks tumour invasion and metastatic colonization. Nat Commun. 2015, 6: 7465. doi: 10.1038/ncomms8465.

[65] Jiu Y, Lehtimäki J, Tojkander S, Cheng F, Jäälinoja H, Liu X, et al. Bidirectional interplay between vimentin intermediate filaments and contractile actin stress fibers. Cell Rep. 2015, 11(10): 1511–1518.

[66] Willis ND, et al. Lamin A/C is a risk biomarker in colorectal cancer. PLoS One. 2008, 3: e2988.

[67] Machiels BM, et al. Abnormal A-type lamin organization in a human lung carcinoma cell line. Eur J Cell Biol. 1995, 67: 328–335.

[68] Wu Z, et al. Reduced expression of lamin A/C correlates with poor histological differentiation and prognosis in primary gastric carcinoma. J Exp Clin Cancer Res. 2009, 28: 8.

[69] Helfand BT, et al. Chromosomal regions associated with prostate cancer risk localize to lamin B deficient microdomains and exhibit reduced gene transcription. J Pathol. 2011, 226(5): 735–45.

[70] Kalluri R, Neilson EG. Epithelial-mesenchymal transition and its implications for fibrosis. J Clin Invest. 2003, 112: 1776–1784.

[71] Thiery JP, Acloque H, Huang RY, Nieto MA. Epithelial-mesenchymal transitions in development and disease. Cell. 2009, 139: 871–890.

[72] Vega FM, Fruhwirth G, Ng T, Ridley AJ. RhoA and RhoC have distinct roles in migration and invasion by acting through different targets. J Cell Biol. 2011, 193(4): 655–665.

[73] Nakashima J, Liao F, Sparks JA, Tang Y, Blancaflor EB. The actin cytoskeleton is a suppressor of the endogenous skewing behaviour of Arabidopsis primary roots in microgravity. Plant Biol Stuttg. 2014, 16(Suppl 1): 142–150.

[74] Anderson TW, Vaughan AN, Cramer LP. Retrograde flow and myosin II activity within the leading cell edge deliver F-actin to the lamella to seed the formation of graded polarity actomyosin II filament bundles in migrating fibroblasts. Mol Biol Cell. 2008, 19: 5006–5018.

[75] Oyanagi J, Ogawa T, Sato H, Higashi S, Miyazaki K. Epithelial-mesenchymal transition stimulates human cancer cells to extend microtubule-based invasive protrusions and suppresses cell growth in collagen gel. PLoS One. 2012, 7: e53209.

[76] Gu S, Liu Y, Zhu B, Ding K, Yao TP, Chen F, et al. Loss of α-tubulin acetylation is associated with TGF-β-induced epithelial-mesenchymal transition. J Biol Chem. 2016, 291(10): 5396–5405.

[77] Sobierajska K, Wieczorek K, Ciszewski WM, Sacewicz-Hofman I, Wawro ME, Wiktorska M, et al. β-III tubulin modulates the behavior of Snail overexpressed during the epithelial-to-mesenchymal transition in colon cancer cells. Biochim Biophys Acta. 2016, 1863(9): 2221–2233.

[78] Whipple RA, Matrone MA, Cho EH, Balzer EM, Vitolo MI, Yoon JR, et al. Epithelial-to-mesenchymal transition promotes tubulin detyrosination and microtentacles that enhance endothelial engagement. Cancer Res. 2010, 70: 8127–8137.

[79] Li WT, Yeh TK, Song JS, Yang YN, Chen TW, Lin CH, et al. BPR0C305, an orally active microtubule-disrupting anticancer agent. Anticancer Drugs. 2013, 24: 1047–1057.

[80] Shin SY, Kim JH, Yoon H, Choi YK, Koh D, Lim Y, et al. Novel antimitotic activity of 2-hydroxy-4-methoxy-2′,3′-benzochalcone (HymnPro) through the inhibition of tubulin polymerization. J Agric Food Chem. 2013, 61: 12588–12597.

[81] Yamasaki T, Seki N, Yamada Y, Yoshino H, Hidaka H, Chiyomaru T, et al. Tumor suppressive microRNA-138 contributes to cell migration and invasion through its targeting of vimentin in renal cell carcinoma. Int J Oncol. 2012, 41: 805–817.

[82] Kim S, Coulombe PA. Intermediate filament scaffolds fulfill mechanical, organizational and signaling functions in the cytoplasm. Genes Dev. 2007, 21: 1581–1597.

[83] Nieminen M, Henttinen T, Merinen M, MarttilaIchihara F, Eriksson JE, Jalkanen S. Vimentin function in lymphocyte adhesion and transcellular migration. Nat Cell Biol. 2006, 8: 156–162.

[84] Kim S, Kellner J, Lee CH, Coulombe PA. Interaction between the keratin cytoskeleton and eEF1Bgamma affects protein synthesis in epithelial cells. Nat Struct Mol Biol. 2007, 14: 982–983.

[85] Eckes B, Colucci-Guyon E, Smola H, Nodder S, Babinet C, Krieg T, et al. Impaired wound healing in embryonic and adult mice lacking vimentin. J Cell Sci. 2000, 113(Pt 13): 2455–2462.

[86] Herrmann H, Bar H, Kreplak L, Strelkov SV, Aebi U. Intermediate filaments: from cell architecture to nanomechanics. Nat Rev Mol Cell Biol. 2007, 8(7): 562–573.

[87] Mendez MG, Kojima S, Goldman RD. Vimentin induces changes in cell shape, motility, and adhesion during the epithelial to mesenchymal transition. FASEB J. 2010, 24: 1838–1851.

[88] Larue L, Bellacosa A. Epithelial-mesenchymal transition in development and cancer: role of phosphatidylinositol 3′ kinase/AKT pathways. Oncogene. 2005, 24(50): 7443–7454.

[89] Knösel T, Emde V, Schlüns K, Schlag PM, Dietel M, Petersen I. Cytokeratin profiles identify diagnostic signatures in colorectal cancer using multiplex analysis of tissue microarrays. Cell Oncol. 2006, 28(4): 167–175.

[90] Qin W, Tsukasaki Y, Dasgupta S, Mukhopadhyay N, Ikebe M, Sauter ER. Exosomes in human breast milk promote EMT. Clin Cancer Res. 2016, pii: clincanres.0135.2016.

[91] Wicki A, Lehembre F, Wick N, Hantusch B, Kerjaschki D, Christofori G. Tumor invasion in the absence of epithelial-mesenchymal transition: podoplanin-mediated remodeling of the actin cytoskeleton. Cancer Cell. 2006, 9: 261–272.

[92] Scita G, Di Fiore PP. The endocytic matrix. Nature. 2010, 463: 464–473.

[93] Cheng YX, Chen GT, Chen C, Zhang QF, Pan F, Hu M, et al. MicroRNA-200b inhibits epithelial-mesenchymal transition and migration of cervical cancer cells by directly targeting RhoE. Mol Med Rep. 2016, 13(4): 3139–3146.

[94] Gottesman MM, Fojo T, Bates SE. Multidrug resistance in cancer: role of ATP-dependent transporters. Nat Rev Cancer. 2002, 2(1): 48–58.

[95] Giannakakou P, Sackett DL, Kang YK, Zhan Z, Buters JT, Fojo T, et al. Paclitaxel-resistant human ovarian cancer cells have mutant beta-tubulins that exhibit impaired paclitaxel-driven polymerization. J Biol Chem. 1997, 272(27): 17118–17125.

[96] Haber M, Burkhart CA, Regl DL, Madafiglio J, Norris MD, Horwitz SB. Altered expression of M beta 2, the class II beta-tubulin isotype, in a murine J774.2 cell line with a high level of taxol resistance. J Biol Chem. 1995, 270(52): 31269–31275.

[97] Kavallaris M, Kuo DY, Burkhart CA, Regl DL, Norris MD, Haber M, et al. Taxol-resistant epithelial ovarian tumors are associated with altered expression of specific beta-tubulin isotypes. J Clin Invest. 1997, 100(5): 1282–1293.

[98] Schiff PB, Fant J, Horwitz SB. Promotion of microtubule assembly in vitro by taxol. Nature. 1979, 277(5698): 665–667.

[99] Schiff PB, Horwitz SB. Taxol stabilizes microtubules in mouse fibroblast cells. Proc Natl Acad Sci USA. 1980, 77(3): 1561–1565.

[100] Vilalta PM, Zhang L, Hamm-Alvarez SF. A novel taxol-induced vimentin phosphorylation and stabilization revealed by studies on stable microtubules and vimentin intermediate filaments. J Cell Sci. 1998, 111(Pt 13): 1841–1852.

[101] Trisdale SK, Schwab NM, Hou X, Davis JS, Townson DH. Molecular manipulation of keratin 8/18 intermediate filaments: modulators of FAS-mediated death signaling in human ovarian granulosa tumor cells. J Ovarian Res. 2016, 9: 8.

[102] Takeshita H, Kusuzaki K, Ashihara T, Gebhardt MC, Mankin HJ, Hirasawa Y. Actin organization associated with the expression of multidrug resistant phenotype in osteosarcoma cells and the effect of actin depolymerization on drug resistance. Cancer Lett. 1998, 126(1): 75–81.

[103] Luciani F, Molinari A, Lozupone F, Calcabrini A, Lugini L, Stringaro A, et al. P-glycoproteinactin association through ERM family proteins: a role in P glycoprotein function in human cells of lymphoid origin. Blood. 2002, 99(2): 641–648.

[104] Wang J, Chan JY, Fong CC, Tzang CH, Fung KP, Yang M. Transcriptional analysis of doxorubicin-induced cytotoxicity and resistance in human hepatocellular carcinoma cell lines. Liver Int. 2009, 29(9): 1338–1347.

[105] Kim MH, Kim J, Hong H, Lee SH, Lee JK, Jung E, et al. Actin remodeling confers BRAF inhibitor resistance to melanoma cells through YAP/TAZ activation. EMBO J. 2016, 35(5): 462–478.

[106] Cheung CH, Wu SY, Lee TR, Chang CY, Wu JS, Hsieh HP, Chang JY. Cancer cells acquire mitotic drug resistance properties through beta I-tubulin mutations and alterations in the expression of beta-tubulin isotypes. PLoS One. 2010, 5(9): e12564. doi: 10.1371/journal.pone.0012564.

[107] Giannakakou P, Gussio R, Nogales E, Downing KH, Zaharevitz D, Bollbuck B, Poy G, Sackett D, Nicolaou KC, Fojo T. A common pharmacophore for epothilone and taxanes: molecular basis for drug resistance conferred by tubulin mutations in human cancer cells. Proc Natl Acad Sci USA. 2000, 97: 2904–2909.

[108] Hari M, Wang Y, Veeraraghavan S, Cabral F. Mutations in alpha- and beta-tubulin that stabilize microtubules and confer resistance to colcemid and vinblastine. Mol Cancer Ther. 2003, 2: 597–605.

[109] Huzil JT, Chen K, Kurgan L, Tuszynski JA. The roles of beta-tubulin mutations and isotype expression in acquired drug resistance. Cancer Inform. 2007, 3: 159–181.

[110] Verdier-Pinard P, Wang F, Martello L, Burd B, Orr GA, Horwitz SB. Analysis of tubulin isotypes and mutations from taxol-resistant cells by combined isoelectrofocusing and mass spectrometry. Biochemistry. 2003, 42: 5349–5357.

[111] Shalli K, Brown I, Heys SD, Schofield AC. Alterations of beta-tubulin isotypes in breast cancer cells resistant to docetaxel. FASEB J. 2005, 19: 1299–1301.

[112] Seve P, Reiman T, Lai R, Hanson J, Santos C, et al. Class III beta tubulin is a marker of paclitaxel resistance in carcinomas of unknown primary site. Cancer Chemother Pharmacol. 2007, 60: 27–34.

[113] Kim Y, Kim H, Jeoung D. Tubulin beta3 serves as a target of HDAC3 and mediates resistance to microtubule-targeting drugs. Mol Cells. 2015, 38(8): 705–714.

[114] Cress AE, Dalton WS. Multiple drug resistance and intermediate filaments. Cancer Metastasis Rev. 1996, 15(4): 499–506.

[115] Fickert P, Fuchsbichler A, Wagner M, Silbert D, Zatloukal K, Denk H, Trauner M. The role of the hepatocyte cytokeratin network in bile formation and resistance to bile acid challenge and cholestasis in mice. Hepatology. 2009, 50(3): 893–899.

[116] Caulin C, Ware CF, Magin TM, Oshima RG. Keratin-dependent, epithelial resistance to tumor necrosis factor-induced apoptosis. J Cell Biol. 2000, 149(1): 17–22.

[117] Wang J, Tai LS, Tzang CH, Fong WF, Guan XY, Yang M. 1p31, 7q21 and 18q21 chromosomal aberrations and candidate genes in acquired vinblastine resistance of human cervical carcinoma KB cells. Oncol Rep. 2008, 19(5): 1155–1164.

Chapter 16

Targeting the Cytoskeleton with Plant-Bioactive Compounds in Cancer Therapy

Anca Hermenean and Aurel Ardelean

Additional information is available at the end of the chapter

http://dx.doi.org/10.5772/66911

Abstract

In this overview we describe the main plant-derived bioactive compounds used in cancer therapy which has the cell cytoskeleton as therapeutic target. Three major classes of these compounds are described: antimitotics with microtubule-destabilizing and—stabilizing effects, plant-bioactive compounds that interact with intermediate filaments/actin, and plant-bioactive compounds that interact with intermediate filaments like keratins and vimentin. We also focus on the molecular aspects of interactions with their cellular targets: microtubules, intermediate filaments, and microfilaments. Some critical aspects of cardiac side effects of cancer chemotherapy are also discussed, focusing on cardiac cytoskeleton and protective effect of plant-derived compounds. The application of plant bioactives in the treatment of cancer has resulted in increased therapeutic efficacy through targeting the cytoskeleton, respectively, prevention of the injury of cytoskeletal components elicited by chemotherapeutics.

Keywords: plant-derived compounds, cancer therapy, microtubules, intermediate filaments, microfilaments

1. Introduction

Chemotherapy is routinely used for cancer treatment. Since tumor cells lose many of the regulatory pathways of the normal cells, they continue to divide without control. Chemotherapeutic drugs try to solve these abnormalities, but sometimes the toxicity of allopathic treatments creates a significant problem.

The cytoskeleton constitutes the supporting framework of the cell, and it is composed of three types of cytosolic filaments: microtubules, intermediate filaments, and microfilaments. The entire cytoskeletal network is a dynamic structure which regulates the cell structure, and it is involved in many cellular functions such as movement, transport, or cell division [1]. The cytoskeleton is one of the main therapeutic targets in cancer cells [2].

Various cancer therapies use plant-derived bioactive products. There are four classes of plant-derived anticancer drugs currently used in oncotherapy: vinca alkaloids (vinblastine, vincristine), epipodophyllotoxins (etoposide and teniposide), taxanes (paclitaxel and docetaxel), and camptothecin derivatives (camptothecin and irinotecan) [3]. To date, new generations of vinca alkaloids, camptothecins, and epothilones as well as a novel class of taxanes have been developed. Some of these are in clinical use, others in clinical trials.

The major inconvenience in using antimicrotubule agents in oncotherapy is that these compounds cause significant side effects such as neutropenia and neurotoxicity and because of their limited efficacy as single agents [3].

This review describes the main natural compounds identified in the last year as potential anticancer agents, which have cell cytoskeleton as therapeutic target. We focus on the interactions of plant-derived anticancer drugs with all three types of cytosolic filaments: microtubules, intermediate filaments, and microfilaments. In addition, we summarize the most recent advances in the understanding of the molecular aspects of these interactions.

Some critical aspects of cardiac side effects of cancer chemotherapy are also discussed, focusing on cardiac cytoskeleton and protective effects of plant-derived compounds.

2. Microtubules as chemotherapeutic targets of plant-derived bioactives

Microtubules are dynamic structures involved in different cellular processes including cell division, where they are the most important constituents of the mitotic spindle apparatus during the M phase of cell division [4]. They are polymers composed of α- and β-tubulin heterodimers, characterized by high dynamics of polymerization/depolymerization, resulting in the elongation or shrinkage of the filaments. Polymerization of microtubules occurs when α- and β-tubulin monomers bind to a GTP at the nucleotide exchangeable site (E-site) in β-tubulin and the non-exchangeable site (N-site) in α-tubulin. Once GTP is hydrolyzed, it becomes non-exchangeable, which matches the addition of the next tubulin dimer to the plus (+) end of the microtubule. Upon depolymerization, the GTP cap is detached, allowing the microtubules depolymerize releasing the α-/β-tubulin heterodimers into the cytoplasm. Subsequently, the GDP attached to another free β-tubulin and can exchange to GTP at the E-site, before another polymerization cycle begins [4, 5].

Dynamic instability is regulated by a number of microtubule-associated proteins (MAPs), which bind to stabilize the microtubules [6]. MAP phosphorylation induces its dissociation leading to microtubule instability. Some cytokines have a critical role in the regulation of MAPs and microtubule dynamics, such as controlling centromere localization Cdc2 kinases, mitogen-activated protein kinases ERK, controlling cell migration JNK, and the main serine/threonine phosphatases, type 1 (PP1) and type 2A (PP2A) [7–10].

The dynamic ability of microtubules to polymerize and depolymerize is essential for cellular division and chromosome segregation during mitosis. Due to their crucial roles in dividing cells, microtubules have been considered a major target for cancer therapy. Microtubule-interacting plant-derived biomolecules, namely, antimitotics, can be classified into two main groups based on their apparent mechanisms of action: microtubule-destabilizing agents act as tubulin polymerization inhibitors, and microtubule-stabilizing agents act as tubulin polymerization promoters [11].

2.1. Microtubule-destabilizing agents

Vinca alkaloids and colchicines prevent the polymerization of tubulin and promote the depolymerization of microtubules.

Vinca alkaloids are a series of biologically active agents isolated from *Catharanthus roseus* (*Vinca rosea*) with a potent antitumor activity, related to their ability to inhibit the polymerization of microtubules and preventing cell division [12]. There are approximately 130 vinca alkaloids distributed in different vegetal tissues: vincristine, vinblastine, and yohimbine in the aerial parts; catharanthine and vindoline in leaves; and almalicine and reserpine in roots [13]. They have demonstrated clinical efficacy in a broad spectrum of cancers, both as single agents and in combination. Vincristine, vinblastine, and vindesine are the first vinca alkaloids used as antitumor drugs. Vinorelbine is the first new second-generation vinca alkaloid, while vinflunine, a bis-fluorinated vinorelbine derivative, was synthesized by superacid chemistry and studied in phase I–III clinical trials [14, 15].

The vinca alkaloids are dimeric compounds consisting of two multi-ringed subunits, vindoline and catharanthine, linked by a carbon-carbon bridge [16]. They act by binding specifically to β-tubulin and block its ability to polymerize with α-tubulin into microtubules, thus disrupting the mitotic spindle. This blocks mitosis and kills actively dividing cells. The results indicate that vinorelbine and vinflunine affect microtubule dynamics differently from vinblastine and proved to be weak binders [17].

Vincristine is used in the treatment of hematological and lymphatic neoplasms, whereas vinblastine in breast cancer, testicular cancer, choriocarcinoma, and vindesine in non-small cell lung cancer or breast cancer. Vinorelbine is useful for the treatment of non-small-cell lung cancer, and vinflunine has been used in the treatment of bladder, non-small-cell lung, and breast cancers [17].

Similar to Vinca alkaloids, colchicine extracted from plants of the genus *Colchicum* (*autumn crocus*) is a microtubule-destabilizing agent at high concentrations and stabilizes microtubule dynamics at low concentrations [18]. It first binds to soluble tubulin, leading to a complex that copolymerizes into the ends of the microtubules and prevents the elongation of the microtubule polymer. It is severely toxic to normal tissues at high dose, which limits its use in cancer therapies [19]. Colchicine showed different antitumoral effects which include inhibition of metastatic potential [20] and angiogenesis [21], cell blebbing through a Rho/Rho effector kinase (ROCK)/ myosin light-chain kinase (MLCK) pathway [22], decrease of ATP influx into mitochondria [23].

Novel microtubule-destabilizing plant-bioactive compounds are summarized in **Table 1**.

Active substance/herbal formulation	Mechanism of action	Therapeutic use	References
Flavonoids isolated from *Tanacetum gracile*	—Modulate microtubule depolymerization by activating mitotic spindle checkpoint —Bind at α- β interfacial site of tubulin	Breast cancer	[24]
Artelastin isolated from the wood bark of *Artocarpus elasticus*	—Radial structure disorganization of the microtubule network —Kinetochores are not affected	Breast cancer	[25]
Podoverine A isolated from *Podophyllum versipelle*	—Mitotic arrest and inhibition of microtubule polymerization by targeting the vinca-binding site on tubulin	Renal cancer Breast cancer	[26]
Plinabulin chemical probe KPU-244-B3	—Binds in the boundary region between α- and β-tubulin near the colchicine-binding site —Induce tubulin depolymerization	Fibrosarcoma	[27]
2′-Hydroxy-2,4,6-trimethoxy-5′,6′-naphthochalcone	—Disruption of microtubular networks by inhibition of tubulin polymerization —Failure of mitotic spindle formation and blocking mitosis at the prometaphase or metaphase-anaphase transition	Colon cancer	[28]
Aqueous extract of ginger	—Disruption of interphase microtubule network of A549 and HeLa cells —Inhibition of temperature-dependent reassembly of cold-treated depolymerized microtubule of A549 and HeLa cells	Cervical carcinoma Lung carcinoma	[29]
Safranal	—Inhibition of tubulin assembly (IC_{50} was obtained at 72.19 μM) —Binds between α- and β-tubulin closer to alpha-tubulin and hydrogen bond with Gly 142 —Hydrophobic interactions play critical roles for safranal molecule stabilization in binding site	Cancer therapy	[30]
Isochaihulactone	—Inhibition of tubulin polymerization in a concentration-dependent manner in A549 non-small-cell lung cancer cells —Cause G2/M phase arrest and apoptosis in a time- and concentration-dependent manner	Lung cancer	[31]
Carnosol	—Modulation of autophagic markers microtubule- associated protein 1A/1B light-chain 3 I (LC3 I) to microtubule-associated protein 1A/1B light-chain 3 II (LC3 II) and p62 in MDA-MB-231 cells	Breast cancer	[32]

Active substance/herbal formulation	Mechanism of action	Therapeutic use	References
Angelica shikokiana methanol extract (AME)	—AME and all isolated compounds inhibited tubulin polymerization —Angelicin and kaempferol-3-O-rutinoside were the most active compounds —Phenolic compounds and furanocoumarins showed binding affinity to colchicine-binding site —Quercetin, kaempferol, luteolin, chlorogenic acid, and methyl chlorogenate exhibited the strongest activity against histone deacetylase 8 (HDAC8) and the highest affinity to trichostatin A-binding site.	Human hepatocellular carcinoma, rhabdomyosarcoma (RD), colorectal carcinoma, human epithelioma, and human breast adenocarcinoma	[33]
Alkaloids from beach spider Lily (*Hymenocallis littoralis*)	—Interrupt polymerization of microtubules in Hep-G2 cells	Hepatocarcinoma	[34]
DYZ-2-90	—Binds to microtubules and rapidly induces tubulin depolymerization	Colorectal cancer	[35]
Indicine N-oxide (INO)	—200 μM induced a mitotic block of about 22% in HeLa cells; 300 μM concentration-induced depolymerization of interphase microtubular network —The effect was similar to the depolymerizing effects of the drugs such as colchicine and vinblastine, although the concentration used here was 1000-fold higher than those drugs —Binds to the tubulin dimer through hydrogen bonds and hydrophobic interactions —INO does not make any interactions with the amino acid residues on the tubulin dimer that were reported to be interacting with the taxol or colchicine, but INO-binding site partially overlaps with the griseofulvin-binding site (docking)	Cervical cancer	[36]

Table 1. Potential plant-bioactive compounds that interact with microtubules as microtubule-destabilizing agents for cancer therapy.

2.2. Microtubule-stabilizing agents

Taxanes are the main class of microtubule-stabilizing agents, which prevent the depolymerization of microtubules and promote the polymerization of tubulin to microtubules.

One of the most important plant compounds in the fight against cancer was discovered in the bark of *Taxus brevifolia*—taxol, now named paclitaxel, which has become one of the most effective drugs against breast and ovarian cancer and has been approved for the clinical treatment of cancer patients. Since the first discovery of paclitaxel in the 1960s, a variety of other microtubule-stabilizing agents have been derived primarily from natural resources [37]. The molecular mechanism includes polymerization of tubulin to stable microtubules and also interacts directly with microtubules, stabilizing them against depolymerization and thereby blocks cells in the G2/M phase of the cell cycle [38]. The binding of taxol to β-tubulin in the polymer results in cold-stable microtubules even in the absence of exogenous GTP. Hydrogen/deuterium exchange (HDX) coupled to liquid chromatography-electrospray ionization MS demonstrated a marked reduction in deuterium incorporation in both β- and α-tubulin in the presence of taxol and contributed to increased rigidity in taxol microtubules and complementary to that due to GTP-induced polymerization [39].

Initially obtained from *Taxus brevifolia* bark, paclitaxel is now a semisynthetic product of 10-deacetylbaccatin III, which is extracted from the needles of the *Taxus baccata*. Similarly, docetaxel, a second-generation taxane, was directly obtained semisynthetically by esterification from the inactive taxane precursor 10-deacetylbaccatin III [40]. Paclitaxel and docetaxel bind to the specific binding sites of tubulin, which is different from the binding site of guanosine triphosphate, vinblastine, colchicine, and podophyllotoxin [41].

Docetaxel has a 1.9-fold higher affinity for the site than paclitaxel and induces tubulin polymerization at a 2.1-fold lower critical tubulin concentration. The effect on the cell cycle is different: paclitaxel inhibits the cell cycle traverse at the G2/M phase junction [42], while docetaxel produces its maximum cell-killing effect against cells in the S phase [43].

To decrease the toxicity and enhance delivery and distribution, new taxane formulations of micelles were investigated, including nanoparticles, emulsions, and liposomes [44]. Compounds such as Abraxane, CT-2103, and docosahexaenoic acid (DHA)-paclitaxel are examples of new taxanes with higher activity than paclitaxel in taxane-resistant cancers, as well as in tumors that have been unresponsive to paclitaxel [16].

Protopine is a benzylisoquinoline alkaloid isolated from *Opium poppy*, *Corydalis tubers*, and *Fumaria officinalis*. It stabilizes tubulin polymerization process but has no affinity to taxol-binding site. It induces a marked increase of tubulin polymerization in a dose-dependent manner in human hormone-refractory prostate cancer (PC-3 cells), similar to paclitaxel. It enhances microtubule assembly and formation of mitotic spindles in PC-3 cells [45].

Taccalonolides are plant steroids possessing a C2–C3 epoxide group and an enol-lactone isolated from *Tacca leontopetaloides*, *Tacca plantaginea*, *Tacca chantrieri*, *Tacca plantaginea*, *Tacca integrifolia*, etc. They act as microtubule stabilizers by binding to another microtubule site than taxol resulting in the formation of microtubule bundles and leading to cell cycle arrest and apoptosis. It is also reported that taccalonolides bind to β-tubulin near the lumen of microtubule, which is different from the taxol-binding site stabilizers which bind to α-tubulin protofilaments [46–49].

Recent study shows that the dietary flavonoid fisetin binds to tubulin and stabilizes microtubules with binding characteristics far superior than paclitaxel. It induces upregulation of microtubule-associated protein (MAP)-2 and microtubule-associated protein (MAP)-4 and increases α-tubulin acetylation, an indicator of microtubule stabilization [50].

3. Microfilaments as chemotherapeutic targets of plant-derived bioactives

Actin filaments are composed of globular actin (G-actin) which polymerizes into filamentous (F) actin and participates in many important cellular processes including cell division and cytokinesis, cell signaling, vesicle and organelle movement, cell junction establishment, and maintenance.

Like microtubules, actin microfilaments can change rapidly their structure in response to external stimuli. Actin polymerization is stimulated by nucleating factors such as the Arp2/3 complex, which mimics a G-actin dimer in order to stimulate actin polymerization [51]. Actin binds ATP to stabilize microfilament formation and hydrolysis [52]. The growth of microfilaments is regulated by thymosin, which binds G-actin to lead the polymerizing process, whereas profilin binds G-actin and catalyzes the exchange of ADP to ATP, promoting monomeric addition to the plus end of F-actin [53].

During cytokinesis, disruption of actin polymerization can effect cellular structure. Cytokinesis inhibitors such as cytochalasin B disrupt the actin cytoskeleton, and the cell is unable to divide [54] but is still able to initiate another mitotic event, continuing to form nuclei and eventually becoming enlarged and multinucleated [55, 56]. Cell lines derived from bladder, kidney, and prostate carcinomas become multinucleated when grown in cytochalasin B-supplemented medium, whereas cells from corresponding normal tissue remain mono- or binucleate under comparable conditions [55]. These particular features make tumor cells ideal targets for chemotherapy, as they have reduced cytoskeletal integrity and multiple nucleation and increased mitochondrial activity [57].

Actin filaments are also of substantial importance to cancer cell migration. Cancer cell migration can convert between mesenchymal and amoeboid types. This latter can occur, e.g., when cells are exposed to protease inhibitors [58] and thereby mesenchymal cancer cell invasion is repressed by specific targeting of protease function. Inhibiting RhoA/ROCK signaling promotes the formation of multiple competing microfilament-derived lamellipodia that suppress amoeboid migration of tumor cells [59]. Tumor cells unable to move through amoeboid migration will switch to mesenchymal migration [60]. However, tumor cells exposed to protease inhibitors will move mainly through amoeboid migration. Using microfilament disrupting RhoA/ROCK inhibitors in combination with protease inhibitors would simultaneously block both types of cell migration.

Phytomedicine developed actin-targeted potential drugs, designed for cancer therapy (**Table 2**).

Active substance/herbal formulation	Mechanism of action	Therapeutic use	References
Resveratrol	−50 μM resveratrol decreases Rac and Cdc42 signaling to the actin cytoskeleton −5 μM resveratrol increases Rac signaling to the actin cytoskeleton	Breast cancer	[61]
Oleuropein	−Disrupt actin filaments in a dose-dependent manner	Sarcoma	[62]
Alkaloid mixture derived from *Senna spectabilis*—cassine and spectaline	−Altered normal distribution pattern of F-actin filaments	Liver cancer	[63]
Deoxyelephantopin (DET)	−Affects the actin cytoskeleton network and downregulates calpain-mediated proteolysis of several actin-associated proteins −Inhibition of proteolysis of actin cytoskeleton-associated proteins identified by differential proteomic profiling	Lung metastasis of mammary adenocarcinoma	[64]
Cucurbitacin E	−Disruption of the F-actin cytoskeleton −Increases the filamentous or polymerized actin fraction	Prostate carcinoma cells	[65]
Cucurbitacin E	−Damaged F-actin without affecting beta-tubulin	95D lung cancer cells	[66]
Cucurbitacin I	−Induced the co-aggregation of actin with phospho-myosin II by stimulation of the RhoA/ROCK pathway and inhibition of LIM-kinase	HeLa cells	[67]
Cucurbitacin B	−Induced rapid and improper polymerization of the F-actin network	Myeloid leukemia cells	[68]
Jasplakinolide (JAS)	−Rearranged the actin cytoskeleton −JAS has a phalloidin-like action −Distribution of actin filaments was different from that induced by cytochalasin D	Cancer cells	[69]
Ganoderma lucidum extracts	−Inhibits growth and induce actin polymerization	Bladder cancer cells	[70]
4-Hydroxycoumarin	−Disorganized the actin cytoskeleton correlated with reductions in cell adhesion to four extracellular matrix proteins and inhibition of random motility	Melanoma cell line B16–F10	[71]

Table 2. Plant-bioactive compounds which interact with actin for cancer therapy.

4. Intermediate filaments as chemotherapeutic targets of plant-derived bioactives

Along with microfilaments and microtubules, intermediate filaments are the other component of the cytoskeleton that can be exploited in the clinical treatment of cancer. All intermediate filaments have a central alpha-helical domain that is composed of four protofibrils separated

http://dx.doi.org/10.5772/66911

by three linker regions [72]. The N- and C-terminus segments of intermediate filaments are non-alpha-helical regions of polypeptide sequences, associated with head to tail into protofilaments that pair up laterally into protofibrils; four of these protofibrils form an intermediate filament.

Whereas microfilaments and microtubules are actin or tubulin polymers, intermediate filaments are composed of 50 different proteins classified into six types based on similarities in amino acid sequence [72]. In regard to potential chemotherapeutic targets, the most promising intermediate filaments are keratins, nestin, and vimentin.

4.1. Anti-keratin agents

Keratin and cytokeratin are intermediate filaments found in the cytoskeleton of epithelial tissue. There are twenty different keratin polypeptides (K1–K20) identified and classified into type I (K9–K20) and type II (K1–K8) intermediate filaments [73]. Keratins of importance to cancer therapy are keratin 8 (K8) and keratin 18 (K18), the most common and characteristic members of intermediate filaments expressed in single-layer epithelial tissues [74, 75]. Oncogenes, which activate Ras signaling, stimulate expression of K18 through transcription factors [76]. However, aberrant K8 and K18 expression has been noticed in particularly invasive carcinomas [77, 78]. K18 was found to be a substrate of the cysteine-aspartic proteases during epithelial apoptosis [77].

Based on aberrant keratin expression found in many cancers, these intermediate filaments present a novel chemotherapeutic target that need to be investigated.

Crude acetone extract of *Bupleurum scorzonerifolium* (AE-BS) showed antiproliferative activity, induced cell arrest in G2/M phase, and apoptosis in A549 human lung cancer cells [79]. In a further study, Chen et al. [73] noticed K8 phosphorylation after AE-BS treatment of A549 cells. The association of ERK1/2 activation with K8 phosphorylation may be related to the apoptotic effect of AE-BS.

4.2. Anti-vimentin agents

Vimentin functions as a regulator in cancer cells undergoing epithelial-mesenchymal transition (EMT), an important change during tumor progression where cells detached from their original tissue become highly motile and invasive. Studies have shown that quercetin prevented epidermal growth factor (EGF)-induced EMT, migration, and invasion of prostate cancer cells by suppressing the expression of vimentin and N-cadherin [80]. Genistein, an isoflavone found in soybeans, fava beans, and lupine, has been shown to downregulate mesenchymal markers ZEB1, slug, and vimentin and therefore cause reversal of EMT in gemcitabine-resistant pancreatic cancer cells [81]. Similarly, this flavonoid was able to decrease protein expression of vimentin, cathepsin D, and MMP-2 and thus suppressed epithelial-mesenchymal transition and migration capacity of BG-1 ovarian cancer cells [82]. Other natural compounds, like silibinin, induced the morphological reversal of mesenchymal phenotype to epithelial phenotype through downregulation of vimentin and MMP-2 and upregulation of cytokeratin-18 [83]. Moreover, silibinin meglumine, a water-soluble form of milk thistle silymarin, impedes the EMT in EGFR-mutant non-small-cell lung carcinoma cells by upregulation of the relative mRNA expression of CDH1 (E-cadherin) accompanied by downregulation of vimentin [84]. Berberine, a plant alkaloid present in various plants like *Berberis,* decreased the expression of the mesenchymal markers vimentin and fibronectin and restored the epithelial marker E-cadherin, thereby contributing to the reversal of EMT [85].

Piplartine, a biologically active component from *Piper* species (Piperaceae), also suppresses tumor progression and migration by disruption of the p120-ctn/vimentin/N-cadherin complex, which plays a critical role in tumor progression and invasion/metastasis [86].

Phenethyl isothiocyanate (PEITC), the main bioactive compound present in cruciferous vegetables, decreases breast and prostate tumor growth inhibition through vimentin suppression [87]. Cucurbitacin E induced disruption of vimentin cytoskeleton in prostate carcinoma cells, while microtubules were unaffected [65]. The natural product withaferin A (WFA) exhibits antitumor activity by binding to vimentin and covalently modifying its cysteine residue, which is present in the highly conserved helical coiled coil 2B domain [88]. Penduletin and casticin, flavonoids from the Brazilian plant *Croton betulaster,* induced changes in the pattern of expression of the cytoskeletal protein vimentin and thereby inhibit the growth of human glioblastoma cells [89].

5. Protective effect of plant-bioactive compounds on anthracycline-induced cardiac cytoskeletal toxicity

Cardiotoxicity is the most serious side effect of antitumoral anthracyclines, which include adriamycin, doxorubicin, mitoxantrone, daunorubicin, or epirubicin [90]. The main cause of toxicity is their effect on the cardiac cytoskeleton, consisting of myofibrils disarray [91], including both structural and functional changes: troponin I and troponin C phosphorylation mediated by a doxorubicin-induced protein kinase C activation [92, 93] and decrease of troponin I, and changes of α-actin, creatine kinase, and myosin light-chain 2 expression [93]. In other studies, degradation of cardiac cytoskeletal proteins, including titin [94] and dystrophin [95], was observed. Recently, changes in the cardiac distribution of desmin have been detected, with areas of decreased expression in the cytoplasm and protein aggregation after mitoxantrone treatment [96, 97]. The use of plant bioactives might protect against the oxidative stress caused by anthracycline drugs, including cytoskeleton injuries. Our group recently demonstrated that the flavonoid chrysin inhibits mitoxantrone-triggered cardiomyocyte apoptosis via multiple pathways, including decrease of the Bax/Bcl-2 ratio and caspase-3 expression along with preservation of the desmin disarray [96].

6. Conclusions

Plant-derived bioactive molecules constitute promising tools for the treatment of cancer. The application of plant bioactives in the treatment of cancer has resulted in increased therapeutic

http://dx.doi.org/10.5772/66911

efficacy through targeting the cytoskeleton and prevention of cytoskeletal injuries due to chemotherapy side effects. Research results testify both the evolution of knowledge coming from pharmacognosy and the great possibilities of future progress by means of a rational approach of natural product-based drug discovery or new pharmaceutical formulations.

Author details

Anca Hermenean[1]* and Aurel Ardelean[2]

*Address all correspondence to: anca.hermenean@gmail.com

1 Department of Histology, Faculty of Medicine, Vasile Goldis Western University of Arad, Arad, Romania

2 Department of Cell and Molecular Biology, Faculty of Medicine, Vasile Goldis Western University of Arad, Arad, Romania

References

[1] Lodish H., Berk A., Zipursky S.L., et al. Molecular Cell Biology. 4th edition. New York: W. H. Freeman. 2000.

[2] Saraf S., Patel D.R., Kaur C.D., Saraf S.. Cytoskeleton analysis as target for bioactives. Trends in Applied Sciences Research. 2001;1–12.

[3] Marzo I., Naval J. Antimitotic drugs in cancer chemotherapy: promises and pitfalls. Biochemical Pharmacology. 2013;8:703–710.

[4] Valiron O., Caudron N., Job D. Microtubule dynamics. Cellular and Molecular Life Sciences CMLS. 2001;58:2069–2084.

[5] Etienne-Manneville S. From signaling pathways to microtubule dynamics: the key players. Current Opinion in Cell Biology. 2010;22:104–111.

[6] Regnard C., Audebert S., Boucher D., Larcher J.C., Edde B., Denoulet P. Microtubules: functional polymorphisms of tubulin and associated proteins (structural and motor MAP's). Comptes Rendus des Seances de la Societe de Biologie et de Ses Filiales. 1996;190:255–268.

[7] Nigg E.A. Mitotic kinases as regulators of cell division and its checkpoints. Nature Reviews Molecular Cell Biology. 2001;2:21–32.

[8] Ma H.T., Poon R.Y.C. How protein kinases co-ordinate mitosis in animal cells. Biochemical Journal. 2011;435:17–31.

[9] Johnson G.L., Lapadat R. Mitogen-activated protein kinase pathways mediated by ERK, JNK, and p38 protein kinases. Science. 2002;298:1911–1912.

[10] Tournebize R., Andersen S.S., Verde F., Doree M., Karsenti E., Hyma A.A. Distinct roles of PP1 and PP2A-like phosphatases in control of microtubule dynamics during mitosis. The EMBO Journal. 1997;16:5537–5549.

[11] Negi A.S., Gautam Y., Alam S., Chanda D., Luqman S., Sarkar J., Khan F., Konwar R. Natural antitubulin agents: importance of 3,4,5-trimethoxyphenyl fragment. Bioorganic & Medicinal Chemistry. 2015;23:373–389.

[12] Himes R.H. Interactions of the catharanthus (Vinca) alkaloids with tubulin and microtubules. Pharmacology & Therapeutics. 1991;51:257–267.

[13] Liu Z., Wu H.L., Li Y., Gu H.W., Yin X.L., Xie L.X., Yu R.L. Rapid and simultaneous determination of five vinca alkaloids in *Catharanthus roseus* and human serum using trilinear component modeling of liquid chromatography–diode array detection data. Journal of Chromatography B. 2016;1026:114–123.

[14] Fahy J. Modifications in the "upper" velbenamine part of the Vinca alkaloids have major implications for tubulin interacting activities. Current Pharmaceutical Design. 2001;7:1181–1197.

[15] Yun-San Yip A., Yuen-Yuen Ong E., Chow L.W. Vinflunine: clinical perspectives of an emerging anticancer agent. Expert Opinion on Investigational Drugs. 2008;17:583–591.

[16] Nobili S., Lippi D., Witort E., Donnini M., Bausi L., Mini E., Capaccioli S. Natural compounds for cancer treatment and prevention. Pharmacological Research. 2009;59:365–378

[17] Ngan V.K., Bellman K., Hill B.T., Wilson L., Jordan M.A. Mechanism of mitotic block and inhibition of cell proliferation by the semisynthetic Vinca alkaloids vinorelbine and its newer derivative vinflunine. Molecular Pharmacology. 2001;60:225–232.

[18] Leung Y.Y., Li L., Hui Y., Kraus V.B. Colchicine—update on mechanisms of action and therapeutic uses. Seminars in Arthritis and Rheumatism. 2015;45:341–350.

[19] Bhattacharyya B., Panda D., Gupta S., Banerjee M. Anti-mitotic activity of colchicine and the structural basis for its interaction with tubulin. Medicinal Research Reviews. 2008;28:155–183.

[20] Charpentier M.S., Whipple R.A., Vitolo M.I., Boggs A.E., Slovic J., Thompson K.N., et al. Curcumin targets breast cancer stem-like cells with microtentacles that persist in mammospheres and promote reattachment. Cancer Research. 2014;74:1250–1260.

[21] Ganguly A., Yang H., Zhang H., Cabral F., Patel K.D. Microtubule dynamics control tail retraction in migrating vascular endothelial cells. Molecular Cancer Therapeutics. 2013;12:2837–2846.

[22] Meshki J., Douglas S.D., Hu M., Leeman S.E., Tuluc F. Substance P induces rapid and transient membrane blebbing in U373MG cells in a p21-activated kinase-dependent manner. PLoS One. 2011;6:e25332.

[23] Maldonado E.N., Patnaik J., Mullins M.R., Lemasters J.J. Free tubulin modulates mitochondrial membrane potential in cancer cells. Cancer Research. 2010;70:10192–10201.

[24] Sinha S., Amin H., Nayak D., Bhatnagar M., Kacker P., Chakraborty S., Kitchlu S., Vishwakarma R., Goswami A., Ghosal S. Assessment of microtubule depolymerization property of flavonoids isolated from Tanacetum gracile in breast cancer cells by biochemical and molecular docking approach. Chemico-Biological Interactions. 2015;239:1–11.

[25] Pedro M., Ferreira M.M., Cidadea H., Kijjo A., Bronze-da-Rocha E., Nascimento M. Artelastin is a cytotoxic prenylated flavone that disturbs microtubules and interferes with DNA replication in MCF-7 human breast cancer cells. Life Sciences. 2005;77:293–311.

[26] Tran T.T., Gerding-Reimers C., Schölermann B., Stanitzki B., Henkel T., Waldmann H., Ziegler S. Podoverine A—a novel microtubule destabilizing natural product from the *Podophyllum* species. Bioorganic & Medicinal Chemistry. 2014;22:5110–5116.

[27] Yamazaki Y., Sumikura M., Hidaka K., Yasui H., Kiso Y., Yakushiji F., Hayashi Y. Anti-microtubule 'plinabulin' chemical probe KPU-244-B3 labeled both alpha- and beta-tubulin. Bioorganic & Medicinal Chemistry. 2010;18:3169–3174.

[28] Lee J.M., Lee M.S., Koh D., Lee Y.H., Lim Y., Shin S.Y. A new synthetic 2′-hydroxy-2,4,6-trimethoxy-5′,6′-naphthochalcone induces G2/M cell cycle arrest and apoptosis by disrupting the microtubular network of human colon cancer cells. Cancer Letters. 2014;354:348–354.

[29] Choudhury D., Das A., Bhattacharya A., Chakrabarti G. Aqueous extract of ginger shows antiproliferative activity through disruption of microtubule network of cancer cells. Food and Chemical Toxicology. 2010;48:2872–2880.

[30] Naghshineh A., Dadras A., Ghalandari B., Riazi G.H., Modaresi M.S., Afrasiabi A., Aslani M.K. Safranal as a novel anti-tubulin binding agent with potential use in cancer therapy: an *in vitro* study. Chemico-Biological Interactions. 2015;238:151–160.

[31] Chen Y.L., Lin S.Z., Chang J.Y., Cheng Y.L., Tsai N.M., Chen S.P., Chang W.L., Harn H.J. *In vitro* and *in vivo* studies of a novel potential anticancer agent of isochaihulactone on human lung cancer A549 cells. Biochemical Pharmacology. 2006;72:308–319.

[32] Dhaheri Y.A., Attoub S., Ramadan G., Arafat K., Bajbouj K., Karuvantevida N., et al. Carnosol induces ROS-mediated beclin1-independent autophagy and apoptosis in triple negative breast cancer. PLoS One. 2014;9:e109630.

[33] Mira A., Shimzu K.. *In vitro* cytotoxic activities and molecular mechanisms of angelica shikokiana extract and its isolated compounds. Pharmacognosy Magazine. 2015;11:S564–S569.

[34] Ji Y.B., Chen N., Zhu H.W., Ling N., Li W.L., Song D.X., Gao S.Y., Zhang W.C., Ma N.N. Alkaloids from beach spider lily (*Hymenocallis littoralis*) induce apoptosis of HepG-2 cells by the fas-signaling pathway. Asian Pacific Journal of Cancer Prevention. 2014;15:9319–9325.

[35] Wang L.T., Pan S.L., Chen T.H., Dong Y.Z., Lee K.H., Teng C.M. DYZ-2-90, a novel neo-tanshinlactone ring-opened compound, induces ERK-mediated mitotic arrest and subsequent apoptosis by activating JNK in human colorectal cancer cells. ChemBioChem. 2012;13:1663–1672.

[36] Appadurai P., Rathinasamy K. Indicine N-oxide binds to tubulin at a distinct site and inhibits the assembly of microtubules: a mechanism for its cytotoxic activity. Toxicology Letters. 2014;225:66–77.

[37] Wani M.C., Horwitz S.B. Nature as a remarkable chemist: a personal story of the discovery and development of taxol. Anti-Cancer Drugs. 2014;25:482–487.

[38] Horwitz S.B.. Taxol (paclitaxel): mechanisms of action. Annals of Oncology. 1994;5 Suppl 6:S3–S6.

[39] Xiao H., Verdier-Pinard P., Fernandez-Fuentes N., Burd B., Angeletti R., Fiser A., Horwitz S.B., Orr G. Insights into the mechanism of microtubule stabilization by Taxol. Proceedings of the National Academy of Sciences of the United States of America. 2006;103:10166–10173.

[40] Ringel I., Horwitz S.B. Studies with RP 56976 (Taxotere). A new semisynthetic analogue of taxol, Journal of the National Cancer Institute. 1989;83:288–291.

[41] Rowinsky E.K. The development and clinical utility of the taxane class of antimicrotubule chemotherapy agents. Annual Review of Medicine. 1997;48:353–374.

[42] Dorr R.T. Pharmacology of the taxanes. Pharmacotherapy. 1997;17:96S–104S.

[43] Hennequin C., Giocanti N., Favaudon V. S-phase specificity of cell killing by docetaxel (Taxotere) in synchronized HeLa cells. British Journal of Cancer. 1995;71:1194–1198.

[44] Hemmenfent K.L., Govindan R. Novel formulations of taxanes: a review. Old wine in a new bottle?. Annals of Oncology. 2006;17:735–749.

[45] Chen C.H., Liao C.H., Chang Y.L., Guh J.H., Pan S.L., Teng C.M. Protopine, a novel microtubule-stabilizing agent, causes mitotic arrest and apoptotic cell death in human hormone-refractory prostate cancer cell lines. Cancer Letters. 2012;315:1–11.

[46] Buey, R.M., Barasoain, I., Jackson, E., Meyer, A., Giannakakou, P., Paterson, I., Mooberry, S., Andreu, J.M., Diaz J.F. Microtubule interactions with chemically diverse stabilizing agents: thermodynamics of binding to the paclitaxel site predicts cytotoxicity. Chemistry and Biology. 2005;12:1269–1279.

[47] Tinley T.L., Randall-Hlubek D.A., Leal R.M., Jackson E.M, Cessac J.W., Quada, J.C., Hemscheidt T.K., Mooberry S.L. Taccalonolides E and A: plant-derived steroids with microtubule-stabilizing activity. Cancer Research. 2003;63:3211–3220.

[48] Li J., Risinger A.L., Peng, J., Chen, Z., Hu L., Mooberry S.L.J. Hydrolysis reactions of the taccalonolides reveal structure activity relationships. Journal of Natural Products. 2013;76:1369–1375.

[49] Peng J., Risinger A.L., Fest G.A., Jackson E.M., Helms G., Polin L.A., Mooberry S.L. Identification and biological activities of new taccalonolide microtubule stabilizers. Journal of Medicinal Chemistry. 2011;54:6117–6124.

[50] Mukhtar E., Adhami V.M., Sechi M., Mukhtar H. Dietary flavonoid fisetin binds to β-tubulin and disrupts microtubule dynamics in prostate cancer cells. Cancer Letters. 2015;367:173–183.

[51] Higgs H.N., Pollard T.D. Regulation of actin filament network formation through ARP2/3 complex: activation by a diverse array of proteins. Annual Review of Biochemistry. 2001;70:649–676.

[52] Dominguez R., Holmes K.C. Actin structure and function. Annual Review of Biophysics. 2011;40:169–186.

[53] dos Remedios C.G., Chhabra D., Kekic M., Dedova I.V., Tsubakihara M., Berry D.A., Nosworthy N.J. Actin binding proteins: regulation of cytoskeletal microfilaments. Physiological Reviews. 2003;83:433–473.

[54] Somers K.D., Murphey M.M. Multinucleation in response to cytochalasin B: a common feature in several human tumor cell lines. Cancer Research. 1982;42:2575–2578.

[55] Somers K.D., Murphey M.M. Cytochalasin B-induced multinucleation of human tumor and normal cell cultures. Cell Biology International Reports. 1980;4:487–495.

[56] Medina D., Oborn C.J., Asch B.B. Distinction between preneoplastic and neoplastic mammary cell populations *in vitro* by cytochalasin B-induced multinucleation. Cancer Research. 1980;40:329–333.

[57] Trendowski M., Yu G., Wong V., Acquafondata C., Christen T., Fondy T.P. The real deal: using cytochalasin B in sonodynamic therapy to preferentially damage leukemia cells. Anticancer Research. 2014;34:2195–2202.

[58] Fayard B., Bianchi F., Dey J., Moreno E., Djaffer S., Hynes N.E., Monard D. The serine protease inhibitor protease nexin-1 controls mammary cancer metastasis through LRP-1-mediated MMP-9 expression. Cancer Research. 2009;69:5690–5698.

[59] Worthylake R.A., Burridge K. RhoA and ROCK promote migration by limiting membrane protrusions. Journal of Biological Chemistry. 2003;278:13578–13584.

[60] Yamazaki D., Kurisu S., Takenawa T. Regulation of cancer cell motility through actin reorganization. Cancer Science. 2005;96:379–386.

[61] Azios N.G., Krishnamoorthy L., Harris M., Cubano L.A., Cammer M., Dharmawardhane S.F. Estrogen and resveratrol regulate rac and Cdc42 signaling to the actin cytoskeleton of metastatic breast cancer cells. Neoplasia. 2007;9:147–158.

[62] Hamdi H.K., Castellon R.. Oleuropein, a non-toxic olive iridoid, is an anti-tumor agent and cytoskeleton disruptor. Biochemical and Biophysical Research Communications. 2005;334:769–778.

[63] Machado P.R., Alvaro F.S., Marcos P., de Avila S.L., da Silva B.V., et al. Alkaloids derived from flowers of Senna spectabilis, (−)-cassine and (−)-spectaline, have antiproliferative

activity on HepG2 cells for inducing cell cycle arrest in G1/S transition through ERK inactivation and downregulation of cyclin D1 expression. Toxicology *In Vitro*. 2016;31:86–92.

[64] Lee W.L., Shyur L.F. Deoxyelephantopin impedes mammary adenocarcinoma cell motility by inhibiting calpain-mediated adhesion dynamics and inducing reactive oxygen species and aggresome formation. Free Radical Biology & Medicine. 2012;52:1423–1436.

[65] Duncan K., Duncan M., Alley M.C., Edward A., Sausville E.A. Cucurbitacin E-induced disruption of the actin and vimentin cytoskeleton in prostate carcinoma cells. Biochemical Pharmacology. 1996;52:1553–1560.

[66] Ma G., Luo W., Lu J., Ma D., Leung C.H., Wang Y., Chen X. Cucurbitacin E induces caspase-dependent apoptosis and protective autophagy mediated by ROS in lung cancer cells. Chemico-Biological Interactions. 2016;253:1e9.

[67] Sari-Hassoun M., Clement M.J., Hamdi I., Bollot G., Bauvais C., Joshi V., Toma F., Burgo A., Cailleret M., Rosales-Hernández M.C., Macias Pérez M.E., Chabane-Sari D., Curmi P.A. Cucurbitacin I elicits the formation of actin/phospho-myosin II co-aggregates by stimulation of the RhoA/ROCK pathway and inhibition of LIM-kinase. Biochemical Pharmacology. 2016;102:45–63.

[68] Haritunians T., Gueller S., Zhang L., Badr R., Yin D., Xing H., Fung M.C., Koeffler H.P. Cucurbitacin B induces differentiation, cell cycle arrest, and actin cytoskeletal alterations in myeloid leukemia cells. Leukemia Research. 2008;32:1366–1373.

[69] Sawitzky H., Liebe S., Willingale-Theune J., Menzel D. The anti-proliferative agent jasplakinolide rearranges the actin cytoskeleton of plant cells. European Journal of Cell Biology. 1999;78:424–433.

[70] Lu Q.Y., Jin Y.S., Zhang Q., Zhang Z., Heber D., Go V.L., Li F.P., Rao J.Y. *Ganoderma lucidum* extracts inhibit growth and induce actin polymerization in bladder cancer cells *in vitro*. Cancer Letters. 2004;216:9–20.

[71] Velasco-Velázquez M.A., Agramonte-Hevia J., Barrera D., Jiménez-Orozco A., García-Mondragón M.J., Mendoza-Patiño N., Landa A., Mandoki J. 4-Hydroxycoumarin disorganizes the actin cytoskeleton in B16–F10 melanoma cells but not in B82 fibroblasts, decreasing their adhesion to extracellular matrix proteins and motility. Cancer Letters. 2003;198:179–186.

[72] Cooper G.M. The Cell: A Molecular Approach, 2nd ed.. Sinauer Associates. 2000.

[73] Chen Y.L., Lin S.Z., Chang W.L., Cheng Y.L., Harn H.J. Requirement for ERK activation in acetone extract identified from Bupleurrum scorzonerifolium induced A549 tumor cell apoptosis and keratin 8 phosphorylation. Life Sciences. 2005;76:2409–2420.

[74] Oshima R.G., Baribault H., Caulín C. Oncogenic regulation and function of keratins 8 and 18. Cancer and Metastasis Reviews. 1996;15:445–471.

[75] Weng Y.R., Cui Y., Fang J.Y. Biological functions of cytokeratin 18 in cancer. Molecular Cancer Research. 2012;10:485–493.

[76] Pankov R., Umezawa A., Maki R., Der C.J., Hauser C.A., Oshima R.G. Oncogene activation of human keratin 18 transcription via the Ras signal transduction pathway. Proceedings of the National Academy of Sciences of the United States of America. 1994;91:873–877.

[77] Oshima R.G. Apoptosis and keratin intermediate filaments. Cell Death and Differentiation. 2002;9:486–492.

[78] Fortier A.M., Asselin E., Cadrin M. Keratin 8 and 18 loss in epithelial cancer cells increases collective cell migration and cisplatin sensitivity through claudin1 upregulation. Journal of Biological Chemistry. 2013;288:11555–11571.

[79] Cheng Y.L., Chang W.L., Lee S.C., Liu Y.G., Lin H.C., Chen C.J., Yen C.Y., Yu D.S., Lin S.Z., Harn H.J. Acetone extract of *Bupleurum scorzonerifolium* inhibits proliferation of A549 human lung cancer cells via inducing apoptosis and suppressing telomerase activity. Life Sciences. 2003;73:2383–2394.

[80] Bhat F.A., Sharmila G., Balakrishnan S., Arunkumar R., Elumalai P., Suganya S., Singh R.P., Srinivasan N., Arunakaran J. Quercetin reverses EGF-induced epithelial to mesenchymal transition and invasiveness in prostate cancer (PC-3) cell line via EGFR/PI3K/Akt pathway. Journal of Nutritional Biochemistry. 2014;25:1132–1139.

[81] Li Y., Vanden Boom T.G., Kong D., Wang Z., Ali S., Philip P.A., et al. Up-regulation of miR-200 and let-7 by natural agents leads to the reversal of epithelial-to-mesenchymal transition in gemcitabine-resistant pancreatic cancer cells. Cancer Research. 2009;69:6704–6712.

[82] Kim Y.S., Choi K.C., Hwang K.A. Genistein suppressed epithelial–mesenchymal transition and migration efficacies of BG-1 ovarian cancer cells activated by estrogenic chemicals via estrogen receptor pathway and downregulation of TGF-β signaling pathway. Phytomedicine. 2015;22:993–999.

[83] Polachi N., Bai G., Li T., Chu Y., Wang X., Li S., Gu N., Wu J., Li W., Zhang Y., Zhou S., Sun H., Liu C. Modulatory effects of silibinin in various cell signaling pathways against liver disorders and cancer—a comprehensive review. European Journal of Medicinal Chemistry. 2016;123:577–595.

[84] Cufí S., Bonavia R., Vazquez-Martin A., Corominas-Faja B., Oliveras-Ferraros C., Cuyàs E., Martin-Castillo B., Barrajón-Catalán E., Visa J., Segura-Carretero A., Bosch-Barrera J., Joven J., Micol V., Menendez J.A. Silibinin meglumine, a water-soluble form of milk thistle silymarin, is an orally active anti-cancer agent that impedes the epithelial-to-mesenchymal transition (EMT) in EGFR-mutant non-small-cell lung carcinoma cells. Food and Chemical Toxicology. 2013;60:360–368.

[85] Naveen C.R., Gaikwad S., Agrawal-Rajput R. Berberine induces neuronal differentiation through inhibition of cancer stemness and epithelial-mesenchymal transition in neuroblastoma cells. Phytomedicine. 2016;23:736–744.

[86] Lee S.W., Mandinova A. Patent application title: methods for the treatment of cancer using piperlongumine and piperlongumine. Analogs 2009. WO 20090312373.

[87] Gupta P., Wright S.E., Kim S.H., Srivastava S.K. Phenethyl isothiocyanate: a comprehensive review of anti-cancer mechanisms Biochimica et Biophysica Acta. 2014;1846:405–424.

[88] Bargagna-Mohan P., Hamza A., Kim Y, Ho K.H., Mor-Vaknin N., Wendschlag N., Liu J., Evans R.M., Markovitz, D.M., Zhan C.G., Kim K.B., Wendschlag N., Liu J., Evans R.M., Markovitz D.M., Zhan C.G., Kim K.B., Mohan R. The tumor inhibitor and antiangiogenic agent withaferin A targets the intermediate filament protein vimentin. Chemistry & Biology. 2007;14:623–634.

[89] Cerqueira Coelho P.L., Villas-Boas de Freitas S.R., Seara Pitanga B.P., da Silva V.D.A., Oliveira M.N., Grangeiro M.S., dos Santos Souza C., dos Santos El-Bachá R., Costa M.D., Barbosa P.R., de Oliveira Nascimento I.L., Lima Costa S. Flavonoids from the Brazilian plant croton betulaster inhibit the growth of human glioblastoma cells and induce apoptosis. Revista Brasileira de Farmacognosia. 2016;26:34–43.

[90] Monsuez J.J., Charniot J.C., Vignat N., Artigou J.Y. Cardiac side-effects of cancer chemotherapy. International Journal of Cardiology. 2010;144:3–15.

[91] Hsiu-Chuan Y., Oberley T.D., Vichitbandha S., Ye-Shhi H.O., St Clair D.K. The protective role of manganese superoxide dismutase against adriamycin-induced acute cardiac toxicity in transgenic mice. Journal of Clinical Investigation. 1996;98:1253–1260.

[92] Ito H., Miller S.C., Billingham M.E., Akimoto H., Torti S.V., Wade R. Doxorubicin selectively inhibits muscle gene expression in cardiac muscles cells *in vivo* and *in vitro*. Proceedings of the National Academy of Sciences of the United States of America. 1990;87:4275–4279.

[93] Sussman M.A., Hamm-Alvarez S.F., Vilalta P.M., Welch S., Kedes L. Involvement of phosphorylation in doxorubicin-mediated myofibril degeneration. Circulation Research. 1997;80:52–61.

[94] Lim C.C., Zuppinger C., Guo X., Kuster G.M., Helmes M., Eppenberger H.M., Suter T.M. et al. Anthracyclines induce calpain-dependent titin proteolysis and necrosis in cardiomyocytes. Journal of Biological Chemistry. 2004;279:8290–8299.

[95] Chen Y., Daosukho C., Opii W.O., Turner D.M., Pierce W.M., Klein J.B., Vore M. et al. Redox proteomic identification of oxidized cardiac proteins in adriamycin-treated mice. Free Radical Biology and Medicine. 2006;41:1470–1477.

[96] Anghel N., Cotoraci C., Ivan A., Suciu M., Herman H., Balta C., Nicolescu L., Olariu T., Galajda Z., Ardelean A., Hermenean A. Chrysin attenuates cardiomyocyte apoptosis and loss of intermediate filaments in a mouse model of mitoxantrone cardiotoxicity. Histology and Histopathology. 2015;30:1465–1475.

[97] Fisher P.W., Salloum F., Das A., Hyder H., Kukreja R.C. Phosphodiesterase-5 inhibition with sildenafil attenuates cardiomyocyte apoptosis and left ventricular dysfunction in a chronic model of doxorubicin cardiotoxicity. Circulation Journal. 2005;111:1601–1610.